Hydrology and Environmental Engineering

Hydrology and Environmental Engineering

Edited by Cohen Foster

SYRAWOOD
PUBLISHING HOUSE
New York

Published by Syrawood Publishing House,
750 Third Avenue, 9th Floor,
New York, NY 10017, USA
www.syrawoodpublishinghouse.com

Hydrology and Environmental Engineering
Edited by Cohen Foster

International Standard Book Number: 978-1-64740-430-7 (Hardback)

Cataloging-in-publication Data

Hydrology and environmental engineering / edited by Cohen Foster.
p. cm.
Includes bibliographical references and index.
ISBN 978-1-64740-430-7
1. Environmental engineering. 2. Hydrology. 3. Environmental hydraulics. I. Foster, Cohen.
TA170 .H93 2023
628--dc23

TABLE OF CONTENTS

PREFACE

The purpose of the book is to provide a glimpse into the dynamics and to present opinions and studies of some of the scientists engaged in the development of new ideas in the field from very different standpoints. This book will prove useful to students and researchers owing to its high content quality.

Environmental engineering is the creation of infrastructure and procedures for waste disposal, water supply, and pollution control. It is a broad field that incorporates hydrology, geology, microbiology, chemistry, mathematics, ecology, hydraulics and economics. Hydrology refers to the study of water cycling through various reservoirs on the Earth. It focuses on the effects of climate on the water cycle and the distribution of water in the atmosphere, subsurface and surface. It also studies the chemistry of water from different sources. The vast majority of hydrologic applications involve surface water systems such as estuaries, rivers, and open-channels. Geographically, the study of hydrology is divided into three areas, namely, surface water hydrology, groundwater hydrology (hydrogeology), and marine hydrology. Some of the major areas of study within hydrology are hydrogeology, hydrometeorology, drainage-basin management, water quality and surface hydrology. This book contains some path-breaking studies in the field of hydrology and environmental engineering. It consists of contributions made by international experts. Scientists and students actively engaged in these fields will find this book full of crucial and unexplored concepts.

At the end, I would like to appreciate all the efforts made by the authors in completing their chapters professionally. I express my deepest gratitude to all of them for contributing to this book by sharing their valuable works. A special thanks to my family and friends for their constant support in this journey.

Editor

1

Modeling River Runoff Temporal Behavior through a Hybrid Causal–Hydrological (HCH) Method

Santiago Zazo [1], José-Luis Molina [1,*], Verónica Ruiz-Ortiz [2], Mercedes Vélez-Nicolás [3] and Santiago García-López [3]

1 IGA Research Group, Higher Polytechnic School of Ávila, University of Salamanca, Av. de los Hornos Caleros, 50, 05003 Ávila, Spain; szazo@usal.es
2 Department of Industrial Engineering and Civil Engineering, University of Cadiz, Campus Bay of Algeciras, Avda. Ramon Puyol s/n, 11202 Algeciras, Spain; veronica.ruiz@uca.es
3 Department of Earth Sciences, University of Cadiz, Campus Rio San Pedro, s/n, 11510 Puerto Real, Spain; mercedes.velez@uca.es (M.V.-N.); santiago.garcia@uca.es (S.G.-L.)
* Correspondence: jlmolina@usal.es

Abstract: The uncertainty in traditional hydrological modeling is a challenge that has not yet been overcome. This research aimed to provide a new method called the hybrid causal–hydrological (HCH) method, which consists of the combination of traditional rainfall–runoff models with novel hydrological approaches based on artificial intelligence, called Bayesian causal modeling (BCM). This was implemented by building nine causal models for three sub-basins of the Barbate River Basin (SW Spain). The models were populated by gauging (observing) short runoff series and from long and short hydrological runoff series obtained from the Témez rainfall–runoff model (T-RRM). To enrich the data, all series were synthetically replicated using an ARMA model. Regarding the results, on the one hand differences in the dependence intensities between the long and short series were displayed in the dependence mitigation graphs (DMGs), which were attributable to the insufficient amount of data available from the hydrological records and to climate change processes. The similarities in the temporal dependence propagation (basin memory) and in the symmetry of DMGs validate the reliability of the hybrid methodology, as well as the results generated in this study. Consequently, water planning and management can be substantially improved with this approach.

Keywords: Bayesian causal modeling; HCH method; hydrological modeling; deterministic and stochastic modeling; rainfall–runoff modeling; temporal dependence; basin memory

1. Introduction

Currently, extreme events such as heavy rainfall, droughts, and floods are becoming more common and less exceptional [1–5], even within a particular territory [6,7], resulting in growing variability in certain hydrological processes [8,9], in addition to pressure on, and uncertainty regarding, water resources [10–12], which is especially noticeable using runoffs series [13]. In this sense, the frequency of droughts in Southern Europe is increasing significantly [14,15]. This is making it necessary to improve the knowledge on the temporal behavior of rivers [16–18], as not all the reasons that explain this increasing variability are new [16,18].

As such, there is a strong need to develop new analytical approaches capable of capturing the induced and widespread effects that these new hydrological phenomena are causing on water resource availability [19]. This is essential not only for planning and development of effective water resource management strategies [18], but also in terms of optimal dimensioning of hydraulic infrastructure, such as reservoirs [13,20].

The temporal behavior of rivers has traditionally received special attention from the scientific and engineering communities [16,20–22]. This topic has been developed through previous studies and approaches that are currently the benchmarks and through persistence, and is strongly related to the measurement of the long-term memory of time series using the Hurst coefficient [23], as well as storage and drought statistics [24]. This issue has mainly been addressed by models of: (1) the interactions between multiple physical factors, such as meteorological, geological, and hydrogeological factors [13]; and (2) the analysis of hydrological records [24]. Furthermore, it should firstly be noted that any model is an abstraction of reality [25], which is only partially known, and secondly, as is well-known, the scarcity of data has been a constant issue in hydrological research [26].

This has led to diverse points of view [22] on how to address these issues, which are essentially classified into two main approaches, deterministic and stochastic [27,28], as was detailed by Molina et al. [16]. Moreover, the growing global demand for water resources [29,30], negative scientific predictions of their availability due to climate change [31,32], and the partial knowledge of the underlying relationships in complex natural systems such as water systems [33–36] give this research special relevance.

Deterministic models, which are often complex, are characterized by (1) being accurate, (2) needing large amounts of data and even inputs from other models, (3) and by the data processing being time consuming and frequently based on a unique time series [16]. The above advantages allow a better understanding of the modeling process through the abstraction (i.e., simplification) of complex simulated natural phenomena [16]. This is especially useful in the case of controlled and gauged river basins that are barely affected by climate change [27]. Moreover, this general method is currently applied using software such as HEC-HMS [37] (which is widely accepted in the field of hydrological engineering [16]) by most water administrations [38]. However, the main weakness of these models is their limited capacity for analysis of certain relevant aspects, such as the temporal dependence or persistence of the basin´s memory [16,27], in addition to not being viable for basins with limited data [39].

In recent decades, stochastic-model-based approaches have emerged in the development of hydrological models as a result of advances in computer science [16,20,22,36,40]. These models are fundamentally different from the previous ones in how they deal with the uncertainty inherent in any hydrological process [27,41]. This distinctive feature makes them suitable for hydrological modeling [41]. Some relevant examples are autoregressive moving average (ARMA) and autoregressive integrated moving average (ARIMA) models, multivariate adaptive regression splines (MARS), and causal reasoning (CR). The theoretical backgrounds of all these approaches were covered in depth by Molina et al. [16,20].

In this sense, CR, as an artificial intelligence (AI) technique, is a powerful stochastic approach and is especially relevant to this research for the following reasons: (1) it uses raw data directly [42]; (2) it does not require *"a priori"* knowledge of the process [43]; (3) it handles a large amount of information from dynamic and non-linear systems [36]; (4) it is able to define sophisticated relationships in complex natural systems [44]; (5) it is a powerful tool for discovering causal structures in raw statistical data [45].

Additionally, as shown by the studies by Molina et al. [20] and Zazo [22], the potential for using causality to deal with temporal river behavior has recently begun to be explored using CR. This involves coupling traditional and novel methodologies through an ARMA model and Bayesian causal modeling (BCM). This has given rise to an active research area focused on increasing the knowledge of water resources and based on extracting the hidden logical time dependency structure that inherently underlies hydrological series. A comprehensive and in-depth theoretical and mathematical background and the main hydrological contributions to this research area are covered in [13,17,18,28].

The sustainability of a hydrological system as assessed through the characterization of basin memory is very relevant, especially in the short term. Consequently, this research work aims to capture that temporal signature through a hybrid causal framework, which combines the advantages of deterministic and stochastic models. This is mainly highlighted through the relationship between an input signal (rainfall) and an output signal (runoff).

Additionally, this work aims to overcome the lack of suitable and available hydrological data for CR analysis, as causal models need to be populated by as much information as possible to be representative [16]. This scarcity is largely due to several reservoirs having been built in the study basin (Almodovar, Celemín, and Barbate reservoirs) at the end of the 20th century in order to solve the problems related to recurrent droughts, floods, and water availability in this study zone.

In order to solve these challenges, a hybrid framework called the hybrid causal–hydrological (HCH) approach is developed in this study. This is done by hybridizing a BCM (stochastic platform) with a rainfall–runoff model (RRM) (deterministic module), which is the data source for the BCM platform.

Here, the RRM is the Témez RRM model (henceforth referred as T-RRM) [46–48]. This is an aggregate and semi-distributed model of parameters that is suitable for homogeneous basins with a reduced amount of data. The T-RRM model is well-established in hydrological engineering, particularly in Spain [15,49–53], and is supported here by EvalHid software [54]. The T-RRM can provide reliable and long-term natural regime runoff time series and is used here to generate synthetic series using a parsimonious and unconditioned ARMA model (first stochastic approach). Then, these equiprobable time series are used for causal reasoning (second stochastic series) and the results are analyzed in depth.

After this introductory section, this manuscript is organized as follows. A case study, dataset description, and the applied methodology are discussed in Section 2. Section 3 provides the main experimental results from the research. In Section 4, the results are discussed in detail. Lastly, Section 4 is devoted to the general conclusions drawn from the study.

2. Materials and Methods

2.1. Case Study

The Barbate River Basin is located in the province of Cadiz, southern Spain, near the Strait of Gibraltar. It covers an approximate area of 1330 km^2, which includes a significant extension of Los Alcornocales Natural Park and a small sector at the SW of Breña and Marismas in the Barbate Natural Park. The basin has a smooth orography, with more than the 70% of its area ranging between 0 and 100 m; however, in its northeast sector, there are mountainous reliefs that reach 1090 m above sea level (m a.s.l.; Figure 1a–c) [55].

The Köppen–Geiger classification describes the climate of this area as temperate, with hot, dry summers [56,57]. The basin is characterized by a Mediterranean climate with a strong oceanic influence, which is bolstered by its orography, in addition to a marked seasonal rainfall regime, with average annual values of 835 mm [55]. Additionally, the general rainfall pattern is characterized by strong interannual irregularity, with values ranging between 365 (dry years) and 1606 mm/year (wet years). The precipitation is mainly concentrated in autumn and winter months, with practically no precipitation in the summer period. The average annual temperature is 18 °C [55], with milder values in mountainous areas. In this sense, it should be noted that the study area is repeatedly affected by intense and persistent warm–dry wind periods, commonly known as Levante winds, which can last for 7 consecutive days, with an average velocity of 50 km/h and maximum velocities of up to 100 km/h [58,59]. This wind plays an important role in evaporation processes [55].

Geologically, the basin comprises a wide variety of materials. The oldest ones are Triassic Keuper facies (pre-orogenic), which are composed of multicolored gypsum-rich clays, sands, and limestones [60]. On these, tecto-sedimentary materials such as cemented siliceous sandstones and clays are overlaid, corresponding to turbiditic deposits [61]. Post-orogenic materials were subsequently deposited, filling the tertiary basins with sediments that were deposited in marine and coastal environments during the Upper Miocene and Pliocene (marls, biocalcarenites, and sands). Lastly, continental materials from the Quaternary age were deposited, with these being associated with recent sedimentary processes, aeolian and beach sands, alluvial material linked to river dynamics, and slope deposits [55].

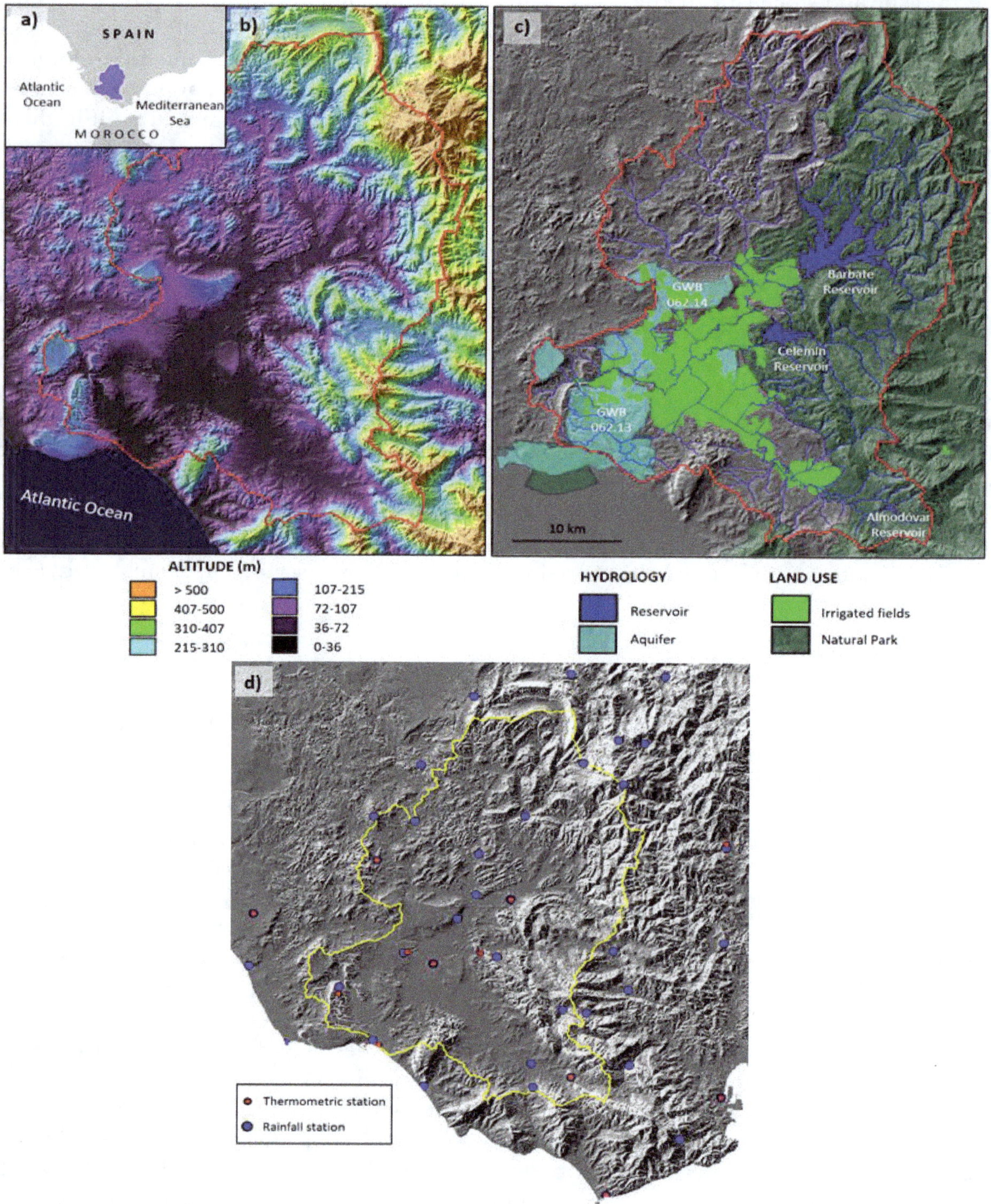

Figure 1. (**a**) Situation map. (**b**) Barbate River Basin orography map. (**c**) Land occupation and hydrological elements. (**d**) Rainfall and thermometric stations map [6].

In hydrogeological terms, two systems of a certain significance are located in the study area, named the Benalup aquifer (groundwater body GWB 062.014, of 33 km^2) and the Barbate aquifer (GWB 062.013, of 93 km^2). Both systems, which are constituted by biocalcarenites and sands, are characterized by an intergranular porosity that is locally increased by dissolution processes [62].

Regarding hydrological aspects, the main water course in the zone is the Barbate river, which is regulated by the Barbate dam (storage capacity of 228 Hm^3) and its tributaries, the Álamo, Celemín, and Almodóvar rivers. The latter two are also regulated, in this case by the Celemín (45 Hm^3) and

Almodóvar (5.7 Hm^3) dams, respectively (Figure 1c). The water resources for the Celemín and Barbate reservoirs are jointly used to irrigate an area measuring 122.3 Km^2, whereas the Almodóvar reservoir is exploited for both irrigation (3.63 Km^2) and for supply of the municipality of Tarifa. Finally, the basin's water resources were estimated to be 247 Hm^3/year for the period 1999–2016 [55], with an average annual demand of 102 Hm^3 [63].

2.2. Dataset Description

The analyzed period comprises 68 years, ranging from 1951 to 2017. Monthly rainfall and temperature data were initially obtained from 67 and 35 stations, respectively. These stations were distributed throughout the Barbate river basin and adjacent catchment areas (Figure 1d), with all of them belonging to the Spanish Official Network of Hydrometeorological Records [64,65].

Given that the available time series displayed several recording periods and numerous data gaps, it was necessary to homogenize them. This was done using the following methodological processes: (1) the neighboring stations (less than 1 km) were completed directly by substitution [66] or by linear regression when the correlation coefficient between them was greater than 0.95; (2) time series with less than 180 monthly data points were removed (equivalent to 15 years record) [67,68]; (3) the resulting series were systematically completed through multivariate regression models involving previous monthly stationarization [69,70] using the free software CHAC [71].

Likewise, in order to guarantee the robustness of the data and reliability of the results, the following quality criteria were applied: (i) the prioritization exponent (0.1), which weighs the importance of the number of common data among the series used [71]; and (ii) the prioritization threshold (ranging between 0.80 to 0.64), which is the minimum value of multiple correlation coefficients and is used to establish pairs of stations and complete the gaps [72]. According to the aforementioned criteria, a total of 43 rainfall and 16 thermometric stations were subsequently considered in the T-RRM.

The existence of three reservoirs in the study area affected the availability of the runoff time series under natural regimes (Figure 1c). These data were provided by the Guadalete–Barbate River Basin Authority [73] and only cover a limited period of time ranging from October 1999 to June 2016.

Finally, given the limited data available under the natural regime, two types of input series were considered for each reservoir in this research: first, a long series ranging from 1951 to 2017 obtained from the T-RRM, comprising 67 records; and second, a short series with a length of 16 years, which was obtained from observed records from 2000 to 2015.

2.3. Justification of HCH Method Utility

The methodological framework proposed here combines a traditional approach (rainfall–runoff and ARMA models) with new strategies based on AI [20]. In this case, the AI technique comprises a Bayesian network (BN), in line with the recent research trends in the field of stochastic hydrology [22]. This is mainly due to the fact that this union of methodologies between classical and novel approaches is able to provide more accurate outcomes in the modeling of complex natural processes [43,74–76], such as water resources.

Furthermore, the usage of a specific RRM model represents an improvement in this research line, which is focused on causality, because it overcomes the traditional limitations related to the availability of hydrological runoff data in natural regimes [16,26]. Moreover, it is also worth highlighting the fact that causal models improve and increase the knowledge on the temporal behavior of a basin's water resources [13,17,18,27]. For example, causal models are able to provide a specific temporal horizon of dependence, whereas correlogram outcomes are much more limited. This implies that a novel approach should be used to assess the time order of dependence of a given time series. Additionally, a new qualitative approach is provided for assessing temporal dependence by means of DMGs, taking a geometrical perspective.

Additionally, the HCH method takes advantage of the following features, among others: (1) the potential for dynamic analysis and omnidirectional step-by-step analysis along the whole network [20];

(2) the simplicity in the definition of relationships in complex natural systems [44]; and (3) the capacity to identify causal structures in raw statistical data supported by the BNs [77]. The aim of this is to study the relationships of dependence between objective variables and the previous variables. It should be noted that under this approach, every relationship between variables is a temporal relationship [13]. In this sense, the variable relationship strength is omnidirectionally quantified through its p-value, which measures the strength of evidence against the dependence relationship [20,22]. Therefore, the inverse value (1-p-value) represents the strength of the dependence relationship. Consequently, a value equal to 1 indicates strong evidence of total dependence, in contrast to 0, which represents little evidence of dependence [18].

Conceptually, BCM, as a form of Causality, is characterized by searching reasoning patterns in a hierarchical way, from top to bottom. This is supported by queries in form of conditional probability to assess the influence among the variables, which are instances of CR. This type of analysis is focused on the cause, and the objective comprises the prediction of the "downstream" consequences [77,78]. Figure 2 shows the conceptual scheme of developed causal models.

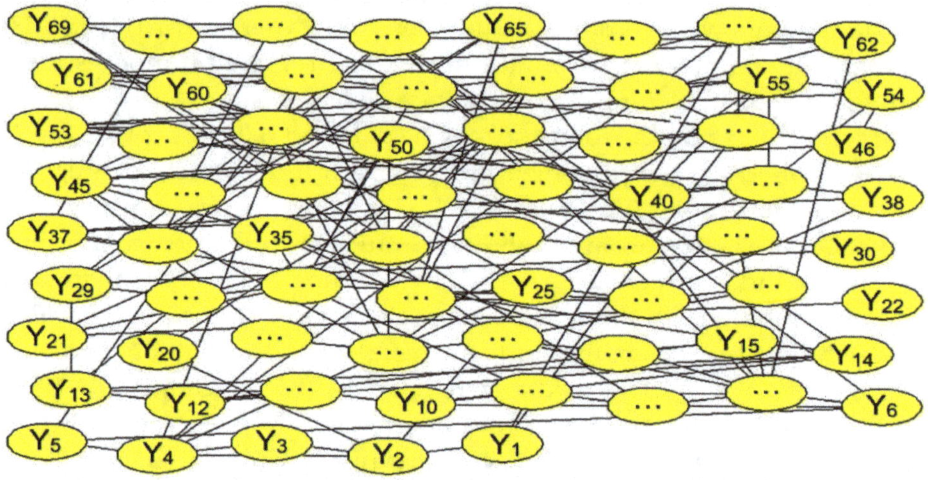

Figure 2. Causal model. Conceptual scheme for the case of a runoff time series of 69 years (from the initial year, Y_1, to the final year, Y_{69}). The visible threshold of independence between variables is 5% (0.05), which means that up to 95% (0.95) of the dependence relationships between the variables are displayed. Each link between variables (consecutive or not, trivial or not) denotes a relationship.

2.4. General Methodology

In this study, we develop a hybrid causal–hydrological (HCH) framework based on two modules and four sequential phases (Figure 3). The first module, a deterministic module, comprises a long-duration rainfall–runoff model definition in a natural regime (addressed entirely in phase 1) to overcome the limitation of the availability of hydrological data for posterior causal analysis. The second module, which is purely stochastic, includes all the other phases. In phase 2, a BCM design is developed, supported by the two subphases. In subphase 2a, a parsimonious and unconditioned ARMA model is defined. Next, subphase 2b is entirely devoted to the definition of BCM. After this, phase 3 is the core of this research study. This encompasses the advantages of a deterministic model with the analytical capabilities that causality offers through causal reasoning, supported by Bayesian modeling (using AI techniques). Through the hybridization of both modules, the real temporal dependence structure is revealed. This is a hidden, logical, and non-trivial interdependence structure that inherently underlies runoff data and that can be discovered, extracted, and modeled over time using BCM. Finally, in phase 4, the basin's hydrological memory is dynamically characterized, focusing on a short time series. This is performed through an analysis of the temporal dependence strength over time through a dependence mitigation graph (DMG). This a novel and dynamic qualitative approach for assessing the temporal dependence of hydrological series from a geometrical perspective [17].

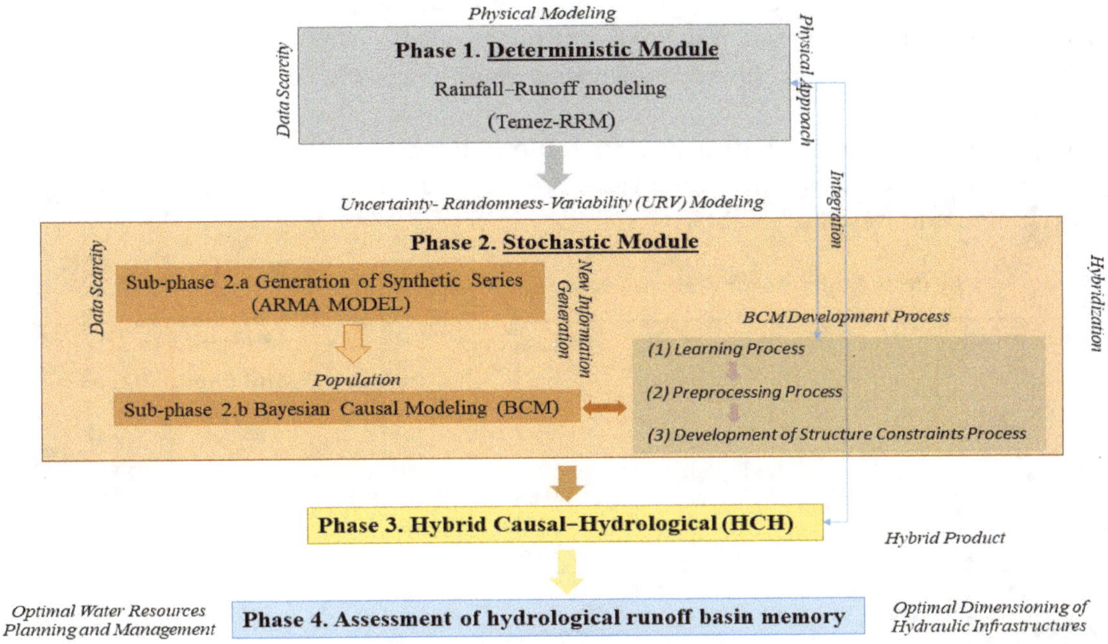

Figure 3. Methodological framework.

2.4.1. Phase 1 Deterministic Module: Rainfall–Runoff Model Definition

In order to estimate the monthly runoff series, and given the lack of complete and reliable runoff data available for causal analysis, a specific rainfall–runoff model was applied. In this case, the T-RRM was selected [46] according to criteria related to gauging data availability. This is an aggregate model of a semi-distributed application, whose use is restricted to homogeneous basins without exhaustive information [50,79]. This rainfall–runoff model is well-established in the field of hydrological engineering, particularly in Spain [15,49–53], and here it is supported by EvalHid software [54]. The T-RRM conceptual framework is shown in Figure 4a.

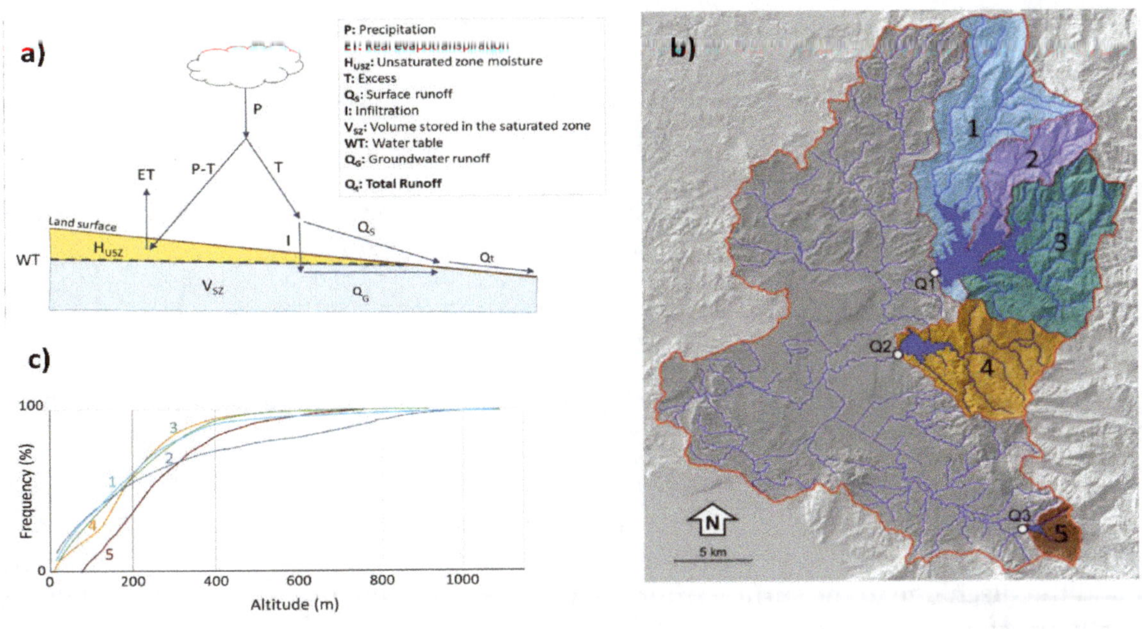

Figure 4. (**a**) The T-RRM conceptual scheme. (**b**) Barbate River Basin. Sub-basin scheme associated with the availability of gauging data (Qi): Barbate reservoir (Q1, sub-basins 1, 2, and 3); Celemín reservoir (Q2, sub-basin 4); Almodóvar reservoir (Q3, sub-basin 5). (**c**) Hypsometric curves of the 5 sub-basins.

In this case, three gauging control points were considered, located in the three reservoirs of the study areas: Barbate (Q1), Celemín (Q2), and Almodóvar (Q3). Therefore, according to the availability of gauging data and the applied homogeneity criteria, five (5) sub-basins were finally selected (Figure 4b).

Regarding the 5 sub-basins, Figure 4c and Table 1 summarize the key factors considered, such as the hypsometric curves, geological materials, and sub-basin extension, as well as the maximum, minimum, average, and median heights.

Table 1. Main morphometric characteristics of the studied sub-basins.

Sub-Basins/ Code Number		Q_i [1]	Geological Materials	Extension (Km^2)	Elevation (m) Maximum	Minimum	Mean	Median
Barbate	/1	Q1	[2] Flysch. Trias, Jurassic	149.7	1087.4	11.3	209.1	181.2
	/2		Flysch	59.7	1089.6	18.0	290.2	198.9
	/3		Flysch	141.8	911.8	18.0	202.2	180.9
Celemín	/4	Q2	Flysch	94.1	630.3	19.4	196.7	185.1
Almodóvar	/5	Q3	Flysch	16.6	763.4	98.0	283.2	248.4

[1] Gauging control points (see Figure 4b). [2] Triassic-Jurassic period.

In the T-RRM, the calculation process is controlled by the hydraulic principle of continuity and the hydrological principle of mass balance. The calculation of the excess (T_i) is a function of the precipitation (P_i), soil moisture deficit (H_{max}–H_{i-1}), evapotranspiration (ET_i), and the surplus parameter (C):

$$T_i = \frac{(P_i - P_0)^2}{P_i - \delta_t - 2 \cdot P_0} \rightarrow if: \; P_i > P_0 \; \rightarrow if \; not: \; T_i = 0, \tag{1}$$

where:

$$\delta = H_{max} - H_{i-1} + PET_i, \tag{2}$$

and

$$P_0 = C \cdot (H_{max} - H_{i-1}), \tag{3}$$

The soil moisture (H_i) and real evapotranspiration (ET_i) can be obtained from the following expressions:

$$H_i = \max(0, \; H_{i-1} + P_i - T_i - PET_i), \tag{4}$$

$$ET_i = \min(H_{i-1} + P_i - T_i, PET_i), \tag{5}$$

The infiltration to the aquifer, which corresponds to the recharge by direct precipitation (I_i), is a function of the excess (T_i) and maximum infiltration parameter (I_{max}):

$$I_i = I_{max} \cdot \frac{T_i}{T_i + I_{max}}, \tag{6}$$

The excess (Q_S) that does not infiltrate to the aquifer becomes runoff and is obtained through the following expression:

$$Q_S = T_i - I_i, \tag{7}$$

For the saturated zone, the T-RRM uses a single-cell aggregated model, whose defining parameter is the aquifer depletion curve (α, expressed as $month^{-1}$). The evolution of the volume stored in the aquifer and the discharge to the surface drainage network or to the sea can be obtained through the following expressions:

$$V_i = V_{i-1} - V_{i-1} \cdot e^{-\alpha \cdot \Delta t} \cdot R_i \cdot \left(\frac{1 - e^{-\alpha \cdot \Delta t}}{\alpha} \right), \tag{8}$$

$$Q_G = V_{i-1} - V_i \cdot R_i, \tag{9}$$

The total contribution (Q_t) corresponds to the excess water that does not infiltrate ($T - I$) plus the groundwater contribution:

$$Q_t = T_i - I_i + Q_G = Q_S + Q_T, \quad (10)$$

In the T-RRM, the input data for each sub-basin were series of monthly precipitation and potential evapotranspiration data. The first series was obtained through weighted averages of the precipitation series from the rainfall stations belonging to or surrounding each sub-basin after filling their data gaps, as indicated in the previous section. Weighting for each sub-basin was carried out according to the station influence–area criterion (Thiessen polygons) using GIS software. Since potential evapotranspiration (*ETP*) cannot be determined instrumentally, an empirical estimation method based on temperature was applied; more concretely, the Thornthwaite framework was used, a well-established and widely accepted method in the fields of hydrological engineering and soil science [80]:

$$ETP_C = ETP_{SC} \cdot \frac{d}{30} \cdot \frac{h}{12} = 16 \cdot \left(\frac{10 \cdot t}{I}\right)^a \cdot \frac{d}{30} \cdot \frac{h}{12}, \quad (11)$$

where $ETPS_C$ is the unadjusted monthly evapotranspiration value (mm/month), d is the number of days in a month, h is the light hours depending on the latitude, t is the monthly average temperature (°C), I is the annual heat index (mm/year), i is the monthly heat index (mm/month), and a is the dimensionless parameter. For each sub-basin, monthly temperature series were obtained from the thermometric records belonging to nearby stations, following a similar procedure to that used for precipitation. In this case, however, it was necessary to apply a correction coefficient based on the altitude–temperature gradient in order to compensate for the discrepancies between a sub-basin's average altitude and the average altitude of the thermometric stations considered.

Finally, the model was calibrated and validated by comparing the output values obtained using EvalHid software with the standard values from the gauging control points. As in the Barbate basin, there was only one control point available; basins 1, 2, and 3 were grouped after the calibration process. Subsequently, a backward validation technique was applied [63,81]. In this technique, the most recent periods (2000–2015) are used to calibrate the model, while the oldest (1950–1999) are used for validation. The calibration parameters of the T-RRM were the maximum moisture (H_{max}), coefficient of runoff (C), maximum infiltration (I_{max}), and aquifer depletion curve (α). These parameters depend mainly on the land uses, geology, lithology, and morphology of the terrain [82,83].

2.4.2. Phase 2 Stochastic Module

This phase comprised two other subphases. On the one hand, subphase 2a produced a equiprobable synthetic runoff series using a parsimonious and unconditioned ARMA model (the first stochastic approach). This information was exclusively generated from the T-RRM runoff outcomes. On the other hand, in subphase 2b, the BCM design was developed (second stochastic approach).

Subphase 2a ARMA Model Definition

In order to generate an ARMA model, first it is necessary to determine the main statistical parameters of the runoff time series. This was addressed by using a traditional statistical analysis approach based on the mean, standard deviation, and variation and skewness coefficients [24]. Additionally, the measure of persistence for the runoff time series (inherent statistical property of time series) was evaluated using the Hurst coefficient, which is expressed as [23]:

$$(R/S)_n = c \cdot n^H, \quad (12)$$

where the term (R/S) is the rescaled range (with zero average, expressed in terms of standard deviation), n is the number of data points per interval, c is a constant of proportionality, and H is the Hurst coefficient. Accordingly, H values in the range of [0, 0.5) indicate a negative autocorrelation (switching between high and low values in the future); H values in the range of (0.5, 1] indicate a positive autocorrelation (persistent series). In contrast, $H = 0.5$ indicates a series with independent data (without memory), and finally $H = 1$ implies a deterministic series. Furthermore, the Hurst coefficient, also known as the index of dependence, is a classical temporal indicator that is widely accepted in the hydrology community for assessing the memory of a basin [26,84]. This was performed following the general framework proposed by Salas et al. [24].

Subsequently, a set of 200 equiprobable synthetic runoffs were generated exclusively from T-RRM runoff outcomes, supported by a parsimonious and unconditioned ARMA (*1,1*). The procedure was implemented according to the method used by Molina et al. [20]. The synthetic series were obtained based on the classical and well-established scheme shown in the paper by Salas et al. [24]. In addition, the ARMA (*1,1*) model confers a high level of freedom for subsequent analysis by maintaining the *"a priori"* structure of consecutive relationships between years. This is extremely useful in seeking out non-trivial relationships [13,20,22]. Furthermore, according to Molina et al. [20] and Molina and Zazo [13,28], the "boundary effects" were avoided by using a warm-up period of 20 years. Under these circumstances, the whole process of generation of synthetic series was "non-conditioned".

Although ARMA models are widely used in hydrological modeling, a brief description of these models seems appropriate. It is well-known that an autoregressive moving average (ARMA) (*p,q*) model is the result of the union of two components or models. The first model, the autoregressive (AR) model ("*p*" part), is the temporal correlation of a time series [24], representing the temporal dependence delay within a series. The second component, the moving average (MA) model ("*q*" part), uses delays of the forecast errors to improve the process [20]. Mathematically speaking, an ARMA (*p,q*) has the following formulation [85]:

$$X_t = a_t + \sum_{j=1}^{p}\left(\varnothing_j \cdot X_{t-j}\right) - \sum_{j=1}^{q}\left(\theta_j \cdot a_{t-j}\right) , \tag{13}$$

where X_t is the value of the variable at a certain time step t; p is the number of autoregressive parameters; q is the number of moving average parameters; $\varnothing_j$ and θ_j are the coefficient of autoregressive and moving average model, respectively; and a_t is a random variable that represents the historical residuals (error term).

Subphase 2b Bayesian Causal Modeling (BCM) Design

Firstly, it should be pointed out that BCM is a type of probabilistic graphical model based on a direct acyclic graph (DAG) that implements the theorem of Bayes throughout a network (named a Bayesian network), whereby conditional probability is propagated omnidirectionally [22]. In this sense, BNs, which were first proposed by Pearl [86], are types of probabilistic reasoning networks based on Bayesian conditional probability and graph theory [87]. Furthermore, they offer a compact representation of the joint probability distribution over sets of random variables [88,89]. Indeed, they are the most commonly used types of PGMs [90]. Formally, a BN $N = (G, P)$ defines a joint probability distribution $P(V)$ over a set of nodes V (or variables) present in the graph. For each variable $X \in V$, there is a conditional probability distribution $P(X|pa(X))$ that belongs to a set of probability distributions. According to Madsen [91], this is expressed as:

$$P(V) = \prod_{X \in V} P(X|pa(X)) , \tag{14}$$

On the other hand, BCM is addressed here through a causal reasoning (CR) framework, which is characterized by searching reasoning patterns focused on the cause and where the objective is the prediction of the effect [78]. This kind of methodological approach is based on conditional probability queries, which are performed hierarchically from top to bottom [18]. In this way, BCM is performed over a set of random decision variables, known as "nodes", which are consecutively interconnected by means of "links", in addition to a set of conditional probability tables between the decision variables [86,92]. One of the main advantages of this methodological strategy is that it computes inference omnidirectionally throughout the BN [22]. Therefore, given an observation with any type of evidence for any of the network's nodes (or a subset of nodes), BNs have the ability to compute the posterior probabilities of all other nodes in the network, regardless of the arc direction, through observational inference [93]. This enables "dynamic quantification" of the strength of the relationships between these decision variables over time by assessing both the "*a priori*" and "*a posteriori*" probability distribution for each of them [88]. On this point, it is important to note that decision variables in this study are represented by hydrological years, as well as noting that each variable can influence the previous and following ones (trivial and natural relationships) through their dependence strength [18]. In essence, and according to Bayes´ theorem, this is formally expressed as [94]:

$$P(B|A) = \frac{P(A|B) \cdot P(B)}{P(A)}, \tag{15}$$

where $P(B|A)$ is the conditional probability of B for a given state of variable A; $P(A|B)$ is the inverse conditional probability; $P(B)$ is the probability of B; and $P(A)$ is the probability of A.

On the other hand, the BCM design, based on the design used by Madsen [91], involved the following key substeps: (1) the learning process involved 200 synthetic runoff series obtained from ARMA models; (2) the preprocessing process involved discretization of the synthetic data into discrete probability distributions by means of five intervals of same length; (3) the development of a structure constraints process, considering that the main "*a priori*" relationship among variables is natural, i.e., between consecutive years (hydrological years in this case). This method was performed using HUGIN Expert software version 7.3 [95]. In this way, the causal design respects the inherent temporal nature of the income information from the ARMA model, as well as its high degree of freedom.

2.4.3. Phase 3 Hybrid Causal–Hydrological (HCH) Modeling

This is the key phase in this study, as it combines the advantages offered by deterministic models with the analytical power from AI techniques based on PGMs. Furthermore, this highlights the suitability of the BCM for the dynamic and multitemporal analysis of the behavior of the water resources of a basin, which is especially useful in a context of lack of reliable observed data.

From the BCM design obtained in the previous phase, a structure learning process was developed, which is the real core of this methodological approach. This allows for discovering, extracting, and modeling the hidden, non-trivial, and logical interdependence temporal structures with time lag > 1 [22], which inherently underlies the hydrological series [18]. That hidden structure reveals the real and general temporal behavior of water resources in the recorded point.

The structure learning process was supported by a necessary path condition (NPC) algorithm. This is an improvement over the PC algorithm [91], which is similar to the inductive causation (IC) algorithm [96] proposed by Bonissone et al. [97]. In this regard, it should be noted that the PC algorithm is based on conditional independence decisions [98]. Conceptually, and according to Madsen [91], in a DAG dataset, the NPC is based on the principle that "in order for two variables X and Y to be independently conditional on a set S and no subset S′ ⊂ S, a path must exist between X and every Z ∈ S (not crossing Y) and between Y and every Z ∈ S (not crossing X). Otherwise, the inclusion of each Z in S is unexplained". This is explained here by the fact that all variables are considered a-priori-dependent in this research. The structure learning process significance level was established at a value of 0.05. Given that an exhaustive explanation of the NPC algorithm is not the objective of this research, please refer to Madsen [91] for a complete explanation.

Afterwards, a temporal dependence process was carried out through a dependence analysis of the relative percentage of change in runoff that a time lag produces on the following variables (throughout the network). According to Molina et al. [20], this can be done by means of the maximization of the highest interval of the "*a priori*" discrete probability distribution of the analyzed decision variable and the observation of its impact across the BN. According to this approach, large modifications are associated with strong dependence between variables, even among non-consecutive ones (non-trivial relationships; in other words, time lag > 1), while slight modifications reveal weak relationships. In this way, for each decision variable a propagation curve (mitigation) of temporal dependence across the BN [17] can be obtained. The resulting set of curves was encompassed by two dependence propagation wrap-around functions, one positive (named "W-MAX") and one negative (named "W-MIN"), which were fitted to mathematical functions. Both final functions (W-MAX and W-MIN jointly) define the dependence mitigation graph (DMG), which summarizes the general temporal behavior of water resources at the analyzed point. In this sense, DMG is a novel qualitative approach for assessing the time dependence from a geometrical perspective, in a dynamic and continuous manner, against the classical, static, and punctual analysis that the correlogram technique offers. The analysis of the symmetry of the whole graph indicates dependent temporal behavior in the case of an asymmetric graph. In contrast, an independent behavior is indicated by symmetric graphs [17]. More details on the theoretical foundations of DMG can be found in [13,17,18,20,28].

On the other hand, it is also worth noting the number of causal models built in this research. This was mainly done in concordance with the input series considered for each reservoir (long and short ones), which define the multitemporal and homogeneous analysis. However, regarding the backward validation technique applied to calibrate the T-RRM (see Section 2.3), two causal models were built, which were focused on the time period ranging from 2000 to 2015. In this way, it was possible to validate both the reliability of the T-RRM and the suitability of the hybrid causal–hydrological (HCH) approach in order to overcome the well-known limitation regarding the availability of hydrological data. Therefore, nine (9) causal models were defined—three (3) for each sub-basin (reservoir), one over a long series (from 1951 to 2017) and the other two over a short series (from 2000 to 2015).

2.4.4. Phase 4 Assessment of the Basin's Hydrological Runoff Memory

This is the last phase of the research. It comprises the assessment of the hydrological memory, given the particular relevance of this aspect to the sustainability of the hydrological system, especially in the short term.

The hydrological memory was characterized in detail through an analysis of the temporal dependence strength over time using DMG and according to the temporal horizon of the behavior revealed through BCM. In this sense, DMG is a novel qualitative approach for assessing the temporal dependence of a hydrological series based exclusively on a geometrical perspective, in a dynamic and continuous manner against the classical, static, and punctual analysis that the correlogram technique offers [17].

Using the DMG analysis technique, the symmetry between the W-MAX and W-MIN functions leads to independent temporal behavior, while the asymmetric graph displays temporal dependence behavior, which is greater if the graph is more asymmetrical. Furthermore, DMG presents other important advantages compared to classical approaches such as correlograms by: (1) providing a clear temporal horizon of dependence defined by the convergence of W-MAX and W-MIN to 0 on X axis (time lags); (2) providing a joint analysis of short-, medium-, and long-term dependence; (3) highlighting the specific region of temporal behavior based on a detailed analysis of the W-MAX and W-MIN slope functions (which is especially useful in the case of temporal dependence basins). For a complete definition of the DMG approach, the reader is invited to refer to Molina et al. [18,20], Molina and Zazo [13], and Zazo et al. [17].

3. Results

3.1. *T-RRM Outputs*

The T-RRM model obtained the monthly runoff in natural regime of the studied sub-basin (Barbate, Celemín, and Almodóvar) reservoirs for the study period (1951–2017). Figure 5 shows the results of the backward validation technique and Table 2 displays the results of the calibration process for the maximum moisture (H_{max}), coefficient of runoff (C), maximum infiltration (I_{max}), and aquifer depletion curve (α) parameters. The modeled runoff fits very well with the observed data for the validation period (Figures 5 and 6).

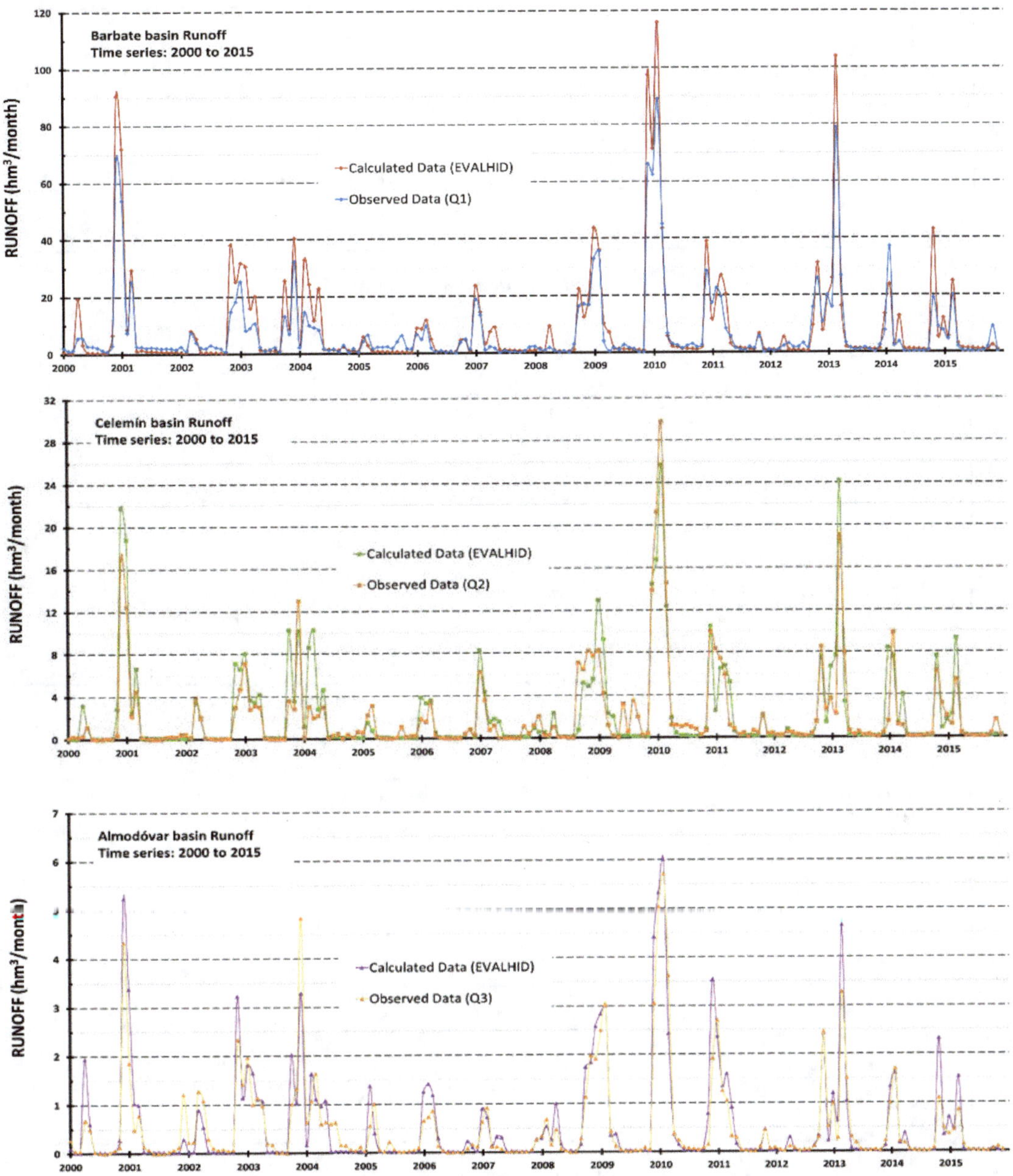

Figure 5. Flow hydrographs of calculated and observed contributions for the period 2000–2015.

Table 2. Parameters after T-RRM calibration.

Sub-Basin [1]/Code Number	H_{max}	C	I_{max}	α (Month^{-1})
/1	350 mm	0.4	20 mm	0.1
Barbate/2 Q1	280 mm	0.3	3 mm	0.1
/3	280 mm	0.3	3 mm	0.1
Celemín/4 Q2	280 mm	0.3	3 mm	0.1
Almodóvar/5 Q3	280 mm	0.3	3 mm	0.1

[1] Gauging control points (see Figure 4b).

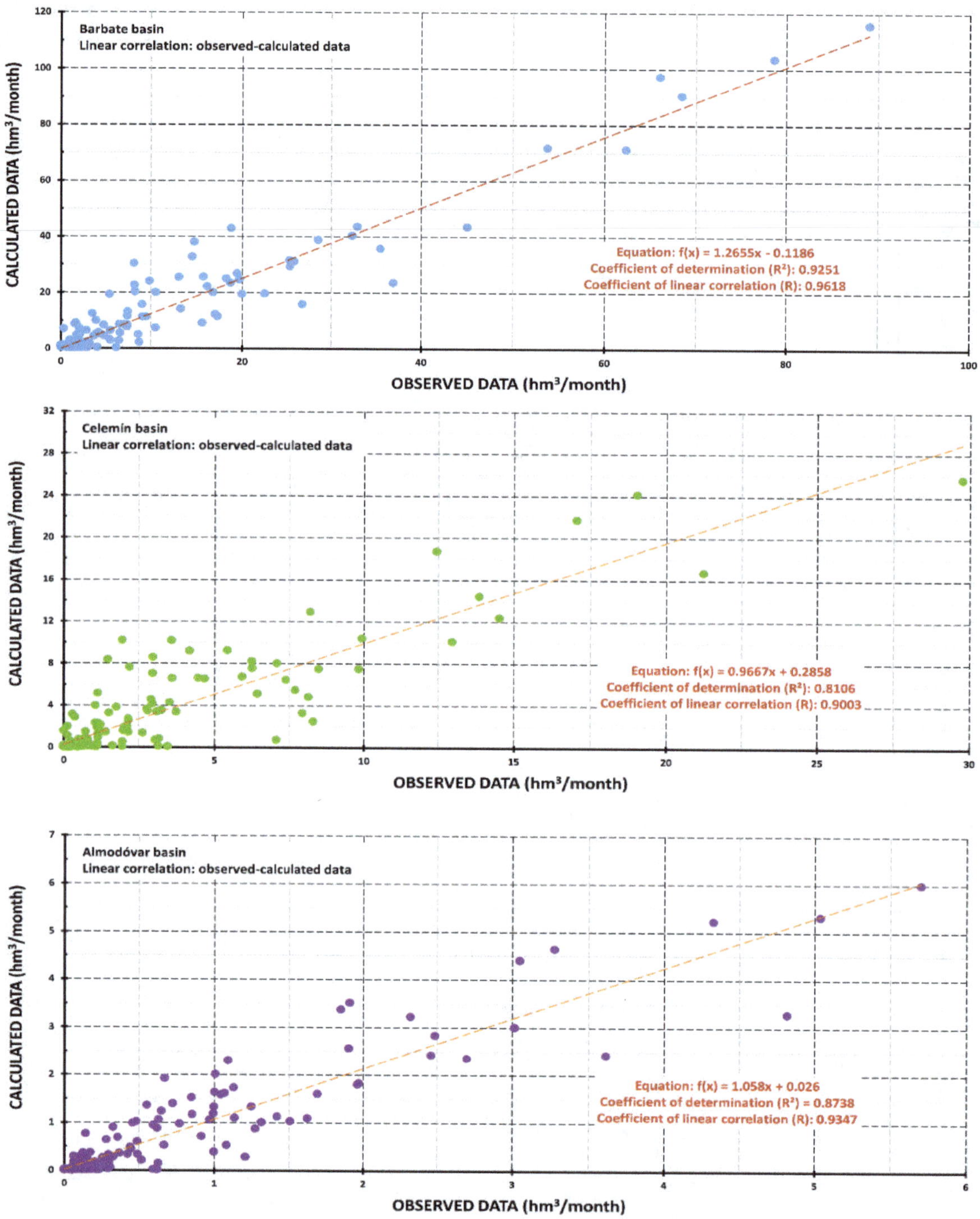

Figure 6. Linear correlation between calculated and observed data for the period 2000–2015.

An average difference between the estimated and the observed contributions of 3.3 Hm^3/month was detected in the case of the Barbate reservoir, of 1 Hm^3/month for Celemín, and of 0.2 Hm^3/month for the Almodóvar reservoir. Considering the maximum contribution of each reservoir, the average deviation was 2.9% for Barbate, 3.9% for Celemín, and 3.5% for Almodóvar.

Table 3 shows that the monthly average contribution values ranged between 0.5 and 10.8 Hm^3/month, implying annual values of between 6 and 130 Hm^3. The greatest contributions are from the Barbate sub-basin, owing to its physiography and more abundant rainfall, which enable greater surface runoff. On the contrary, the Almodóvar basin gave the lowest contributions, mainly due to the reduced surface area of its basin (16.6 Km^2, see Table 1).

Table 3. Average monthly and annual runoff values for the Barbate, Celemín, and Almodóvar reservoirs (period: 1951–2017).

Sub-Basins	Month												Average (Hm^3)	
	January	February	March	April	May	June	July	August	September	October	November	December	Monthly	Annual
Barbate	25.0	24.5	18.8	10.5	5.5	1.2	1.0	0.9	0.9	4.2	11.3	25.2	10.8	129.1
Celemín	6.7	6.3	5.0	2.3	1.1	0.2	0.1	0.1	0.1	0.8	2.8	6.3	2.7	31.8
Almodóvar	1.1	1.2	0.9	0.5	0.2	0.0	0.0	0.0	0.0	0.2	0.6	1.3	0.5	6.2

Figure 6 shows the regression adjustments and provides linear correlation coefficients (R) ranging between 0.90003 and 0.9618, with determination coefficients (R^2) ranging between 0.8106 and 0.92514.

Finally, Figure 7 displays the annual flow hydrographs for the three studied sub-basins. The storage capacity of the Barbate reservoir is 228 Hm^3, for Celemín it is 45 Hm^3, and for the Almodóvar reservoir it is 5.7 Hm^3. Considering the average contributions, it can be deduced that the only infrastructure with multiannual regulation capacity within the basin (approximately 2 years of storage) is the Barbate reservoir. Moreover, for 11 of the 68 years of study (16%), the inputs exceeded the storage capacity. This situation accounts for 26% of the years in the case of Celemín and 46% for Almodóvar, showing the reduced capacity of the latter. In addition, the hydrographs also show pronounced dry periods. The first took place in the 1950s, while the two most extreme periods in terms of length and volume took place in the 1990s and in the 2000s. During these decades, there were also contribution peaks, which evidence greater irregularity in the precipitation events during the last decades.

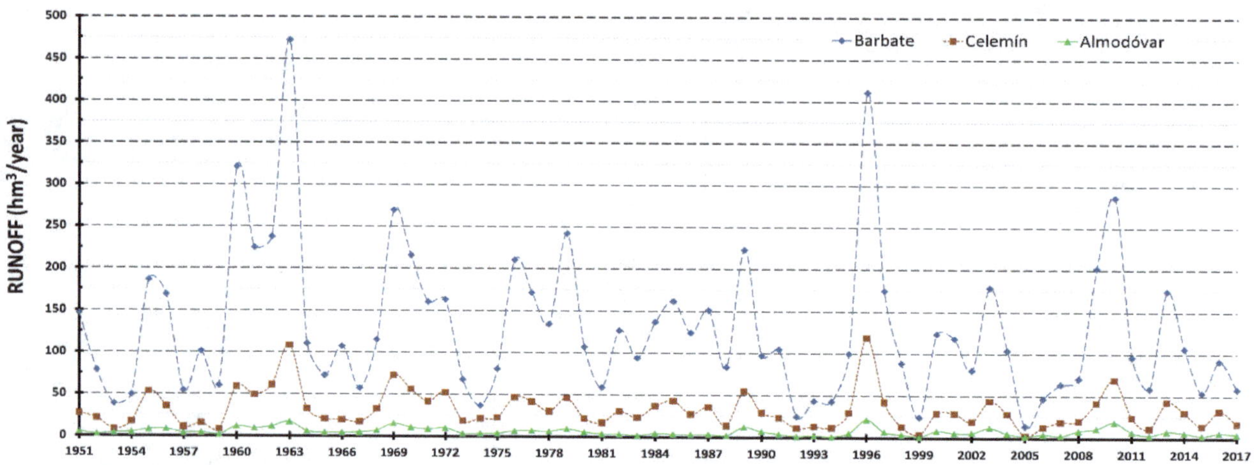

Figure 7. Annual runoff time series from the T-RRM (time period ranging from 1951 to 2017).

3.2. Stochastic Module: Statistical Parameters and Design of Bayesian Causal Modeling (BCM)

Table 4 shows the main statistical parameters of the considered time series (long and short series), as well as the Hurst coefficient as a dependence indicator. It can be clearly observed that the ARMA (*1,1*) model is able to preserve the main statistical parameters of historical time series. Equally remarkable are the differences in the statistical parameters of short series (obtained from the T-RRM and observed records) in the case of the Barbate sub-basin (mean: 110.98 Hm^3 and 88.82 Hm^3; standard deviation: 69.92 Hm^3 and 56.08 Hm^3), which is in agreement with the results shown in the previous section.

Regarding the Hurst coefficient obtained from the T-RRM time series, in the case of the long series (1951–2017), all sub-basins had a value of 0.66, while the short ones (2000–2015), for which the results were less homogeneous, were in the range of [0.61,0.69] (0.65 Barbate/Q1; 0.61 Celemín/Q2; 0.69 Almodóvar/Q3). However, the average value ($H = 0.65$) was practically the same. In contrast, this observed trend was not observed for the gauging data (short series exclusively). In this case, the variability of result was larger ([0.57, 0.77]; 0.74 Barbate/Q1; 0.77 Celemín/Q2; 0.57 Almodóvar/Q3), although the average value was similar to the other two ($H = 0.69$ versus 0.66 and 0.65, respectively). Furthermore, it should be noted that all H coefficients display values greater than 0.50, implying a positive correlation and a persistent trend (long-term memory).

On the other hand, Figure 8 summarizes the conceptual scheme of the design of BCM using HUGIN© software, which may also be seen as a result in itself. On the left side, the learning and preprocessing processes are shown. Here, the synthetic data were discretized into five intervals of the same length. On the other hand, the right side shows the developed hierarchical structure from top to bottom (initial to final year). Here, each decision variable is connected in such a way that it can influence the previous and following one in a natural way (trivial relationships). This defines the structure constraints process, in which the main *"a priori"* relationship among variables is considered as the natural behavior, i.e., between consecutive years. Subsequently, and by means of the analysis of the *"a posteriori"* probability distributions, non-trivial dependence relationships (time lag > 1) were extracted, owing to the power of analysis that CR supported by DMG offers. This information is implicitly present in hydrological data.

Table 4. Main statistical parameters applied in the development of the 9 causal models.

	Sub-Basins [1]	**Parameters**	**T-RRM Results**	**Average SS [3]**
Long series: From 1951 to 2017	Barbate/Q1	Mean:	129.10 Hm^3	129.71 Hm^3
		Standard deviation:	87.78 Hm^3	93.87 Hm^3
		Skewness coefficient [2]:	1.64	1.83
		Variation coefficient:	68%	72%
		Hurst coefficient:	0.66	——
	Celemín/Q2	Mean:	31.77 Hm^3	31.89 Hm^3
		Standard deviation:	21.47 Hm^3	22.39 Hm^3
		Skewness coefficient [2]:	1.84	1.73
		Variation coefficient:	68%	70%
		Hurst coefficient:	0.66	——
	Almodóvar/Q3	Mean:	6.22 Hm^3	6.16 Hm^3
		Standard deviation:	4.14 Hm^3	3.92 Hm^3
		Skewness coefficient [2]:	1.54	1.42
		Variation coefficient:	67%	64%
		Hurst coefficient:	0.66	——
	Sub-basins [1]	**Parameters**	**T-RRM results**	**Average SS [3]**
Short series: From 2000 to 2015	Barbate/Q1	Mean:	110.98 Hm^3	111.71 Hm^3
		Standard deviation:	69.62 Hm^3	62.46 Hm^3
		Skewness coefficient [2]:	1.13	0.58
		Variation coefficient:	63%	56%
		Hurst coefficient:	0.66	——
	Celemín/Q2	Mean:	27.50 Hm^3	27.76 Hm^3
		Standard deviation:	16.10 Hm^3	15.62 Hm^3
		Skewness coefficient [2]:	1.06	0.70
		Variation coefficient:	59%	56%
		Hurst coefficient:	0.61	——
	Almodóvar/Q3	Mean:	7.05 Hm^3	7.25 Hm^3
		Standard deviation:	4.24 Hm^3	4.23 Hm^3
		Skewness coefficient [2]:	1.45	1.09
		Variation coefficient:	60%	58%
		Hurst coefficient:	0.69	——
	Sub-basins [1]	**Parameters**	**Gauging data**	**Average SS [3]**
Short series (Observed data): From 2000 to 2015	Barbate/Q1	Mean:	88.82 Hm^3	87.83 Hm^3
		Standard deviation:	56.08 Hm^3	52.16 Hm^3
		Skewness coefficient [2]:	1.69	1.28
		Variation coefficient:	63%	59%
			0.74	——
	Celemín/Q2	Mean:	26.58 Hm^3	28.08 Hm^3
		Standard deviation:	18.54 Hm^3	20.48 Hm^3
		Skewness coefficient [2]:	1.89	1.54
		Variation coefficient:	70%	73%
			0.77	——
	Almodóvar/Q3	Mean:	6.71 Hm^3	6.53 Hm^3
		Standard deviation:	4.30 Hm^3	4.13 Hm^3
		Skewness coefficient [2]:	1.14	1.11
		Variation coefficient:	64%	63%
		Hurst coefficient:	0.57	——

[1] Gauging Control Point (see Figure 4b). [2] Classic skewness coefficient. [3] Synthetic series from ARMA (1,1).

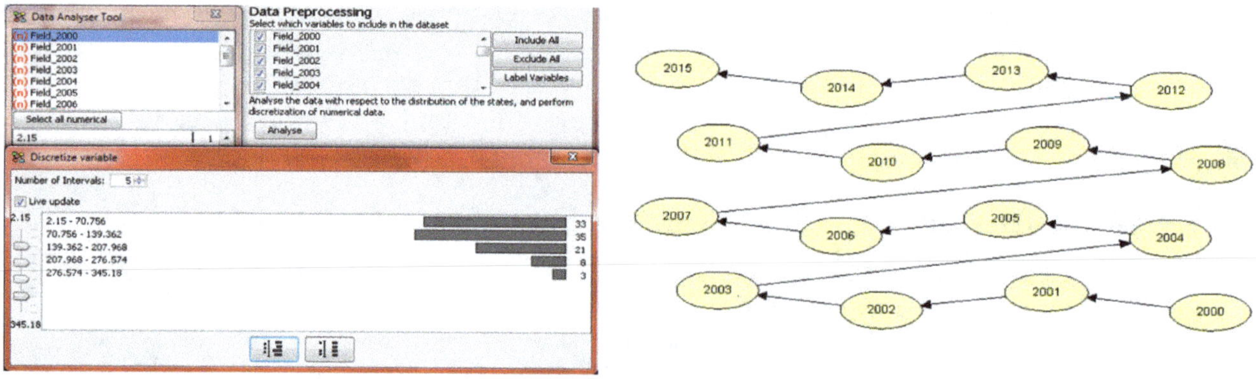

Figure 8. Conceptual design of BCM. (**Left**) Schemes of learning, preprocessing, and structure constraints processes. (**Right**) Hierarchical structure from top to bottom (from initial year to final one). Note: This scheme was replicated in each of the 9 causal models. For this reason, just one causal model is shown (gauging records for Barbate Q1 from 2000 to 2015).

3.3. Runoff Basin Memory Assessment through Hybrid Causal–Hydrological (HCH) Modeling

Given the availability of both the T-RRM results and gauging data and inspired by the backward validation technique, the temporal behavior was validated. This was possible thanks to the qualitative approach that DMGs offer and was focused on the short series (time period ranging from 2000 to 2015). Furthermore, both a reliability analysis of the T-RRM results and a suitability analysis of the hybrid causal–hydrological (HCH) approach were performed. Figure 9 shows a comparative analysis of the results based on both DMGs. In this sense, it is worth noting that the determination coefficients (R^2) of the resulting mathematical functions were almost 1.00 (0.99 in all cases), demonstrating the robustness of the adjustment process.

In general, all the graphs present important asymmetry results, with the minimum result being obtained for the Celemín sub-basin from the T-RRM (time lag = 0 127.27 versus −161.72; see Figure 9b), which is highlighted through a dominant wrap-around function (basically W-MAX). In addition, the behavior trends are maintained (convergence to 0 on X axis-temporal horizon), with a practically equal dependence propagation range. The difference in the Almodóvar sub-basin (see Figure 9c, T-RRM [0, 3] versus gauging [0, 4] ranges) is not significant, because the relative percentage of change for time lag 3 is practically 0 in both cases.

Although the DMGs show differences in absolute values, (1) the relationship between W-MAX and W-MIN, (2) the temporal horizons, and (3) the range of relative percentage change are homogeneous and essentially maintain the observed trends. In particular, in the case of the Barbate sub-basin (Figure 9a), the relationship between W-MAX and W-MIN remains practically constant (3.3/1 versus 3.1/1; T-RRM = [+500.21, −150.90]; gauging records = [+861.50, −276.89]), being equal to the temporal horizon ([0, 4]). In the Celemín sub-basin (Figure 9b), the dependence propagation range is the same ([0, 3]), as well as the pattern of mitigation, with a strong dependence on the interval [0, 1] and symmetry between W-MAX and W-MIN in the remaining time lags. Finally, the Almodóvar sub-basin (Figure 9c) is the case with the lowest difference. Here, even the range is roughly the same (T-RRM: 740.99 versus 686.72 from gauging records).

Therefore, a general analysis shows a more than reasonable concordance between the resulting DMGs (Figure 9), which validates both the T-RRM results and the suitability of HCH approach.

Figure 10 shows the results based on long series. As in the case of the short series, all DMGs display asymmetric graphs with distinctly dominant wrap-around functions (basically W-MAX versus W-MIN), except in the case of the Celemin sub-basin, where this general trend is inversed and the graph shows a certain symmetry (see Figure 9b), thus evidencing clearly dependent behavior. In addition, here all determination coefficients (R^2) for W-MAX and W-MIN mathematical functions (polynomial in all cases) show values close to 1.00, with 0.98 being the lowest value (Figure 10a), demonstrating the excellent fit of the mathematical functions.

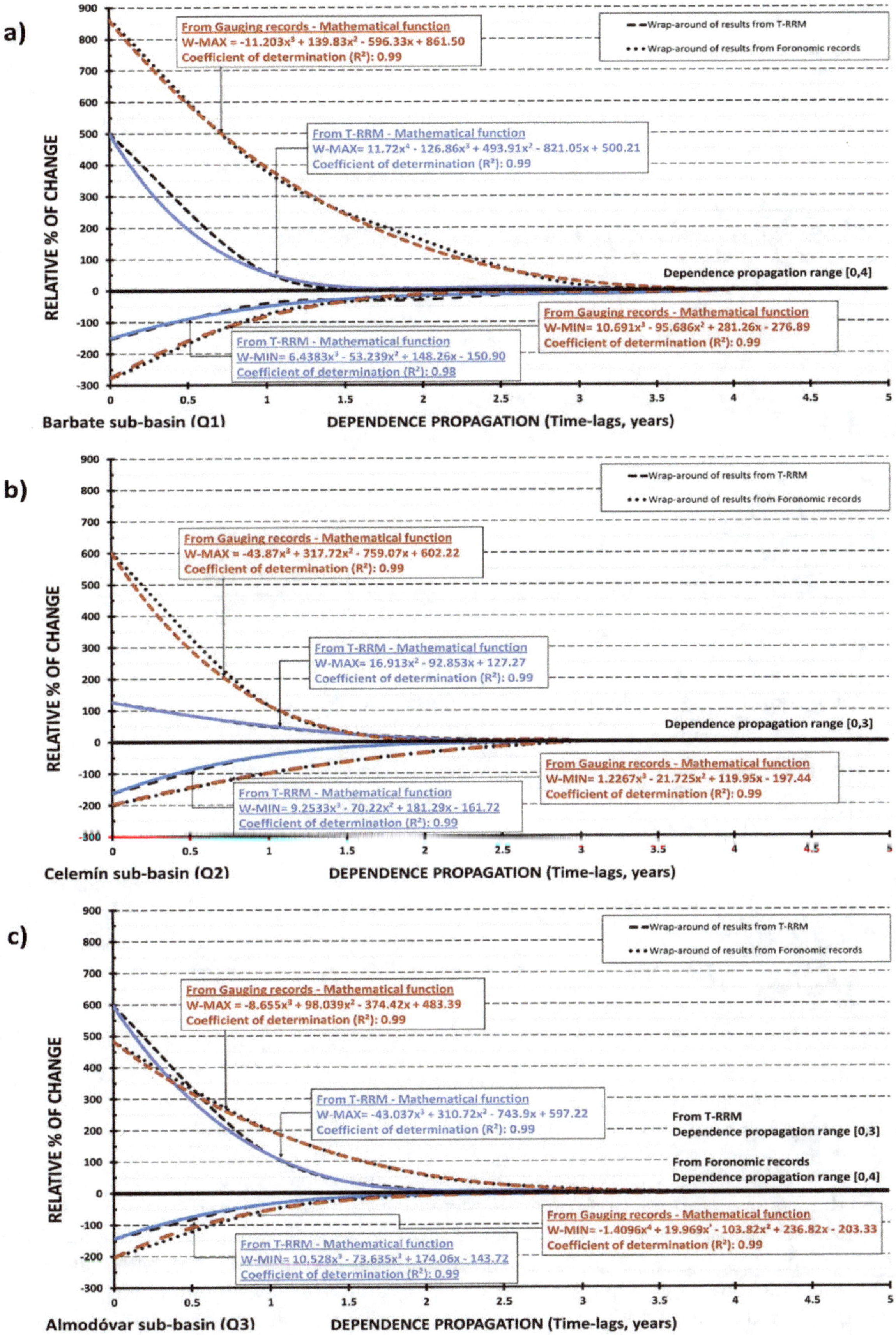

Figure 9. Dependence mitigation graphs (DMGs) for a short series (2000–2015). Comparison of causal model results from T-RRM versus gauging records: (**a**) Barbate Q1 sub-basin; (**b**) Celemín Q2 sub-basin; (**c**) Almodóvar Q3 sub-basin).

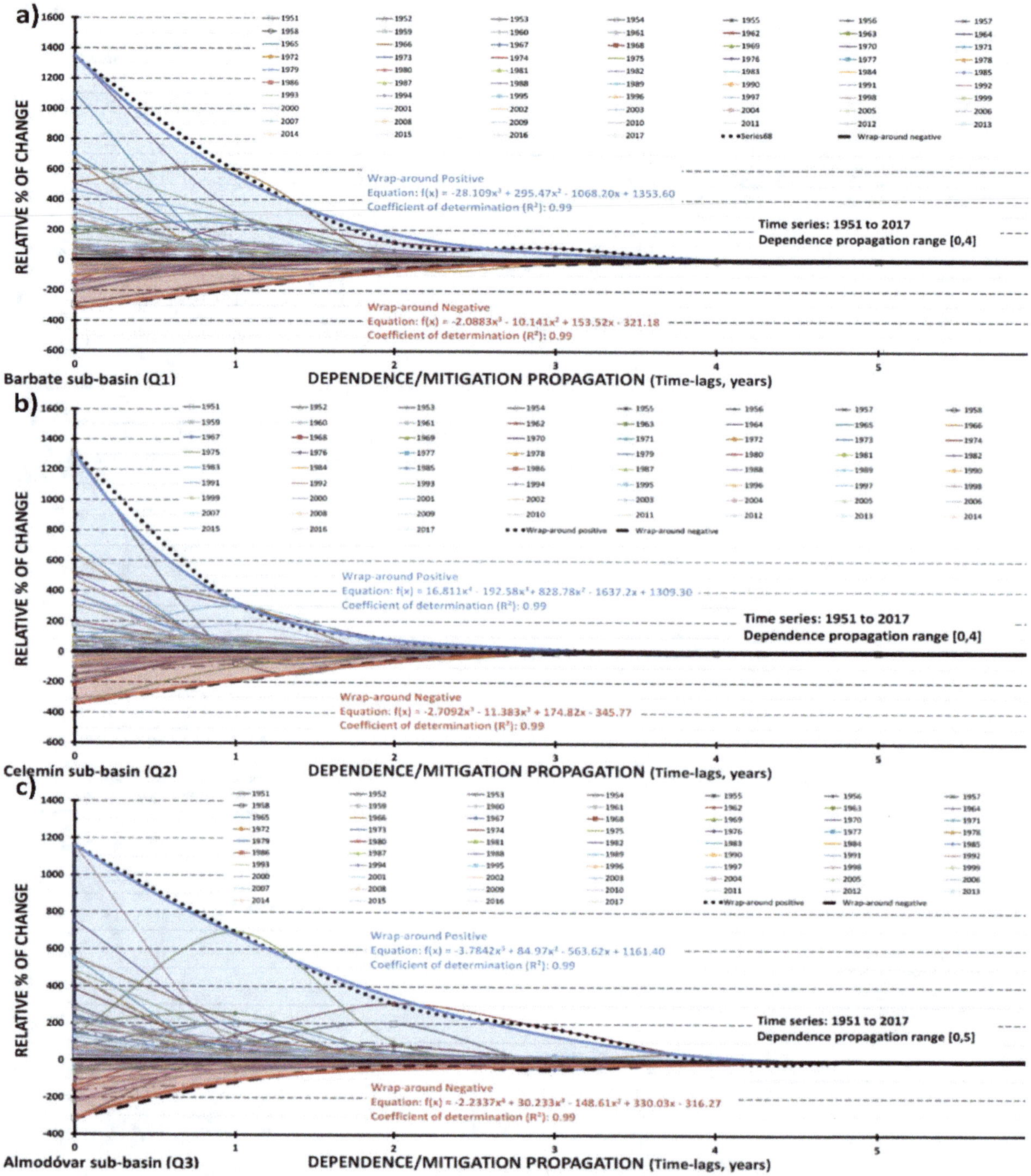

Figure 10. Dependence mitigation graphs (DMGs) for a long series (1951–2017) using the T-RRM: (**a**) Barbate Q1 sub-basin; (**b**) Celemín Q2 sub-basin; (**c**) Almodóvar Q3 sub-basin.

As mentioned above, the convergence of all series and W-MAX and W-MIN to 0 on the X axis (time lags) defines the temporal horizon of the dependence influence. In the case of the long series, the temporal horizons are mainly focused in the short term, with dependence propagation values ranging between [0, 4] (Figure 10a,b) and [0, 5] (Figure 10c). In this case, all W-MIN values present a practically constant convergence trend, whilst W-MAX values display different behavior.

Similarly, based on the DMG gradient, two regions can be clearly observed—the first region, where there is rapid mitigation (high gradient) and therefore greater dependence; and the second region, in which the gradient is lower, characterized by a gradual dissipation or mitigation of the dependence up to a relative percentage of change of 0.

Barbate Q1 and Celemín Q2 sub-basins (Figure 10a,b) present a similar pattern of behavior, both with a temporal horizon of four (4) years (time lags), whereby the first region of rapid mitigation is located in time lags [0, 1] and the second between [1,4]. In contrast, the Almodóvar Q3 sub-basin (Figure 10c) displays slightly different behavior, with a temporal horizon of up to five (5) years, a high-gradient region situated in the interval [0, 2], and a low-gradient region between [2,5].

Generally, the temporal patterns in all cases are clearly dependent; in line with the values of H coefficients, which reveal a high degree of dependence. This is mainly focused on the [0, 1] interval, which has a similar range for the relative percentage of change (maximum 1353.60 and minimum −345.77) and an average relationship that is four times greater for W-MAX than W-MIN (relationship 4:1).

Furthermore, the analytical power and suitability of BCM for discovering, extracting, and modeling hidden relationships is highlighted through the two regions of the DMGs (based on an analysis of the gradient). BCM was used to define a short qualitative indicator of behavior (general) and also a very short-term indicator through the greater influence region (particular). This characteristic is especially useful in the current context of the increasing variability of water resources.

4. Discussion and Conclusions

The effective inclusion of uncertainty in hydrological modeling is a challenging topic. There have been attempts at doing this before, but this is still a challenge that must be overcome. In this context, it seems obvious and rational to merge physical rainfall–runoff models with stochastic developments.

In this sense, this paper presents the hybridization process of traditional hydrological rainfall–runoff modeling (T-RRM) with hydrological Bayesian causal modeling (BCM), which is a prominent, innovative, and productive research field. This research work has generated a tool that the authors have termed the hybrid causal–hydrological" (HCH) method, which can provide the hydrological response of a basin and can stochastically characterize the studied hydrological processes.

This research paper has addressed the methodological process involved in this combination of models and produced results through its implementation in the Barbate River Basin, located in SW Spain. Regarding the first point, the aforementioned iterative process was technically satisfactory, given the outcome obtained from the HCH tool. Regarding the results, from a hydrological perspective, in the very short term (within the first year lag), the long series obtained from the T-RRM model had more dependent hydrological behavior, as the relative percentage of change in the DMG was higher than for the short series generated from the T-RRM model and gauging (observed) records. The gauging data had more dependent behavior than the short series produced by the T-RRM, except for in the Almodóvar sub-basin. This can be attributed to several factors, such as the insufficient amount of consistent data in the hydrological records and climate change processes, which might alter the general hydrological behavior of a basin when a long record period is analyzed. The results for the temporal propagation of the dependence (hydrological memory) are quite similar for the three analyses (long series, short series, and gauging series) as well as for the three sub-basins. In addition, the symmetry or asymmetry of the DMGs remains quite constant across the nine causal models. This outcome should be seen as a validation itself of the usefulness and reliability of the T-RRM and the hydrological causal modeling based on BCM, and therefore of the new HCH tool.

Questions remain regarding the optimal dimensioning of hydraulic infrastructure and efficient water resource planning, for which management approaches can successfully assist in providing dynamic and continuous analysis of the temporal dependence of a basin´s runoff.

There are upcoming research topics based on this work, one of which is especially important given the similarity of the topic to the one addressed here. This topic involves the multitemporal analysis of hydrological behavior across river basins. Ingeniería y Gestión del Agua (IGA) Research Group and their associates are currently studying the optimum parameters from hydrological series, such as the length of the analysis period or the volume of data required for the BCM.

Author Contributions: S.Z., S.G.-L., and J.-L.M. conceived and designed the research. S.Z. led the research. V.R.-O. and M.V.-N. performed the T-RRM process. S.G.-L., V.R.-O., and M.V.-N. analyzed the data from deterministic models. S.Z. and J.-L.M. analyzed the data from stochastic models. The Discussion and Conclusions sections were addressed by all authors, and all authors wrote the paper. All authors have read and agreed to the published version of the manuscript.

Acknowledgments: This work was partially developed within the framework of the REMABAR project, supported by the Biodiversity Foundation of the Ministry for Ecological Transition. The authors thank the Universities of Cádiz and Salamanca for the facilities made available in order to jointly undertake this research.

References

1. Botai, J.O.; Botai, C.M.; de Wit, J.P.; Muthoni, M.; Adeola, A.M. Analysis of Drought Progression Physiognomies in South Africa. *Water* **2019**, *11*, 299. [CrossRef]
2. Jyrkama, M.I.; Sykes, J.F. The Impact of Climate Change on Spatially Varying Groundwater Recharge in the Grand River Watershed (Ontario). *J. Hydrol.* **2007**, *338*, 237–250. [CrossRef]
3. Lepori, F.; Pozzoni, M.; Pera, S. What Drives Warming Trends in Streams? A Case Study from the Alpine Foothills. *River Res. Appl.* **2015**, *31*, 663–675. [CrossRef]
4. Praskievicz, S. Impacts of Projected Climate Changes on Streamflow and Sediment Transport for Three Snowmelt-Dominated Rivers in the Interior Pacific Northwest. *River Res. Appl.* **2016**, *32*, 4–17. [CrossRef]
5. Zazo, S.; Molina, J.L.; Rodriguez-Gonzalvez, P. Analysis of Flood Modeling through Innovative Geomatic Methods. *J. Hydrol.* **2015**, *524*, 522–537. [CrossRef]
6. Berg, P.; Moseley, C.; Haerter, J.O. Strong Increase in Convective Precipitation in Response to Higher Temperatures. *Nat. Geosci.* **2013**, *6*, 181–185. [CrossRef]
7. Trenberth, K.E. Changes in Precipitation with Climate Change. *Clim. Res.* **2011**, *47*, 123–138. [CrossRef]
8. O'Gorman, P.A. Precipitation Extremes under Climate Change. *Curr. Clim. Chang. Rep.* **2015**, *1*, 49–59. [CrossRef]
9. Pfahl, S.; O'Gorman, P.A.; Fischer, E.M. Understanding the Regional Pattern of Projected Future Changes in Extreme Precipitation. *Nat. Clim. Chang.* **2017**, *7*, 423–427. [CrossRef]
10. Chávez Jiménez, A. Propuesta Metodológica Para la Identificación de Medidas de Adaptación al Cambio Climático en Sistemas de Recursos Hídricos. Ph.D. Thesis, Universidad Politécnica de Madrid, Madrid, Spain, 2013; 232p.
11. Merritt, W.S.; Alila, Y.; Barton, M.; Taylor, B.; Cohen, S.; Neilsen, D. Hydrologic Response to Scenarios of Climate Change in Sub Watersheds of the Okanagan Basin, British Columbia. *J. Hydrol.* **2006**, *326*, 79–108. [CrossRef]
12. Pulido-Velazquez, D.; Luis Garcia-Arostegui, J.; Molina, J.; Pulido-Velazquez, M. Assessment of Future Groundwater Recharge in Semi-Arid Regions under Climate Change Scenarios (Serral-Salinas Aquifer, SE Spain). Could increased Rainfall Variability Increase the Recharge Rate? *Hydrol. Process.* **2015**, *29*, 828–844. [CrossRef]
13. Molina, J.L.; Zazo, S. Assessment of Temporally Conditioned Runoff Fractions in Unregulated Rivers. *J. Hydrol. Eng.* **2018**, *23*, 04018015. [CrossRef]
14. Voss, R.; May, W.; Roeckner, E. Enhanced Resolution Modelling Study on Anthropogenic Climate Change: Changes in Extremes of the Hydrological Cycle. *Int. J. Clim.* **2002**, *22*, 755–777. [CrossRef]
15. Marcos, P.; Lopez-Nicolas, A.; Pulido-Velazquez, M. Analysis of Climate Change Impact on Meteorological and Hydrological Droughts through Relative Standardized Indices. In Proceedings of the 19th EGU General Assembly, Vienna, Austria, 23–28 April 2017; p. 1391.
16. Molina, J.L.; Zazo, S.; Martín-Casado, A.; Patino-Alonso, M. Rivers' Temporal Sustainability through the Evaluation of Predictive Runoff Methods. *Sustainability* **2020**, *12*, 1720. [CrossRef]
17. Zazo, S.; Macian-Sorribes, H.; Sena-Fael, C.M.; Martín-Casado, A.M.; Molina, J.L.; Pulido-Velazquez, M. Qualitative Approach for Assessing Runoff Temporal Dependence through Geometrical Symmetry. In Proceedings of the ICEUBI 2019 International Congress on Engineering, Covilhã, Portugal, 27–29 November 2019.
18. Molina, J.L.; Zazo, S.; Martín-Casado, A.M. Causal Reasoning: Towards Dynamic Predictive Models for Runoff Temporal Behavior of High Dependence Rivers. *Water* **2019**, *11*, 877. [CrossRef]
19. Vogel, K.; Weise, L.; Schroeter, K.; Thieken, A.H. Identifying Driving Factors in Flood-Damaging Processes using Graphical Models. *Water Resour. Res.* **2018**, *54*, 8864–8889. [CrossRef]

20. Molina, J.L.; Zazo, S.; Rodriguez-Gonzalvez, P.; Gonzalez-Aguilera, D. Innovative Analysis of Runoff Temporal Behavior through Bayesian Networks. *Water* **2016**, *8*, 484. [CrossRef]
21. Molina, J.L.; Garcia-Arostegui, J.L.; Bromley, J.; Benavente, J. Integrated Assessment of the European WFD Implementation in Extremely Overexploited Aquifers through Participatory Modelling. *Water Resour. Manag.* **2011**, *25*, 3343–3370. [CrossRef]
22. Zazo, S. Analysis of the Hydrodynamic Fluvial Behaviour through Causal Reasoning and Artificial Vision. Ph.D. Thesis, University of Salamanca, Ávila, Spain, 12 May 2017.
23. Hurst, H.E. Long-Term Storage Capacity of Reservoirs. *Trans. Am. Soc. Civ. Eng.* **1951**, *116*, 770–808.
24. Salas, J.; Delleur, J.; Yevjevich, V.; Lane, W.L. *Applied Modeling of Hydrologic Time Series*, 1st ed.; Water Resources Publications: Littleton, CO, USA, 1980; p. 484.
25. Díaz Caballero, F.F. *Selección de Modelos Mediante Criterios de Información en Análisis Factorial: Aspectos Teóricos Y Computacionales*; Editorial de la Universidad de Granada: Granada, Spain, 2011.
26. Tyralis, H.; Koutsoyiannis, D. Simultaneous Estimation of the Parameters of the Hurst-Kolmogorov Stochastic Process. *Stoch. Environ. Res. Risk Assess.* **2011**, *25*, 21–33. [CrossRef]
27. Farmer, W.H.; Vogel, R.M. On the Deterministic and Stochastic use of Hydrologic Models. *Water Resour. Res.* **2016**, *52*, 5619–5633. [CrossRef]
28. Molina, J.L.; Zazo, S. Causal Reasoning for the Analysis of Rivers Runoff Temporal Behavior. *Water Resour. Manag.* **2017**, *31*, 4669–4681. [CrossRef]
29. Koutroumanidis, T.; Sylaios, G.; Zafeiriou, E.; Tsihrintzis, V.A. Genetic Modeling for the Optimal Forecasting of Hydrologic Time-Series: Application in Nestos River. *J. Hydrol.* **2009**, *368*, 156–164. [CrossRef]
30. Molina, J.L.; Bromley, J.; Garcia-Arostegui, J.L.; Sullivan, C.; Benavente, J. Integrated Water Resources Management of Overexploited Hydrogeological Systems using Object-Oriented Bayesian Networks. *Environ. Model. Softw.* **2010**, *25*, 383–397. [CrossRef]
31. Zhang, H.; Wang, B.; Liu, D.L.; Zhang, M.; Feng, P.; Cheng, L.; Yu, Q.; Eamus, D. Impacts of Future Climate Change on Water Resource Availability of Eastern Australia: A Case Study of the Manning River Basin. *J. Hydrol.* **2019**, *573*, 49–59. [CrossRef]
32. Versini, P.-A.; Pouget, L.; McEnnis, S.; Custodio, E.; Escaler, I. Climate Change Impact on Water Resources Availability: Case Study of the Llobregat River Basin (Spain). *Hydrol. Sci. J.* **2016**, *61*, 2496–2508. [CrossRef]
33. Giupponi, C.; Sgobbi, A. Decision Support Systems for Water Resources Management in Developing Countries: Learning from Experiences in Africa. *Water* **2013**, *5*, 798–818. [CrossRef]
34. Sivapalan, M.; Savenije, H.H.G.; Bloeschl, G. Socio-Hydrology: A New Science of People and Water. *Hydrol. Process.* **2012**, *26*, 1270–1276. [CrossRef]
35. Liu, D.; Tian, F.; Lin, M.; Sivapalan, M. A Conceptual Socio-Hydrological Model of the Co-Evolution of Humans and Water: Case Study of the Tarim River Basin, Western China. *Hydrol. Earth Syst. Sci.* **2015**, *19*, 1035–1054. [CrossRef]
36. Aqil, M.; Kita, I.; Yano, A.; Nishiyama, S. A Comparative Study of Artificial Neural Networks and Neuro-Fuzzy in Continuous Modeling of the Daily and Hourly Behaviour of Runoff. *J. Hydrol.* **2007**, *337*, 22–34. [CrossRef]
37. USACE. US Army Corps of Engineers. Hydrologic Engineering Center. Available online: https://www.hec.usace.army.mil/software/hec-hms/ (accessed on 10 August 2020).
38. MITECO Ministerio para la Transición Ecológica. Modelo SIMPA 2019. Periodo de Simulación:1940/41 a 2017/18. Available online: https://www.miteco.gob.es/es/agua/temas/evaluacion-de-los-recursos-hidricos/evaluacion-recursos-hidricos-regimen-natural/ (accessed on 26 September 2019).
39. Graf, R.; Zhu, S.; Sivakumar, B. Forecasting River Water Temperature Time Series using a Wavelet-Neural Network Hybrid Modelling Approach. *J. Hydrol.* **2019**, *578*, 124115. [CrossRef]
40. Jayakrishnan, R.; Srinivasan, R.; Santhi, C.; Arnold, J.G. Advances in the Application of the SWAT Model for Water Resources Management. *Hydrol. Process.* **2005**, *19*, 749–762. [CrossRef]
41. Koutsoyiannis, D. Hydrology, Society, Change and Uncertainty. *Geophys. Res.* **2014**, *16*, 1–31. [CrossRef]
42. Zounemat-Kermani, M.; Teshnehlab, M. Using Adaptive Neuro-Fuzzy Inference System for Hydrological Time Series Prediction. *Appl. Soft Comput.* **2008**, *8*, 928–936. [CrossRef]
43. Nourani, V.; Kisi, O.; Komasi, M. Two Hybrid Artificial Intelligence Approaches for Modeling Rainfall-Runoff Process. *J. Hydrol.* **2011**, *402*, 41–59. [CrossRef]
44. Molina, J.L.; Pulido-Velazquez, D.; Garcia-Arostegui, J.-L.; Pulido-Velazquez, M. Dynamic Bayesian Networks as a Decision Support Tool for Assessing Climate Change Impacts on Highly Stressed Groundwater Systems. *J. Hydrol.* **2013**, *479*, 113–129. [CrossRef]

45. Spirtes, P. Introduction to Causal Inference. *J. Mach. Learn. Res.* **2010**, *11*, 1643–1662.
46. Témez, J.R. *Modelo Matemático de Trasformación "Precipitación-Escorrentía". Comisión de "Explotación y Garantía". Grupo de Trabajo de Predicción de Precipitaciones y Relación Entre Precipitaciones y Caudales*; Asociación de Investigación Industrial Eléctrica (ASINTEL): Madrid, Spain, 1977.
47. Témez, J.R. *Cálculo Hidrometeorológico de Caudales Máximos en Pequeñas Cuencas Naturales*; Ministerio de Obras Públicas y Urbanismo; Gobierno de España: Madrid, Spain, 1978.
48. Témez, J. Extended and Improved Rational Method. Version of the Highways Administration of Spain. In Proceedings of the XXIV Congress IAHR, Madrid, Spain, 9–13 September 1991; pp. 33–40.
49. Escriva-Bou, A.; Pulido-Velazquez, M.; Pulido-Velazquez, D. Economic Value of Climate Change Adaptation Strategies for Water Management in Spain's Jucar Basin. J. *Water Resour. Plann. Manag.* **2017**, *143*, 04017005. [CrossRef]
50. Estrela Monreal, T.; Cabezas Calvo-Rubio, F.; Estrada Lorenzo, F. La Evaluación De Los Recursos Hídricos En El Libro Blanco Del Agua En España. *Ingeniería Agua* **1999**, *6*, 125–138. [CrossRef]
51. Perez-Sanchez, J.; Senent-Aparicio, J.; Segura-Mendez, F.; Pulido-Velazquez, D.; Srinivasan, R. Evaluating Hydrological Models for Deriving Water Resources in Peninsular Spain. *Sustainability* **2019**, *11*, 2872. [CrossRef]
52. Rico, M.T.; Benito, G. Estimación de caudales de crecida en pequeñas cuencas de montaña: Revisión Metodológica y Aplicación a la cuenca de Montardit (Pirineos Centrales, España). *Revista Cuaternario & Geomorfología* **2002**, *16*, 127–138.
53. Murillo, J.M.; Navarro, J.A. Aplicación Del Modelo De Témez a La Determinación De La Aportación Superficial Y Subterránea Del Sistema Hidrológico Cornisa-Vega De Granada Para Su Implementación En Un Modelo De Uso Conjunto. *Boletín Geológico Minero* **2011**, *122*, 363–388.
54. Paredes-Arquiola, J.; Solera, A.; Álvarez, J.; Elvira, N. *Manual Técnico Herramienta EvalHid Para la Evaluación de Recursos Hídricos Versión 1.1*; 3 De Octubre 2017; Universidad Politécnica de Valencia: Valencia, Spain, 2017.
55. UCA. Universidad de Cadiz. Departamento de Ciencias de la Tierra. Proyecto REMABAR: Análisis de Estrategias para reducir las pérdidas por evaporación y mejora del estado de los acuíferos bajo un contexto de Cambio Climático en la cuenca del río Barbate. In *Informe de Modelización y Simulación de la Cuenca del río Barbate*; Reina Sofía Cultural Center: Cádiz, Spain, 2019; pp. 2–3.
56. Kottek, M.; Grieser, J.; Beck, C.; Rudolf, B.; Rubel, F. World Map of the Köppen-Geiger Climate Classification Updated. *Meteorol. Z.* **2006**, *15*, 259–263. [CrossRef]
57. Peel, M.C.; Finlayson, B.L.; McMahon, T.A. Updated World Map of the Köppen-Geiger Climate Classification. *Hydrol. Earth Syst. Sci.* **2007**, *11*, 1633–1644. [CrossRef]
58. Lobo, F.J.; Hernandez-Molina, F.J.; Somoza, L.; Rodero, J.; Maldonado, A.; Barnolas, A. Patterns of Bottom Current Flow Deduced from Dune Asymmetries Over the Gulf of Cadiz Shelf (Southwest Spain). *Mar. Geol.* **2000**, *164*, 91–117. [CrossRef]
59. Gomez-Pina, G.; Munoz-Perez, J.J.; Ramirez, J.L.; Ley, C. Sand Dune Management Problems and Techniques, Spain. *J. Coast. Res.* **2002**, *36*, 325–332. [CrossRef]
60. Perez-Lopez, A.; Perez-Valera, F. Palaeogeography, Facies and Nomenclature of the Triassic Units in the Different Domains of the Betic Cordillera (S Spain). *Palaeogeogr. Palaeoclimatol. Palaeoecol.* **2007**, *254*, 606–626. [CrossRef]
61. Luján, M.; Crespo-Blanc, A.; Balanyá, J.C. The Flysch Trough Thrust Imbricate (Betic Cordillera): A Key Element of the Gibraltar Arc Orogenic Wedge. *Tectonics* **2006**, 25. [CrossRef]
62. Vélez-Nicolás, M.; García-López, S.; Ruiz-Ortiz, V.; Sánchez-Bellón, Á. Towards a Sustainable and Adaptive Groundwater Management: Lessons from the Benalup Aquifer (Southern Spain). *Sustainability* **2020**, *12*, 5215. [CrossRef]
63. Ruiz-Ortiz, V.; Garcia-Lopez, S.; Solera, A.; Paredes, J. Contribution of Decision Support Systems to Water Management Improvement in Basins with High Evaporation in Mediterranean Climates. *Hydrol. Res.* **2019**, *50*, 1020–1036. [CrossRef]
64. Junta de Andalucia. Estaciones Agroclimáticas. Available online: https://www.juntadeandalucia.es/agriculturaypesca/ifapa/ria/servlet/FrontController (accessed on 10 January 2019).
65. MITECO. Agencia Estatal de Meteorología. Estaciones Climáticas. Available online: https://www.miteco.gob.es/es/cartografia-y-sig/ide/descargas/otros/default.aspx (accessed on 10 January 2019).
66. Karl, T.R.; Williams, C.N. An Approach to Adjusting Climatological Time-Series for Discontinuous Inhomogeneities. *J. Clim. Appl. Meteorol.* **1987**, *26*, 1744–1763. [CrossRef]

67. Akbar, R.; Gianotti, D.S.; McColl, K.A.; Haghighi, E.; Salvucci, G.D.; Entekhabi, D. Hydrological Storage Length Scales Represented by Remote Sensing Estimates of Soil Moisture and Precipitation. *Water Resour. Res.* **2018**, *54*, 1476–1492. [CrossRef]
68. Zhang, C.; Zhou, X.; Lei, W. Necessary Length of Daily Precipitation Time Series for Different Entropy Measures. *Earth Sci. Inform.* **2019**, *12*, 475–487. [CrossRef]
69. Hidalgo, J.C.G.; de Luis Arrillaga, M.; Stepánek, P.; Bonvehi, J.R.; Prats, J.M.C. Reconstrucción, Estabilidad y Proceso de Homogeneizado de Series de Precipitación en ambientes de elevada Variabilidad Pluvial. In *La Información Climática como herramienta de gestión ambiental Bases de Datos y Tratamiento de Series Climatólogicas: Reunión Nacional De Climatología (7th 2002 Albarracín, España)*; Universidad de Zaragoza: Zaragoza, Spain, 2002; pp. 47–58. ISBN 9788495480699.
70. Squintu, A.A.; van der Schrier, G.; van den Besselaar, E.J.M.; Cornes, R.C.; Tank, A.M.G.K. Building Long Homogeneous Temperature Series Across Europe: A New Approach for the Blending of Neighboring Series. *J. Appl. Meteorol. Climatol.* **2020**, *59*, 175–189. [CrossRef]
71. CEDEX. Centro de Estudios y Experimentación de Obras Públicas. Cálculo Hidrometeorológico De Aportaciones Y Crecidas (CHAC). Ministerio De Fomento Y Ministerio Para La Transición Ecológica. Gobierno De España. 2013. Available online: https://ceh.cedex.es/chac/ (accessed on 12 February 2019).
72. Barrera Escoda, A. Técnicas de completado de series mensuales y aplicación al estudio de la influencia de la NAO en la distribución de la precipitación en España. In *Trabajo para la obtención del Diploma de Estudios Avanzados (DEA). Programa de doctorado de Astronomía y Meteorología (Bienio 2002–2004)*; Universidad de Barcelona, Departamento de Astronomía y Meteorología: Barcelona, Spain, 2002; p. 96.
73. HIDROSUR. Junta de Andalucía. Consejería de Agricultura, Ganadería, Pesca y Desarrollo Sostenible. Dirección General de Infraestructuras del Agua. Available online: http://www.redhidrosurmedioambiente.es/saih/ (accessed on 13 January 2019).
74. Sang, Y.; Shang, L.; Wang, Z.; Liu, C.; Yang, M. Bayesian-Combined Wavelet Regressive Modeling for Hydrologic Time Series Forecasting. *Chin. Sci. Bull.* **2013**, *58*, 3796–3805. [CrossRef]
75. See, L.; Openshaw, S. A Hybrid Multi-Model Approach to River Level Forecasting. *Hydrol. Sci. J.* **2000**, *45*, 523–536. [CrossRef]
76. Jain, A.; Kumar, A.M. Hybrid Neural Network Models for Hydrologic Time Series Forecasting. *Appl. Soft Comput.* **2007**, *7*, 585–592. [CrossRef]
77. Pearl, J. Graphical models for probabilistic and causal reasoning. In *Computing Handbook*, 3rd ed.; Tucker, A., Gonzalez, T., Topi, H., Diaz-Herrera, J., Eds.; Computer Science and Software Engineering: Boca Raton, FL, USA, 2014; Volume 1.
78. Pearl, J. *Causality: Models, Reasoning, and Inference*, 2nd ed.; Cambridge University Press: New York, NY, USA, 2009; p. 484. ISBN 978-0-521-89560-6.
79. Garcia, A.; Sainz, A.; Revilla, J.A.; Alvarez, C.; Juanes, J.A.; Puente, A. Surface Water Resources Assessment in Scarcely Gauged Basins in the North of Spain. *J. Hydrol.* **2008**, *356*, 312–326. [CrossRef]
80. Rao, K.H.V.D.; Rao, V.V.; Dadhwal, V.K. Improvement to the Thornthwaite Method to Study the Runoff at a Basin Scale using Temporal Remote Sensing Data. *Water Resour. Manag.* **2014**, *28*, 1567–1578.
81. Paredes, J.; Andreu, J.; Solera, A. A Decision Support System for Water Quality Issues in the Manzanares River (Madrid, Spain). *Sci. Total Environ.* **2010**, *408*, 2576–2589. [CrossRef] [PubMed]
82. Moghadam, E.; Boroumand, N.; Jalali, V.; Sanjari, S. Evaluation the Effect of Different Land use and Soil Characteristics on Saturated Hydraulic Conductivity. *J. Water Soil* **2017**, *31*, 1302–1312.
83. Welde, K.; Gebremariam, B. Effect of Land use Land Cover Dynamics on Hydrological Response of Watershed: Case Study of Tekeze Dam Watershed, Northern Ethiopia. *Int. Soil Water Conserv. Res.* **2017**, *5*, 1–16. [CrossRef]
84. Rea, W.; Oxley, L.; Reale, M.; Brown, J. Estimators for Long Range Dependence: An Empirical Study. *arXiv* **2009**, arXiv:0901.0762.
85. TRASERO. *Tratamiento y Gestión de Series Temporales Hidrológicas*; Diputación Provincial de Alicante-Departamento de Ciclo Hídrico: Alicante, Spain, 2015; pp. 28–30. ISBN 978-84-15327-28-8.
86. Pearl, J. *Probabilistic Reasoning in Intelligent Systems: Networks of Plausible Inference*; Morgan Kaufmann: San Francisco, CA, USA, 1988; p. 552.
87. Hanea, A.M.; Gheorghe, M.; Hanea, R.; Ababei, D. Non-Parametric Bayesian Networks for Parameter Estimation in Reservoir Simulation: A Graphical Take on the Ensemble Kalman Filter (Part I). *Comput. Geosci.* **2013**, *17*, 929–949. [CrossRef]

88. Castelletti, A.; Soncini-Sessa, R. Bayesian Networks and Participatory Modelling in Water Resource Management. *Environ. Model. Softw.* **2007**, *22*, 1075–1088. [CrossRef]
89. Said, A. The Implementation of a Bayesian Network for Watershed Management Decisions. *Water Resour. Manag.* **2006**, *20*, 591–605. [CrossRef]
90. Cabañas de Paz, R. New Methods and Data Structures for Evaluating Influence Diagrams. Ph.D. Thesis, University of Granada, Granada, Spain, 2017; 360p.
91. Madsen, A.L.; Lang, M.; Kjærulff, U.B.; Jensen, F. The Hugin Tool for Learning Bayesian Networks. In *European Conference on Symbolic and Quantitative Approaches to Reasoning and Uncertainty*; Springer: Berlin/Heidelberg, Germany, 2003; pp. 594–605.
92. Jensen, F.V. *An Introduction to Bayesian Networks*; UCL Press: London, UK, 1996.
93. Cain, J. *Planning Improvements in Natural Resources Management*; Centre for Ecology and Hydrology: Wallingford, UK, 2001; Volume 124, pp. 1–123.
94. Bayes, T. LII. An Essay towards Solving a Problem in the Doctrine of Chances. By the Late Rev. Mr. Bayes, FRS Communicated by Mr. Price, in a Letter to John Canton, AMFR S. *Philos. Trans. R. Soc. Lond.* **1763**, *53*, 370–418.
95. HUGIN. Hugin Expert A/S. 7.3. 2010. Available online: http://www.hugin.com (accessed on 23 April 2020).
96. Verma, T.; Pearl, J. An Algorithm for Deciding if a Set of Observed Independencies has a Causal Explanation. In Proceedings of the 8th Conference on Uncertainty in Artificial Intelligence, Stanford, CA, USA, 17–19 July 1992; pp. 323–330.
97. Bonissone, P.; Henrion, M.; Kanal, L.; Lemmer, J. Proceedings of the Sixth Conference on Uncertainty in Artificial Intelligence, 1990. *arXiv* **2013**, arXiv:1304.3854.
98. Le, T.; Hoang, T.; Li, J.; Liu, L.; Liu, H.; Hu, S. A Fast PC Algorithm for High Dimensional Causal Discovery with Multi-Core PCs. *IEEE/ACM Trans. Comput. Biol. Bioinform.* **2019**, *16*, 1483–1495. [CrossRef]

2

Selection of the Best Alternative of Heating System by Environmental Impact Assessment

Vlasta Ondrejka Harbulakova [1,*], Martina Zelenakova [1], Pavol Purcz [2] and Adrian Olejnik [1]

1 Department of Environmental Engineering, Faculty of Civil Engineering, Technical University of Kosice, Vysokoskolska 4, 042 00 Kosice, Slovakia; martina.zelenakova@tuke.sk (M.Z.); adrianolejnik91@gmail.com (A.O.)
2 Department of Applied Mathematics, Faculty of Civil Engineering, Technical University of Kosice, Vysokoskolska 4, 042 00 Kosice, Slovakia; pavol.purcz@tuke.sk
* Correspondence: vlasta.harbulakova@tuke.sk

Abstract: The Framework Directive 2009/28/EC on the promotion of the use of energy from renewable sources aims at developing the local and regional electricity market in order to reduce greenhouse gas emissions. A comparison study of the proposed activities of construction of a new biomass-fired power plant or reconstruction of an old one-gas power plant in town located in eastern Slovakia is presented in this paper. The method of the index coefficient was used for choosing the best alternatives. Multicriteria analysis proved that the construction of biomass-fired power plant is the most suitable solution chosen from three assessed variants (no activity is implemented, biomass power plant and modernized gas boiler).

Keywords: biomass-fired power plant; ecological criteria; economic criteria; multicriteria analysis; technical criteria and technological criteria

---

1. Introduction

The Cardiff summit in 1989 created the platform of coordinated action aimed at protecting the environment. The European Commission has progressively focused its attention on the development and integration of environmental aspects into the sectoral policies of transport, energy, industry, agriculture, industry, internal economic policy and fisheries. The first step was decision-making of the first integration strategy in the energy sector adoption in 1999, which was modified in 2001 and presented in Gothenburg, Sweden, before the European Council [1,2].

Directive 2001/77/EC [3] of the European Parliament and of the Council of 27 September 2001 on the promotion of electricity produced from renewable energy sources in the internal electricity market and Directive 2003/54/EC [4] of the European Parliament and of the Council of 26 June 2003 have stability rules commune in terms of obtaining and distributing electricity. Another important document presented by the European Commission was "Green Paper on a secure, competitive and sustainable energy for Europe" [5], which was released in 2006. The aim of the EC was to create an integrated energy policy in Europe. The Commission Communication [6] of 10 January 2007 entitled "Roadmap for renewable energy—Renewable energies in the 21st century: building a more sustainable future" has demonstrated that a 20% target for the global share of renewable energy would be achievable targets and that a framework that includes binding targets should provide the business community with the long-term stability needed to make rational, sustainable renewable energy investments to reduce dependence on imported fossil fuels and increase the use of new energy technologies.

In December 2008, a wide range of measures in the European Union (EU) was adopted, which were aimed at reducing the impact of the EU states' activities on global warming and also reducing negative effects on the global climate, while ensuring adequate and reliable energy supply. The Framework Directive 2009/28/EC [7] on the promotion of the use of energy from renewable sources aims at developing the local and regional electricity market in order to reduce greenhouse gas emissions.

Energy is one of the sectors that pollute and harm the environment the most. The alignment of energy and environmental conditions is at the moment one of the most serious strategic challenges to addressing global environmental problems. Therefore, the development of energy must strictly respect the principles of sustainable development. With a view to the future, the reduction of the negative effects of energy on the environment in Slovakia can be achieved by promoting the use of renewable energy sources as well as by promoting energy-saving measures [3].

From the point of view of sustainable development, the transition from the use of non-renewable energy sources to the use of renewable energy sources is necessary. To achieve this goal, it is necessary to change habits, practices and technologies not only in production but also in consumption. Total energy consumption is one of the main determinants of the impact of energy on the environment. Therefore, it is necessary to harmonize the relationship between energy and the environment through the introduction of technologies using renewable energy sources. All energy sources must be used not only with respect to the environment but also with respect to human health [8]. There are many Environmental Impact Assessments (EIA) of the thermal power plant worldwide. The energy strategy of the EU has become one of the most important factors influencing the development not only in the polity of the member states. Different mitigation measures for the control of pollution caused by thermal power as well as some new technologies are described in the paper by authors from India [9]. The Environmental Impact Assessment Report for the Tanda thermal power plant (in New Delhi) presents that the adverse environmental impacts due to construction and operation of thermal power can be mitigated to an acceptable level by implementation of various measures for mitigation [10]. In a paper by Pokale [11], the EIA of the thermal power plant and its effect, as well as a cost-benefit analysis, are presented and discussed. Studies not only from Italy [12], Bangladesh [13], China [14,15], Spain [16] or Turkey [17] confirm that the countries try to make every effort to respect the principles of sustainable development. Moridi et al. (2015) [18] presented a Multi-Criteria Decision-Making analysis method (MCDM) with the aim to select and prioritize appropriate technologies based on key criteria including cost, design, maintainability, and size and filtration efficiency in their paper. They were searched for optimal technologies for the treatment of air pollutants emitting from industrial complexes, especially petrochemical industries. EIA is a necessary step also during the early planning stages of environmental structures in order to gain clear insights into the structures' probable impacts with respect to the different components of the total environment. Likewise, the use of appropriate EIA techniques can aid decision-makers in formulating appropriate actions based on informed decisions in light of project urgency and limited resources, which are common constraints in developing countries [19].

This paper is based on legislative and methodological documents relating to the assessment process of environmental impact—Directive 2014/52/EU of the European Parliament and of the Council of 16 April 2014 [20] on the assessment of the effects of certain public and private projects on the environment (only Act in the following) (the Environmental Impact Assessment, or EIA Directive) [20] and exactly Act No. 24/2006 Coll. on Environmental Impact Assessment as amended [21]—national Law in Slovakia (only Act in the following). The purpose of this Act is to ensure the procedure for the overall expert and public assessment of construction, and other activities determined under the Act (see Act Annex) prior to the decisions on the permission thereof under special provisions, and also for the assessment of proposals for certain development policies and generally binding legal directives from the point of view of their presumed effect on the environment.

The main steps of the EIA process in Slovakia according to National Council of the Slovak Republic 2005 are presented in Figure 1 [22].

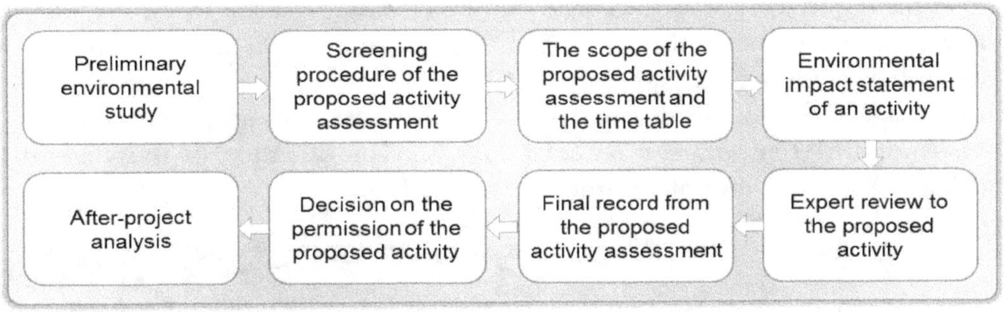

Figure 1. Main steps of the Environmental Impact Assessments (EIA) process in Slovakia (National Council of the Slovak Republic 2005) [22].

The 2006 EIA Act introduced no major changes in EIA procedures; it tightened certain procedural time limits and better delineated EIA responsibilities between the Ministry of the Environment and the regional and district environment offices. The Act was amended many times to respect all requirements of European Union; the last amendment was in June 2017, which was devoted to public participation in the EIA process [22].

There is a direct proportion of energy production and impact of the assessed activity on the environment. Effective and rational use of energy seems to be the best solution how to decrease the negative impact [8].

In the Act, the term "activity" is defined as an operation (structure, facility or others) that by its properties, localization or cooperation with other factors can affect the environment and cultural heritage. To implement such an activity, the permission (approval) or other decision according to specific regulations is necessary.

The Act consists of six parts and sixteen annexes. Annex 8 presents a list of activities subject to environmental impact assessment. It is divided into part A, presenting activities subject to obligatory evaluation, and part B, presenting activities subject to screening.

Variant solutions which are subject to assessment are mostly different in terms of capacity, technology, time of realization, construction costs and maintenance costs. When comparing variants, the result should be in the order of suitability for implementation with regard to human health and the environment. The economic and technical aspects of implementation are the most distinct. The economic aspect is not required by Act 24/2006 [21] to be considered, therefore, in many intentions it is only advisory. Comparing of variations mostly only provides information on the impact of individual variants on the environment and human health [6].

The proposed activity is the construction of a new biomass-fired power plant or reconstruction of an old one-gas power plant in the Trebišov town district in Slovakia. The result is a comparison of the proposed activities with the current state of the area using index coefficient method and finally the selection of the best alternatives of the heating system of Trebišov city.

2. Materials and Methods

2.1. Description of the Study Area

Impact assessment on the environmental compounds of the proposed power plant for heating in Trebišov city in Slovakia (Figure 2) was done according to Act 24/2006 Coll. as amended [21]. Trebišov is situated in the southeast part of Slovakia. It is surrounded by the Slanské Mountains that affect the air circulation. Southwest to northwest winds prevail in this territory. Average annual wind speed in the lowlands of Slovakia is relatively homogeneous. Minimum wind speed is usually connected with situations of temperature inversion, especially in the winter season.

Trebišov district is separated into three climatic areas. Part of the district lies in warm and moderately warm climatic areas and one part lies in cold climatic areas (higher than 800 m a.s.l.) [23].

The study area according to temperature is characterized by two areas. Specifically, part of the district is situated in a warm and dry area with cold winter (January ≤ -3 °C) and part belongs to warm and mildly dry area with cold winter (January ≤ -3 °C).

Long-time average air temperature measured at the Milhostov weather station is 9 °C, and during the vegetative season, the temperature reaches 16.5 °C. The duration of the vegetation period is 200–220 days. The average number of summer days is 67 [23,24].

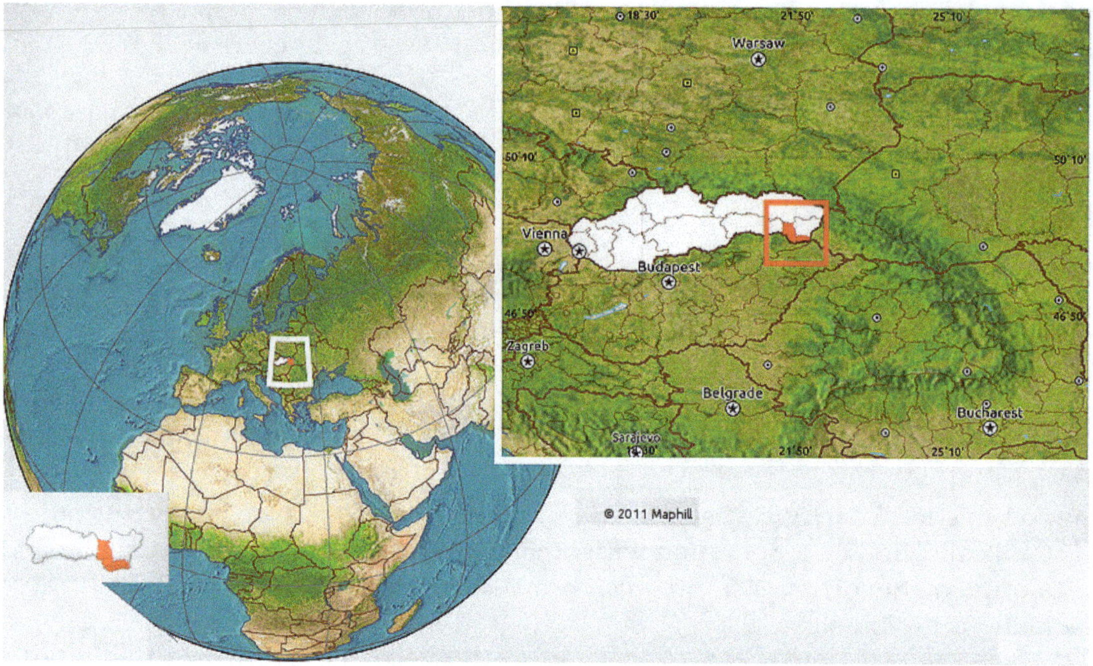

Figure 2. Location of Slovakia and Trebišov district within the European Union (adapted from maphill.com) [25].

The rivers in the Trebišov district are the Ondava River and Latorica River. There are no areas of water bodies which are under the protection. Very important water sources are the Ondava River and Trnávka River, as well as Ruskovský Pond situated in this study area [26]. From the geomorphological division, the study area is located in the Matransko-Slanská area (The Western Carpathians) and in the Eastern Slovakian Lowlands. There are no more significant deposits of ore minerals in the vicinity of Trebišov. There are mainly non-toxic raw materials here as andesite (mainly), which is used primarily as a building material, gravel or paving stone. In particular, the Slanské Hills are covered with andesite volcanoes and stratovolcano.

In the cadastral area of the Trebišov town, the following soils are found: black, brownish and bottomland occurring as subtypes of the black soils and brownish soils [27]. From the agricultural point of view, the most valued are brownish soils. They are more harvested and used for cereal growing. The acidity of the soil is weakly alkaline and has a pH value of 7.3 [27].

2.2. Description of the Project

A brief description of the technical and technological solution is as follows. The central energy source of heating for Trebišov town has two possibilities:

- biomass-fired power plant (alternative 1; wood chips and straw);
- modernization of existing natural gas boiler (alternative 2).

A technological solution of biomass-fired power plant consists of the boiler house (SO 01), a handy storage of straw (SO 02) and technical annex (SO 03). Individual objects are connected structurally and technologically. Objects of power supply are located in the northern part of the plot.

The boiler construction consists of a steel hall with three boilers, one for combustion of wood and two for burning straw. The building object SO 02—a straw storeroom—consists of a single store. There is a firewall designed between the boiler room and the straw storeroom. The building object SO 03—the technical annex—consists of two above-ground floors. Construction of the technical pavilion is built of Porotherm walls. In the technical pavilion, there is the operator, the equipment and the administration of the boiler room [23,24].

2.2.1. Biomass-Fired Power Plant

Wood Chips Combustion

Uncontaminated wood chips will be used as fuel. The preparation of fuel is ensured by cleaving of waste wood (wood-based residue from wood extraction) [23,24]. The fuel will be transported to the boiler room by means of a lorry with a trailer on which two containers are attached with volumes of $2 \times 35\ m^3$. The cars enter via gate No. 1 to the building and the fuel is emptied to a biomass-wood storage site (building SO 04). It is an uncovered, three-sided (height of 2.5 m) concrete walls bounded space. The entrance of fuel into the landfill, as well as its storage, is illustrated by the red arrows, and the transfer of fuel from the biomass landfill to the charging press is illustrated by the blue arrows in Figure 3. Ash removal from the burning process is illustrated by the green arrows. The task of the fuel store exchanger is to gradually add the accumulated quantity of feedstock (wood chips) to the feed press and through it to the inlet part of the boiler in order to ensure the continuous supply of the boiler to the fuel.

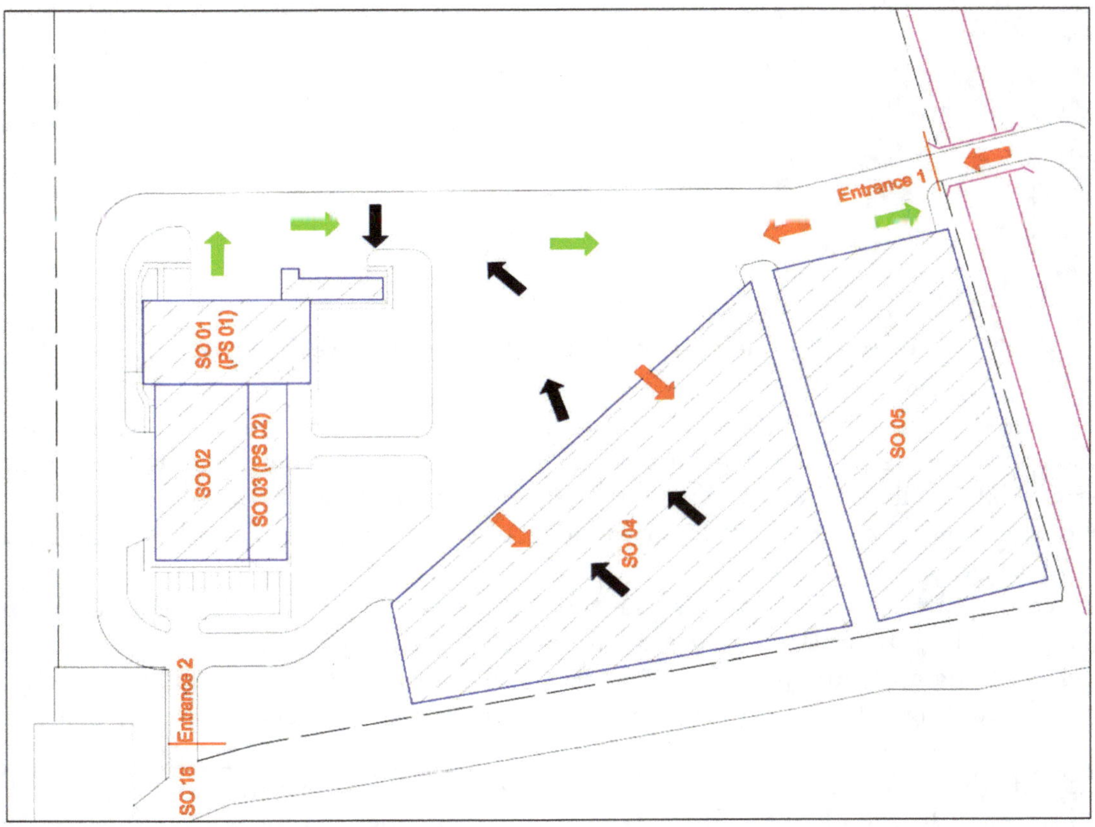

Figure 3. The direction of the fuel handling (source: by authors).

Straw Combustion

As a fuel, the straw used in agriculture as cereal crops will be used. Fuel is provided by packing into large straw bales in the fields.

The fuel will be transported to the boiler room by means of a lorry on which straw bales are stored. The straw is stored on the car; in one layer there are 10 parcels and, in the case of an assumed package with a width of 1200 mm, height 900 mm, length 2200 mm, weight 261 kg, can be placed in three layers. One trailer car carries 60 straw bales for a total weight of 15.66 t [23]. The cars enter via gate No. 1 and the fuel is transported either directly to the straw storeroom (SO 02) or to the straw landfill (SO 05). The placement decision will be based on the fulfilment of both stocks. The unloading is carried out using a forklift truck. The transport of fuel to the boiler house and to the warehouses is illustrated by the red arrows in Figure 4.

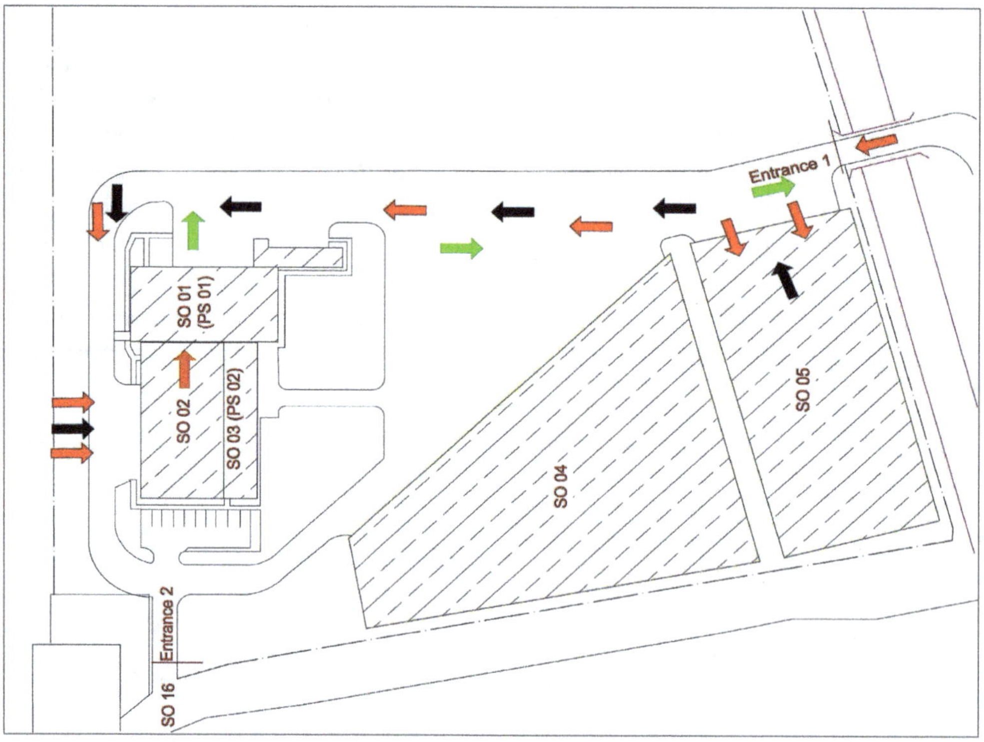

Figure 4. The direction of the straw handling (source: by authors).

The covered straw landfill (SO 05) only serves as a temporary storage to overlap a failure in the fuel supply logistics to the boiler house. In the case of emptying the straw storeroom (object SO 02), straw bales will be taken from the covered straw landfill (object SO 05) using a forklift truck. The transport of fuel from the covered straw landfill to the straw storeroom is illustrated by the blue arrows in Figure 4.

In the case of direct straw removal in the straw storeroom (object SO 02), the packages are stacked on top of a precisely drawn place on the store floor. Storing exactly in place is important for handling straw bundles using an automatic crane. The bridge crane located on the crane track, which will be located on the building of the hall, ensures the transport of the straw package as required by the boiler automation and its placement on the conveyor, which ensures delivery of straw to the boiler. The fuel transportation within the interior spaces is represented by a yellow arrow and ash removal from the burning process is illustrated by the green arrows (Figure 4). Just before entering the straw bales into the boiler, the conveyor ensures that it is rotated to the vertical position. In this position, the straw bundle is cut into three parts by means of the cutting device and transported to the boiler itself, where it burns [23].

2.2.2. Modernization of Existing Natural Gas Boiler

Atmospheric natural gas-fired water tube boilers are used in old central boiler houses. These boilers using the combustion gas transfer the heat to the primary heat exchanger, then it is distributed to the secondary circuit and through pipes to the heated objects. At the secondary circuit is installed a heat meter which measures the heat consumption [23,24,28]. They produce dry combustion products which reach a temperature of 120 °C to 180 °C. Hot combustion products are discharged to the chimney, thereby loss of heat appears. Waste gas contains latent heat which is bound to water vapor resulting from the combustion of natural gas. From the total heat energy gained from natural gas combustion, only 80% is used for water heating. This combustion of natural gas forms a large amount of exhaust gases emitted into the atmosphere through the chimneys. They have a high temperature and steam having a high energy flows through the chimney without further use.

When gas is heated, large amounts of flue gas are generated and released into the atmosphere through the chimneys. They have high temperature and water vapor, which has high energy, escapes the chimney without any further use. With conventional gas boilers, the connection of the condensate hose to the waste is complicated. The hose may clog or freeze, and the condensate can enter the boiler and cause a boiler failure. Sewage water can also be pushed out into the boiler [28,29].

2.3. Application of the Methodology

A lot of methods (tools) have been utilized over the last decades to meet the different actions required in the conduction of impact studies. The objectives of the different actions vary, as do the usable methods for each. These methods are described for example in [30] or [31,32].

Guidelines of European Commission [33] provide information on methods and tools that were selected from case studies and literature research. These generally fall into two groups:

- Scoping and impact identification techniques—these identify how and where an indirect or cumulative impact or impact interaction would occur—Network and analysis; Consultation and questionnaires; Checklists, Matrices; Expert opinion.
- Evaluation techniques—these quantify and predict the magnitude and significance of impacts based on their context and intensity—Matrices; Expert opinion.

It is possible to apply at the outset to define the problem, establish the terms of reference, design the overall EIA process, and set the study boundary. Scoping helps to reform EIA institutional arrangements by focusing on each EIA activity and documents (impact identification and prediction, public and agency consultation and other), as well as subsequent proposal acceptance or rejection [34].

During the EIA process usually, a combination of techniques is used, or approaches are adopted at different stages of the project. Examples of both categories are set out below, in Figure 5.

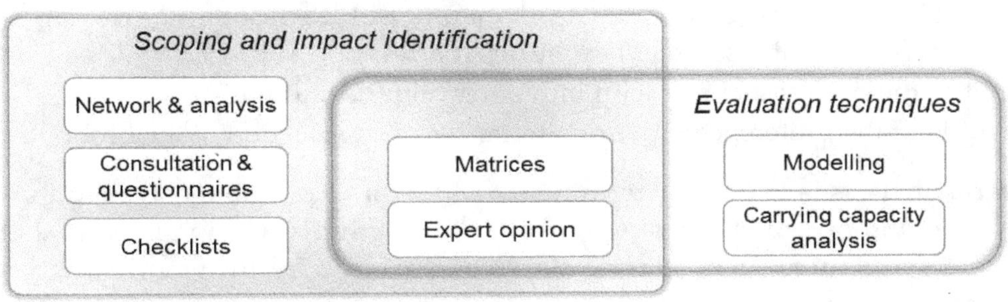

Figure 5. Methods and tools for assessment of indirect and cumulative impact as well as impact interaction (after Walker and Johnston 1999) [22].

The next differentiation of methods in the EIA process is given also in Zeleňáková and Zvijáková (2017) [22]. Table 1 presents the recommended methods suitable for impact assessment in the preliminary impact study, environmental impact statement and methods of multicriteria analysis.

Table 1. Methods for impact assessment in the preliminary impact study, environmental impact statement and methods of multicriteria analysis (after Kozová et al. 1996) [22].

Group of Methods	Methods
Recommended methods suitable for preliminary impact study	Ad hoc methods Checklists and catalogues of criteria Tables and matrices, expressing the causes and the effects Networks and system diagrams Decision trees Overlay mapping
Recommended methods suitable for environmental impact statement	Prognostic methods • Modeling - mathematical models - simulation models - experimental in situ or laboratory models • Comparative methods
Multicriteria analysis	Method of utility function Total Environmental Quality Indicator (TIEQ) method Methods for determining weights of criteria • Ranking method • Allocation method • Grading method • Pairwise comparison method • Dual method ALO-FUL

Multicriteria analysis considers the assessment of the impact on the environment. Objective assessment is based on the use of quantitative factors—objective technical, technological, social and economic units. Qualitative assessment includes [35,36]: indicators; indices; intervals; categories; classes. Indicators can be represented by its description. Indices and intervals can express the magnitude or quality of parameters using the index function. Categories and classes use usually the subjective assessment of the features of interval and/or ratio scales. Formalized workflow involves ensuring that detection of an impact is done using a single method, and prognosis of induced changes is carried out on a scientific basis.

The selection of the optimal alternative is enjoyed by various methods, particularly by multicriteria analysis. The general procedure of multicriteria evaluation of alternatives includes six relatively discrete steps [8]:

- The creation of a purpose-oriented set of evaluation criteria;
- Setting the weights of the evaluation criteria;
- Assessment of the results (consequences, benefits, but also potential damages or losses of alternatives), it is a partial assessment of the alternatives;
- Assessment of the risks associated with implementing the alternatives;
- Determination of the preference order of alternatives and selection of the best option.

Multicriteria analysis is used to determine the value of a comprehensive land use in terms of the quality of the environment affected by humans. Multicriteria method utilizes the catalogue of criteria [8,37,38]. Its structure is hierarchical, adaptive and, basically, the whole society allows you to select the preferred option of a conventional set of alternatives or to give a preferential position of alternatives to a given set of criteria.

In the EIA process, it is always necessary to consider at least two alternatives to the proposed action:

(I) zero alternative—if there is no activity (current state of the environment); and
(II) alternatives of the proposed activity—variants of the activity that usually differ in the locality (site of construction), used technology, time of implementation, etc.

The purpose should be to find the optimal solution, in practice a choice called "preferred option". The selection of the optimal alternative is enjoyed by various methods, particularly by multicriteria analysis [36].

The evaluation presented in a previous paper [39] was done using the method of the Total Environmental Quality Indicator (TIEQ) to calculate the optimal variant. The TIEQ method is a modified utility function. The principle of this method is that the assessed impacts represented by the relevant indicators can be considered from the point of view of quantity and quality and transformed into partial utility functions and these values can already be compared [37]. According to this method alternative of biomass-fired power plant, (A1) is the best one [39]. However, due to the extent of the project, the demand for proof of this fact by another method was required. It makes the calculation more objective and has a higher value. The index coefficient method was chosen as the next method of multicriteria analysis for assessment and for selecting the best option for boiler construction in Trebišov city.

The index coefficient method determines the partial evaluation of the criteria through the calculation of the partial profitability through the index coefficient k_{ij}. The total profitability of the variant U_j (Equation (1)) is determined as the sum of the coefficients of the index coefficients k_{ij} (Equation (2)) and the criteria of the relevant variant and the weight of the i-th criterion [40]:

$$U_j = \sum P_{ij} = \sum k_{ij}.w_i \tag{1}$$

U_j is the overall profitability of the j-th variant; P_{ij} is the i-th criterion of the j-th variant; k_{ij} is the index coefficient of the i-th criterion of the j-th variant; w_i is the weight of the i-th criterion where $\Sigma w_i = 100$. The value of U_j determines the order of the advantage of the variants. There are also different possible methods for determining weights of criteria:

- Ranking method
- Allocation method
- Grading method
- Pairwise comparison method
- Dual method ALO-FUL

Pairwise comparison method by Fuller triangle was used in the presented case. The determined criteria were compared based on a number of preferences to each criterion. The total number of the pairs is n/2(n − 1); in our case, it is 9/2(9 − 1) = 36. The final weight for each criterion is calculated by a number of preferences to criterion divided by 36. The variation that achieves the highest value of overall performance is the best possible [40].

$$k_{ij} = \frac{H}{b_v} \tag{2}$$

The coefficient k_{ij} is determined by comparing the value of the criterion of the variant under consideration (H) with the value of the basic variant (b_v). A basic variant is a fictitious variant whose values of the criteria are theoretically the best (worst) or variants whose values for the criteria are predetermined according to the set goal. The principle of this method is that in a fictitious base variant we replace criterion values by number 1. The profitability of the criteria for the other variants is calculated using a coefficient as the ratio between the value of the calculated variance criteria and the original value of the base criterion. To determine the k_{ij} coefficient, two classes of criteria have to be distinguished. The first class is profit-type criteria, where higher values are preferred to lower performance levels (i.e., the higher the criterion value, the better). The second class is cost-type criteria where lower values are preferred to higher performance levels (i.e., the lower the criterion value, the better this criterion) [40]. Thus, criteria for increasing or decreasing preference are distinguished.

3. Results and Discussion

In the EIA process, considerations of two alternatives of the proposed action in comparison with zero alternative is usually taken in the assessment. In the presented case the study of considered alternatives are:

- Alternative 0—the zero alternative—if no activity is implemented.
- Alternative 1—the biomass-fired power plant in Trebišov district.
- Alternative 2—modernization of existing natural gas boiler.

Comparison of alternatives is done by multicriteria analysis. The beginning of this evaluation is creating a set of criteria and determining their importance (weight). We have chosen nine criteria described in Table 2. For evaluation and comparison, the criteria are divided into qualitative and quantitative ones.

Table 2. Description of criteria.

Aspect	Economic	Technical	Ecological	Social
Criteria (P_i)	the total cost of construction (P1)	time of construction (P3)	waste production (P6)	job opportunities (P8)
	annual operation cost (P2)	land occupation (P4)	emissions production (P7)	extra boiler room construction (P9)
		energy outputs of the power plants (P5)		
Character of assessment	quantitative	quantitative	qualitative	qualitative

The points (0–10) associated with each criterion (Table 3) were stated based on different experts' suggestions [23] (10 experts were involved) by brainstorming method with the aim to get the most objective results. The opinions of experts did not differ widely in most cases.

Table 3. Index coefficient method.

Criteria Pi/ Alternative Ai		P1	P2	P3	P4	P5	P6	P7	P8	P9	Σu_j
b_v		9	8	10	8	9	8	7	9	8	
w_i		3	3	8	3	11	17	19	17	19	**100**
A0	value	0 €	0 €	0 months	0 m^2	0 MW	no	0%	0	yes	
	points	9	8	10	8	0	8	7	0	0	
	k_{i0}	1	1	1	1	0	1	1	0	0	
	$k_{i0}.w_i$	3	3	8	3	0	17	19	0	0	**53**
A1	value	3.8 mil. €	0.54 mil. €	9 months	2000 m^2	14 MW	yes	0%	8	no	
	points	3	6	2	3	9	5	7	9	8	
	k_{i1}	0.33	0.75	0.2	0.375	1	0.625	1	1	1	
	$k_{i1}.w_i$	0.99	2.25	1.6	1.125	11	10.63	19	17	19	**82.6**
A2	value	2.6 mil. €	0.65 mil. €	3 months	1250 m^2	10 MW	yes	6.5%	6	no	
	points	8	5	8	6	8	4	6	7	8	
	k_{i2}	0.88	0.625	0.8	0.75	0.88	0.5	0.86	0.78	1	
	$k_{i2}.w_i$	2.64	1.875	6.4	2.25	9.68	8.5	16.3	13.26	19	**79.9**

Ecological criteria are the most important from an environmental perspective. The evaluation is influenced by subjective opinions of evaluators. The effects of operating variants cannot be measured, assessed and compared. Two criteria are evaluated (P6 and P7). A0 has the best results (no waste production) and A1 and A2 have broadly similar points (some waste production exists). During the combustion processes of biomass, the ash is produced and, in the case of modernized natural gas boiler, condensation and latent heat appear. A0 and A1 produce no emission, which is why they are better than A2.

Economic criteria are important aspects of comparison. In this case, they determine and compare the costs spent on the construction (P1) and annual operation of the alternatives (P2). In our case, the cheapest variant is Alternative A0, while Alternative 1 (A1) is more expensive in the total costs of the construction compared to Alternative 2 (A2), but is more favourable in the annual operating costs.

Technical and technological criteria were evaluated quantitatively. Criterion as the time of construction (P3), land occupation (P4) and energy output (P5) of three boilers belongs here. In two criteria, A0 is more advantageous than A1 and A2, but in the comparison with the third criterion—the output of boilers—Alternative 0 is the worst. Comparing criteria of A1 and A2 at the time of construction and land occupation, A2 is the preferred option, but according to an aspect of energy output, A1 seems to be a better variant.

Social criteria are also qualitatively assessed. The rating is influenced by subjective opinions of evaluators as well. Social impacts directly interfere with the quality of the life of the population, in our case, the alternative which offers more job opportunities (P8) and where it is needed for extra boiler room construction (P9). Since boilers are designed for citizens, they are designed to improve the quality of their lives. In our case, it has been compared which alternative can provide more job opportunities and which alternative contains the residents' need to build their own home boiler room. In both alternatives mentioned above, A0 is the worst. A1 is better than A2 compared to the job opportunities. In terms of the second criterion, both alternatives are able to compensate for the home boiler to a sufficient extent so that residents do not have to build the home boiler rooms.

The second step was to determine the index coefficient k_{ij}. This is determined by evaluating criteria where the value of the highest criterion is replaced by a value of 1. Subsequently, the other values are calculated as the ratio of value 1 to the appropriate variant value. The basic variant in this step is those where number 1 is located. The total profitability was calculated based on Equation (1).

The construction of the biomass-fired power plant is the best of the three compared alternatives based on the selected criteria from the point of view of the environmental impact assessment (Figure 6).

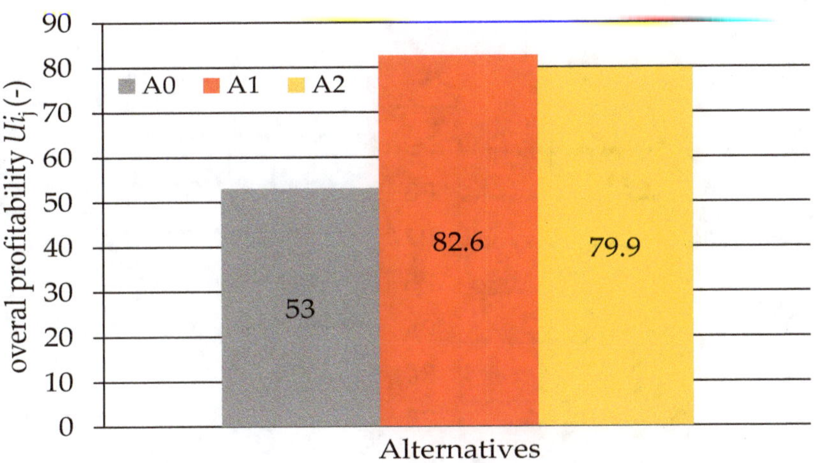

Figure 6. Summary comparison of the assessed alternatives.

The aim of the assessment was to make an overall investigation, description and evaluation of the direct and indirect environmental impacts of the activity on the environment, to determine measures that will prevent or mitigate pollution and damage to the environment, and to explain and compare the advantages and disadvantages of the proposed activity including its variants, in comparison also with the situation that would exist if the activity was not implemented. Biomass-fired power plant seems to be the better alternative (U_2 = 82.6) compared to modernizing the existing natural gas boiler (U_3 = 79.9) and compared to if no activity were implemented (U_0 = 53).

Alternatives 1 and 2 only have 2.7 points of difference. This difference is very low, so we made a sensitivity analysis of the different criteria, which are the criteria that have the real importance in the final decision. Sensitivity analysis of Alternative 0 is presented in Figure 7, Alternative 1 is presented in Figure 8 and Alternative 2 is presented in Figure 9 where criteria: P1—the total cost of construction, P2—annual operation cost, P3—the time of construction, P4—land occupation, P5—energy output, P6—waste production, P7—emissions production, P8—jobs opportunities, P9—extra boiler room construction.

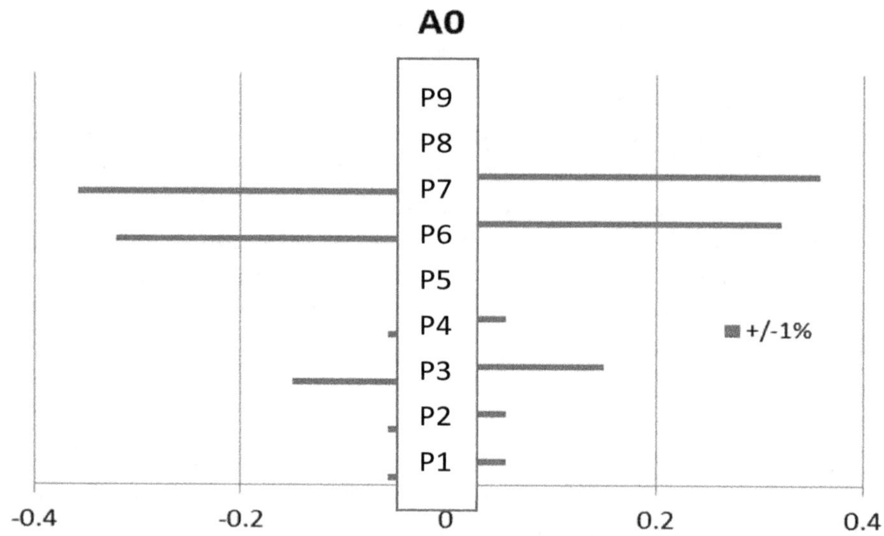

Figure 7. Sensitivity analysis of Alternative 0.

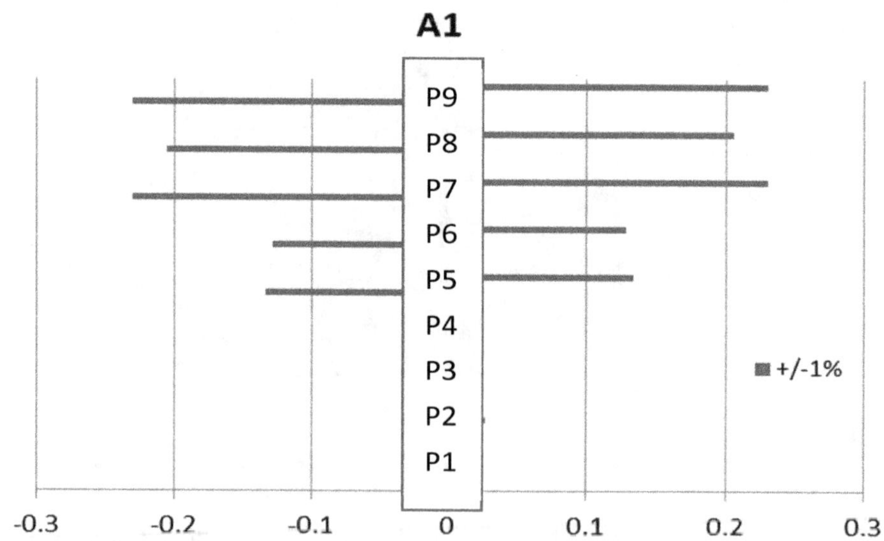

Figure 8. Sensitivity analysis of Alternative 1.

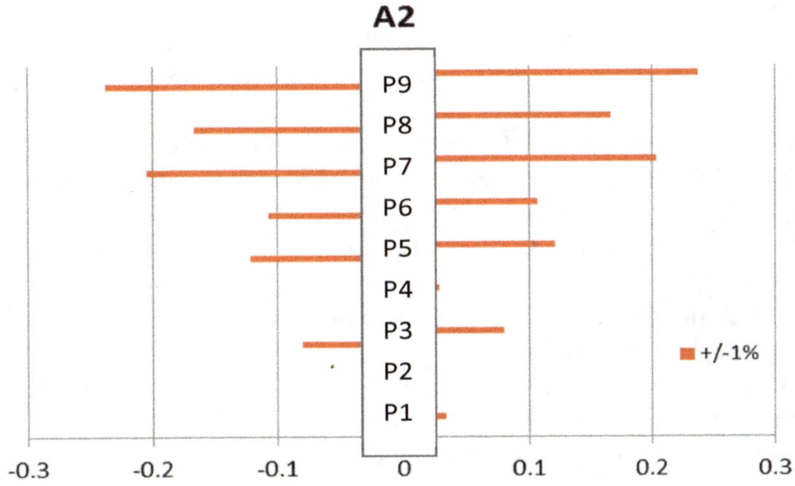

Figure 9. Sensitivity analysis of Alternative 2.

Figures 7–9 show results from sensitivity analysis as to how the model is sensitive to the choice of the criteria. The figures depict the effect on the selected activity (its alternatives) due to the different input data. Criteria: P7—emissions production and P9—extra boiler room construction have major influences on the selection of the best alternative. The assessment shows the highest sensitivity to these criteria. The proposed activity shows little sensitivity to all other criteria P1–P6 and P8.

4. Conclusions

The environmental impact assessment process for plants, structures, facilities and other activities has been applied in developed countries for several decades and is one of the main tools of preventive environmental protection and sustainable development.

The use of biomass in heating systems is beneficial because it uses agricultural, forest, urban and industrial residues and waste to produce heat and electricity with less effect on the environment than fossil fuels.

Coefficient index method as a tool of multicriteria analysis was used for proving that the construction of biomass-fired power plant is the most suitable solution chosen from three assessed variants (no activity is implemented, biomass power plant and modernization of existing gas boiler).

The index coefficient was used to state the weights of criteria. The points (0–10) associated with each criterion were stated based on ten different experts' suggestions with the aim to get the most objective results. In this evaluation, the highest score is the best possible. According to these points, Alternative A2 seems to be the best one (overall profitability of 82.6). Proposals were discussed with professionally qualified persons working in the field of environmental impact assessment as well as civil engineers.

The main contribution of the present paper is using theoretical knowledge of the issue, evaluation on the state of the environment in the area by the multicriteria analyses to select the optimal alternative of the action in the decision-making process in order to preserve environmental quality for further sustainable development of society in the study area. Most important for the assessment is the right selection of the criteria. In our assessment, we found out by a sensitivity analysis that the criteria that have the real importance in the final decision are P7—emissions production and P9—extra boiler room construction.

Author Contributions: Vlasta Ondrejka Harbulakova prepared and edited the manuscript; Martina Zelenakova worked on the materials and methods section—application of the methodology; Pavol Purcz worked on the sensitivity analysis; Adrian Olejnik prepared the expert part—description of the project and the study area.

References

1. Slávka, Štroffeková. Energy and Its Environmental Impact Assessment in the Slovak Republic in 2011. The Indicator Sectoral Report. Available online: https://www.enviroportal.sk/uploads/spravy/sprava-ener-2013-final.pdf (accessed on 25 January 2018). (In Slovak)
2. European Commision. Commission Staff Working Paper on Sustainable Industrial Development. Brussels, 25.10.1999, SEC(1999) 1729. Available online: http://ec.europa.eu/environment/archives/action-programme/pdf/sec991729_en.pdf (accessed on 9 December 2017).
3. Directive 2001/77/EC of the European Parliament and of the Council of 27 September 2001 on the Promotion of Electricity Produced from Renewable Energy Sources in the Internal Electricity Market. Available online: https://publications.europa.eu/en/publication-detail/-/publication/9df8db5d-4776-4fcb-9077-4363e14b836a (accessed on 9 December 2017).
4. Directive 2003/54/EC of the European Parliament and of the Council of 26 June 2003 Concerning Common Rules for the Internal Market in Electricity and Repealing Directive 96/92/EC—Statements Made with Regard to Decommissioning and Waste Management Activities. Available online: https://publications.europa.eu/en/publication-detail/-/publication/caeb5f68-61fd-4ea8-b3b5-00e692b1013c (accessed on 9 December 2017).
5. Impact of Electricity Production on the Environment. Available online: https://online.sse.sk/portal/page/portal/stranka_SSE/spravy/ekologia (accessed on 30 November 2017). (In Slovak)
6. Commission of the European Communities. Commission Staff Working Document. Accompanying Document to the Communication from the Commission to the Council and the European Parliament, Renewable Energy Road Map Renewable Energies in the 21st Century: Building a More Sustainable Future, Impact Assessment. Brussels, SEC (2006) 1719/2. Available online: http://www.ebb-eu.org/legis/renewable%20energy%20roadmap%20full%20impact%20assessment%20100107.pdf (accessed on 9 December 2017).
7. Directive 2009/28/EC of the European Parliament and of the Council of 23 April 2009 on the Promotion of the Use of Energy from Renewable Sources and Amending and Subsequently Repealing Directives 2001/77/EC and 2003/30/EC (Text with EEA Relevance). Available online: http://eur-lex.europa.eu/eli/dir/2009/28/oj (accessed on 9 December 2017).
8. Zvijaková, L.; Zeleňáková, M. *Risk Analysis in Environmental Impact Assessment of Flood Protection Measures*; Leges: Prague, Czech Republic, 2015.
9. Kumar, S.; Katoria, D.; Sehgal, D. Environment Impact Assessment of Thermal Power Plant for Sustainable Development. *Int. J. Environ. Eng. Manag.* **2013**, *4*, 567–572.
10. NTPC Limited. *Environmental Impact Assessment Report for Tanda Thermal Power Project, Stage–II (2 × 660 mw)*; District-Ambedkar Nagar (UP), Document No.: 9562/999/GEG/S/001 Rev. No. 1; NTPC Limited: New Delhi, India, 2009.
11. Pokale, W.K. Effects of thermal power plant on environment. *Sci. Rev. Chem. Commun.* **2012**, *2*, 212–215.
12. Cigognetti, G.; Piccardim, M.; Vital, C. Ministero dell Ambiente e Della Tutela del Territorio e del Mare, Thermal Power Plant of Ponti sul Mincio—Conversion of the Chimney into Lookout Tower and Mincio Park Entrance as a Change to the Prescription of the EIA Screening Determination 2014. Available online: http://www.va.minambiente.it/en-GB/Oggetti/MetadatoDocumento/109647 (accessed on 3 January 2018).
13. Coal Power Generation Comp any of Bangladesh Limited (An Enterprise of the People's Republic of Bangladesh). Report on Environmental Impact Assessment of Construction of Matarbari 600X2 MW. 2013. Available online: https://libportal.jica.go.jp/library/Data/DocforEnvironment/EIA-EPC/EastAsia-SouthwesternAsian/ChittagongACFPPDP/BCEIA.pdf (accessed on 3 January 2018).
14. Bo, X.; Wang, G.; Meng, F.; Wen, R. Air pollution effect of the thermal power plants in Beijing-Tianjin-Hebei region. *China Environ. Sci.* **2015**, *35*, 364–373.
15. Sun, Y.; Liu, Y.; Liu, C. Environmental impact assessment on ambient air fine particulate matter $PM_{2.5}$ role of prevention. *J. Shenyang Jianzhu Univ. (Nat. Sci.)* **2013**, *29*, 1147–1152.
16. Enríquez-de-Salamanca, Á.; Martín-Aranda, R.M.; Díaz-Sierra, R. Consideration of climate change on environmental impact assessment in Spain. *Environ. Impact Assess. Rev.* **2016**, *57*, 31–39.

17. Say, N.P.; Yücel, M.; Yilmazer, M. A computer-based system for environmental impact assessment (EIA) applications to energy power stations in Turkey: ÇEDINFO. *Energy Policy* **2007**, *35*, 6395–6401. [CrossRef]
18. Moridi, F.; Atabi, F.; Nouri, J. Air Pollution Management based on the Selection of Appropriate Technologies for Air Pollutants Filtration using Multi-Criteria Decision-Making. *An. Acad. Bras. Cienc.* **2015**, *87*, 314–323.
19. Shah, A.; Salimullah, K.; Sha, M.H.; Razaulkah, K.; Jan, I.F. Environmental impact assessment (EIA) of infrastructure development projects in developing countries. *OIDA Int. J. Sustain. Dev.* **2010**, *1*, 47–54.
20. Directive 2014/52/EU of the European Parliament and of the Council of 16 April 2014 on the Assessment of the Effects of Certain Public and Private Projects on the Environment. 2014. Available online: http://ec.europa.eu/environment/eia/pdf/EIA_Directive_informal.pdf (accessed on 25 January 2018).
21. National Council of the Slovak Republic 2005 Act of Law No. 24/2006 from December 14th 2005 on Environmental Impact Assessment and on Amendments to Certain Acts. Available online: http://www.ujd.gov.sk/files/legislativa/145_2010_EN.pdf (accessed on 25 January 2018).
22. Zeleňáková, M.; Zvijáková, L. *Using Risk Analysis for Flood Protection Assessment*; Springer: Cham, Switzerlandm, 2017; p. 128.
23. Trebisovksa Energeticka Inc. *Central Energy Source of Trebisov*; TM-P-114.13-B, Technical Report—Summary; Trebisovksa energeticka Inc.: Trebisov, Slovakia, 2013. (In Slovak)
24. Olejník, A. Environmental Impact Assessment of Construction. Master's Thesis, Technical University of Kosice, Košice, Slovakia, 6 May 2016.
25. Maphil.com, Free Satellite Location Map of Rad. Available online: http://www.maphill.com/slovakia/kosice/trebisov/location-maps/satellite-map/ (accessed on 30 November 2017).
26. Slámková, M.; Garčárová, M. Encouraging the Protection of Natura 2000 Sites in Integrating Whole-System of Ecological Stability. Regional Territorial System of Ecological Stability of the Trebišov District. 2012. Available online: www.minv.sk/?verejne-vyhlasky-7&subor=207452 (accessed on 30 November 2017). (In Slovak)
27. Ivaň, M. Mechanisms for Recovery of Wastes of Category "O" Others, Recovery of Construction Debris, Paper, Plastics, Wood," Company Trebišov Town, LAND—Service 2011. Enviroportal.sk, Information Portal of MoE SR. Available online: http://www.enviroportal.sk/sk_SK/eia/detail/ing-michal-ivan-land-servis-trebisov-zariadenie-na-zhodnocovanie-odpad (accessed on 30 November 2017). (In Slovak)
28. Banik and Son Inc. Differences between Traditional and Condensing Boilers. 2015. Available online: http://www.banik.sk/rozdiely-medzi-klasickym-a-kondenzacnym-kotlom/ (accessed on 30 November 2017). (In Slovak)
29. Durikova, K. How Does Condensing Boiler Works? Bratislava: JAGA Group. 2013. Available online: http://mojdom.zoznam.sk/cl/10055/1351440/Oplati-sa-investovat-do-plynoveho-kondenzacneho-kotla- (accessed on 30 November 2017). (In Slovak)
30. Canter, L.W. Methods for Effective Environmental Information Assessment (EIA) Practice. In *Environmental Methods Review: Retooling Impact Assessment for the New Century*; Porter, A.L., Fittipaldi, J.J., Eds.; The Press Club: Fargo, ND, USA, 1998.
31. Canter, L.W. Environmental Impact Assessment. In *Environmental Engineers' Handbook*; Liu, D.H.F., Lipták, B.G., Eds.; CRC Press: Boca Raton, FL, USA, 1999.
32. Kozova, M.; Drdos, J.; Pavlickova, K.; Uradnicek, S. *Environmental Impact Assessment*; Comenius University in Bratislava: Bratislava, Slovakia, 1996.
33. Walker, L.J.; Johnston, J. Guidelines for the Assessment of Indirect and Cumulative Impacts as well as Impact Interactions. 1999. Available online: http://ec.europa.eu/environment/archives/eia/eia-studies-and-reports/pdf/guidel.pdf (accessed on 25 January 2018).
34. Lawrence, D.P. *Environmental Impact Assessment: Practical Solutions to Recurrent Problems*, 2nd ed.; John Wiley & Sons: Hoboken, NJ, USA, 2013.
35. Říha, J. *Environmental Impact Assessment. Methods for Preliminary Decision Analysis*; ČVUT: Praha, Czech Republic, 2001. (In Czech)
36. Říha, J. *Environmental Impact Assessment of Investments*; Multicriteria Analysis and EIA; Academia: Prague, Czech Republic, 1995.
37. Majerník, M.; Húsková, V.; Bosák, M.; Chovancová, J. *Methodology of the Environmental Impact Assessment*; Technical University of Košice (TUKE): Košice, Slovakia, 2008.

38. Gałaś, S.; Krol, E. Indicators for environmental-spatial order assessment on the example of the Busko and Solec Spa communes. *Gospodarka Surowcami Miner.-Miner. Resour. Manag.* **2008**, *24*, 95–115.
39. Zelenakova, M.; Ondrejka Harbulakova, V.; Olejnik, A. Using of Multicriteria Method for Choosing the Best Alternative of the Heating Power Plant, Selected Scientific Papers. *J. Civ. Eng.* **2017**, *12*, 47–56.
40. Kampf, R. Estimation Methods for Weight Criteria. *Sci. Pap. Univ. Pardubice B* **2003**, *9*, 255–261.

3

Combined Forecasting of Rainfall Based on Fuzzy Clustering and Cross Entropy

Baohui Men [1,*], Rishang Long [2], Yangsong Li [1], Huanlong Liu [1], Wei Tian [3,4] and Zhijian Wu [1]

1 Beijing Key Laboratory of Energy Safety and Clean Utilization, North China Electric Power University, Renewable Energy Institute, Beijing 102206, China; lys18811551843@163.com (Y.L.); liuhuanlongHD@163.com (H.L.); wzjdalanqiu@163.com (Z.W.)
2 State Key Laboratory of New Energy Power System, North China Electric Power University, Beijing 102206, China; lrs18810667721@163.com
3 Key Laboratory of Water Cycle and Related Land Surface Process, Institute of Geographic Sciences and Natural Resources Research, Chinese Academy of Sciences, Beijing 100101, China; tianweiBT@163.com
4 University of Chinese Academy of Sciences, Beijing 100049, China
* Correspondence: menbh@ncepu.edu.cn

Abstract: Rainfall is an essential index to measure drought, and it is dependent upon various parameters including geographical environment, air temperature and pressure. The nonlinear nature of climatic variables leads to problems such as poor accuracy and instability in traditional forecasting methods. In this paper, the combined forecasting method based on data mining technology and cross entropy is proposed to forecast the rainfall with full consideration of the time-effectiveness of historical data. In view of the flaws of the fuzzy clustering method which is easy to fall into local optimal solution and low speed of operation, the ant colony algorithm is adopted to overcome these shortcomings and, as a result, refine the model. The method for determining weights is also improved by using the cross entropy. Besides, the forecast is conducted by analyzing the weighted average rainfall based on Thiessen polygon in the Beijing–Tianjin–Hebei region. Since the predictive errors are calculated, the results show that improved ant colony fuzzy clustering can effectively select historical data and enhance the accuracy of prediction so that the damage caused by extreme weather events like droughts and floods can be greatly lessened and even kept at bay.

Keywords: rainfall forecast; cross entropy; ant colony fuzzy clustering; combined forecast

---

1. Introduction

Rainfall forecasts play an important role in agricultural production, urban industry and life. The accurate prediction of rainfall has significant economic and social value. It can provide data support for the relevant departments and help detect droughts and floods and reduce the degree of harm. However, affected by complex factors such as geographical environment, ocean currents, air pressure, temperature, etc. [1], rainfall exhibits strong randomness and nonlinear characteristics that often hamper the forecast of rainfall.

The rainfall prediction methods discussed in this paper are based on mathematical models and algorithms, through the full mining of historical data to establish the forecasting model. At present, a variety of approaches have been applied to predict rainfall at home and abroad and basically they can be classified into five categories: (1) *Numerical prediction model*. This model is based on the physical model of process. Its advantage lies in the fast speed and easy procedure, but due to the impact of the memory required in can only reasonably be used for monthly forecasts and for longer periods (such as an annual forecast) is difficult to use [2,3]; (2) *Time series model exponential smoothing*. The moving average method and Autoregressive Integrated Moving Average Model (ARIMA) [4] belong to this type; they can better describe the linear change process yet sometimes they are not suitable for non-stationary random processes; (3) *Probabilistic simulation methods.* These include the grey model (GM) [5] and Monte Carlo method [6]. The prediction of exponential trends is more accurate with the grey model, but it is only suitable for short and medium term forecasting and the longer the forecast lasts, the larger the errors that may occur. The Monte Carlo method is characterized by describing the random process, but the support of data is also required; (4) *Artificial intelligence methods* such as radical basis function (RBF), genetic algorithm (GA), wavelet analysis (WA) [7–10], can better simulate nonlinear processes with higher prediction accuracy, but may fail due to issues like local optima, overlearning and weak generalization ability; (5) in addition there are some other methods, like Numerical Weather Prediction (NWP) [11,12], R/S analysis [13], trend analysis [14], etc. that can predict rainfall from different angles, but two problems should not be ignored: first, in the long-term forecast, rainfall is a random process hence any single forecasting method cannot ensure the stability during the process of prediction. Additionally, serious errors may occur at certain times resulting in the failure of prediction. Second, obtaining useful information with a single method is one-sided, and overlooks different factors from all perspectives.

For the prediction of rainfall, a large amount of historical data is necessary to ensure the accuracy of the forecast. Nonetheless, historical data inevitably contains some errors or abnormal information and this affects the accuracy of the forecast because rainfall is associated with many factors such as temperature, climate, and human activities. Therefore, the forecast method based on fuzzy clustering prediction method has been applied in recent years [15,16]. However, the traditional fuzzy clustering algorithm approach easily falls into local optimal solutions, and it is difficult to deal with a large number of high-dimensional data from a time performance point of view [17]. In this paper, the ant colony algorithm is proposed to improve the fuzzy clustering. With this method, the reliability and computational efficiency of data filtering and processing are greatly increased.

Bates and Granger established the combined forecasting method based on weights in 1969 [18]. The approach combines the different methods and the features of data to improve the accuracy of forecasting and reduce the risk of failure. The combined forecasting method has been widely used in various fields, including electric power load forecasting, economics, logistics, etc., and facts have proven how effective the method is [19–22]. Nonetheless, as a simple combination of several single methods, the previous method neglects the bias of the selection of a single method. Furthermore, there is no detailed analysis of the time characteristics of historical data. At present, some scholars are paying attention to combined forecasting in the field of rainfall forecasting [23–25]. In Cui's study [23], the wavelet analysis method aims to determine the weight reconstruction of the rainfall forecast, yet the time distribution of historical data is not considered. In Xiong's study [24] and Lu's study [25], real-time river flow or flood forecasting methods have been studied, but they are not suitable for medium and long-term prediction.

The concept of entropy propounded by the German physicist Clausius in 1877 is a function of the state of the system, but the reference value and the variation of entropy are often analyzed and compared. Cross Entropy (CE) is a kind of entropy that reflects the similarity between variables from the perspective of probability. The application of entropy theory in hydrology mainly include the derivation of the distributions and estimation of the corresponding parameters for hydrometeorological variables [26–28], dependence analysis [29] and runoff forecasting [30–34]. The cross entropy is introduced into the combination forecasting by Li et al. [24,25]. Their research put forward a new method of determining weight, which improves the stability of the prediction results. However, the probability density function [24] is not suitable for the prediction of radial flow. The wind power load forecasting method based on normal distribution is proposed by Chen et al. [25]. The time characteristic of historical data is not considered in this method and the solution is too complex to be implemented.

The key of the prediction method based on historical data is not only a prediction model, but also a validity of historical data. And that is the category of data mining. The choice of historical data is fundamentally a clustering process, so clustering method is very important. In terms of the weaknesses of the fuzzy clustering method, the ant colony algorithm attempts to improve the model. Meanwhile, the method for determining weights is also improved by using the cross entropy (CE).

2. Improved Fuzzy Clustering Model

2.1. Research Data

The Beijing–Tianjin–Hebei region is located on the east coast of Eurasia, mid-latitude coastal and inland transfer zone. Influenced by a temperate climate with alternating moist and dry seasons, the annual rainfall in this area ranges from 400 to 800 mm. The study analyses the rainfall from 1969 to 2010 in the Thiessen rainfall station and the forecast is conducted based on the data of rainfall from the Taisen Station in the Beijing–Tianjin–Hebei region. Based on the Taisen Station rainfall data, the forecast is conducted. The results show that improved ant colony fuzzy clustering can effectively select historical data and improve the accuracy of prediction.

On the basis of the data of rainfall from 26 stations, the weight is determined by the Thiessen polygon method, and the weighted average rainfall data sequence is obtained. The Beijing–Tianjin–Hebei administrative divisions and the change of monthly rainfall from 1960 to 2013 in this area are shown in Figure 1.

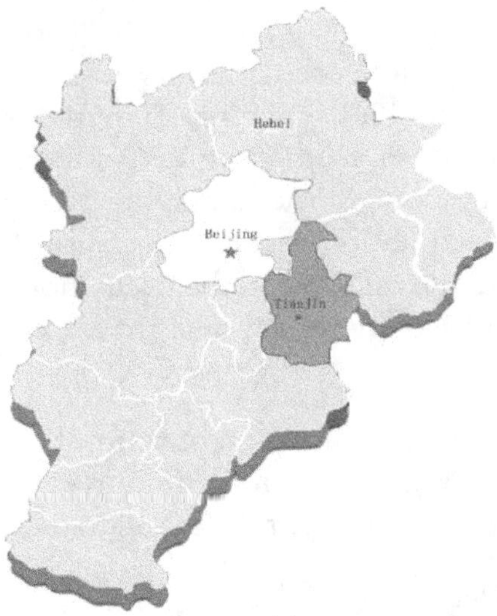

(a)

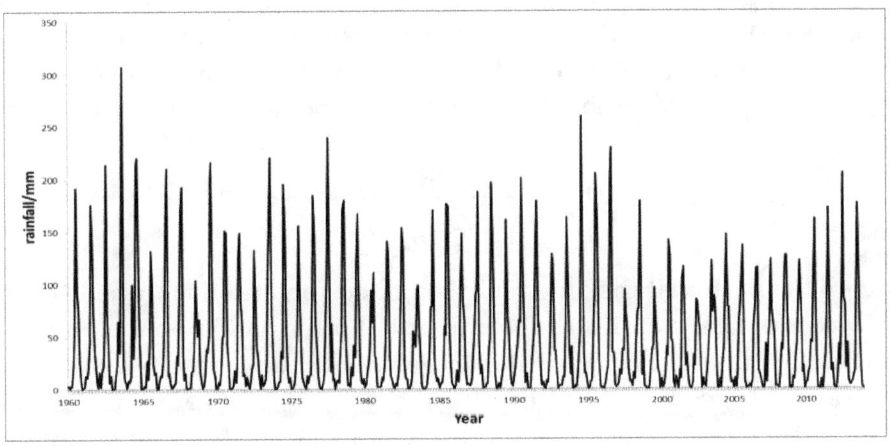

(b)

Figure 1. The Beijing–Tianjin–Hebei administrative divisions and change of monthly rainfall from 1960 to 2013. (**a**) Beijing–Tianjin–Hebei administrative divisions; (**b**) monthly rainfall from 1960 to 2013.

Figure 2 gives the results of the wavelet analysis of the data from 1960–2006. Wavelet analysis is a localized analysis of the time (space) frequency. It multiplies the signal (function) step by step through the telescopic translation operation, finally reaches the time subdivision at high frequency, subdivides the frequency at low frequency, and can automatically adapt to the data analysis.

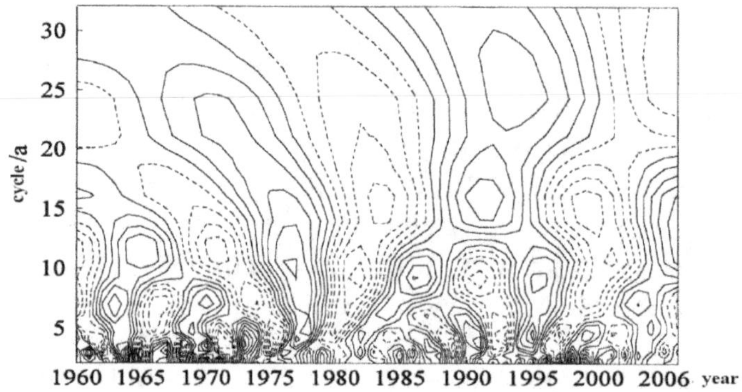

Figure 2. Wavelet analysis of the rainfall data from 1960–2006.

There are four obvious characteristic time scales, namely, 3a, 9a, 14a and 24a, respectively, where the characteristic time scale of 3a is always present from 1960–2006 and the period oscillation is stable. What is more, with 24a time scale cycle time oscillation throughout the study period, the performance is relatively stable. In the middle of the 1960s, Beijing–Tianjin–Hebei area had experienced four dry and wet alternations: from the mid-1960s to the late 1970s, the precipitation was abundant. In the 1980s, the precipitation was relatively low. In the 1990s, the precipitation again entered an abundant period. After the 21st century, the precipitation began to decrease. The characteristics of the two feature scales, 9a and 14a, are similar. Before the mid-1970s, the oscillation of the cycle time was more obvious, and after the rich period of 1970 to 1980, there was a slight increase in the feature scale, respectively, about 10a and about 15a.

The analysis reveals that the periodic variation of rainfall is obvious. Therefore, it is important to predict the future rainfall by grasping the key information of the rainfall in the historical year and using the data mining technology to classify the rainfall-related data reasonably.

2.2. An Introduction of Ant Colony Algorithm

We take the Travelling Salesman Problem (TSP) as an example to illustrate Ant Colony (AC) Algorithm. Suppose there are m cities, d_{ij} is the distance between city i and city j. $\tau_{ij}(t)$ is the amount of information between city i and city j at time t. We use it to simulate the actual ant anterin, set a total of m ants, the term $p_{ij}(t)$ represents the probably of the k-th ant being transferred between city i and city j at time t:

$$p_{ij}(t) = \frac{\tau_{ij}^{a}(t) \cdot \eta_{ij}^{b}}{\sum\limits_{(i,k) \in S, k \notin U} \tau_{ik}^{a}(t) \cdot \eta_{ik}^{b}} \tag{1}$$

where U is the part of the path that the ants have searched for, and S is the set of cities that the next step of the ant k allowed to pass, a indicates the amount of information on the path to the path chosen by the ants, η_{ij} indicates the degree of transfer expectation between city i and j. When $a = 0$, the algorithm is the traditional greedy algorithm; and when $b = 0$, it becomes a pure positive feedback heuristic algorithm. After n moments, the ants can finish all the cities and complete a cycle. In this case, the amount of information on each path is updated according to the following formula:

$$\tau_{ij}^{new} = (1 - \rho)\tau_{ij}^{old} + \Delta\tau_{ij} \tag{2}$$

where $\rho \in (0, 1)$ represents the amount of information that fades with time. The information increment is expressed as:

$$\Delta\tau_{ij} = \sum_{k=1}^{m} \Delta\tau_{ij}^{k} \tag{3}$$

where $\Delta\tau_{ij}^{k}$ is the amount of information left by ant k between city i and j. It can be expressed as:

$$\Delta\tau_{ij}^{k} = \begin{cases} Q/L_k, i, j \in S \\ 0, \text{otherwise} \end{cases} \tag{4}$$

where Q is a constant and L_k is the length of the path traveled by the ant k in this cycle. After several cycles, the calculation can be terminated according to the appropriate stop condition.

2.3. Basic Principles of Fuzzy Clustering

Among the many fuzzy clustering algorithms, the most widely used and successful is Fuzzy C-means (FCM). The FCM algorithm divides n vectors x_k ($k = 1, 2, \ldots, n$) into m fuzzy clusters and obtains the clustering center of each cluster so that the objective function is minimized. The objective function is defined as:

$$J = \sum_{k=1}^{n} \sum_{i=1}^{m} (\mu_{ik})^{h} d(x_k, c_i) \tag{5}$$

where μ_{ij} is the membership function, c_i is the i-th clustering center, h is the fuzzy weight index. $\mu_{ij} \in (0, 1)$ and:

$$\sum_{i=1}^{m} \mu_{ik} = 1 \tag{6}$$

$$d(x_k, c_i) = ||x_k - c_i|| \tag{7}$$

In order to minimize the objective function, the update of the cluster center and membership function is as follows:

$$c_i = \frac{\sum_{k=1}^{n} \mu_{ik}^{h} x_k}{\sum_{k=1}^{n} \mu_{ik}^{h}} \tag{8}$$

$$\mu_{ik} = \frac{1}{\sum_{j}^{m} (d_{ik}/d_{jk})^{1/(h-1)}} \tag{9}$$

Since the solution of a multi-constrained optimization problem is complex, the commonly used method is to fix one of the parameters to optimize the other amounts, and the solution alternates until the difference between two consecutive functions is less than a very small value (the precision requirement).

The flaw of this algorithm is that it needs to be given multiple c values for repeated calculations, and the result is usually a local optimal solution, and the computation time is large because the time required for a matrix multiplication is $O(n^3)$, the time complexity of the first step of the algorithm is reached $O(n^4 \log n)$.

2.4. Improvement of Fuzzy Clustering by Ant Colony Algorithm

One of the keys to improve the speed of fuzzy clustering is to select the initial point of the membership function. If we can get the membership degree approximation result of each parameter point to each cluster, we will improve the speed of the fuzzy clustering algorithm and the ant colony algorithm can achieve this function.

The basic idea is to treat the data as an ant with different attributes. The clustering center is regarded as the "food source" that the ants are looking for, so the data clustering is seen as the process of the ant looking for food sources. The specific process can be described as follows: each ant travels from each cluster center, searches for the next sample point in the entire solution space, and then starts from the cluster center and searches for another sample point in the entire solution space. When the sample point reaches the total number of the original sample points of the cluster, the ant is considered to have completed a search for a path so that the ants do not repeat the same sample point in the search for the same path and set a taboo for each ant $tabu(N)$. If $tabu(N) = 1$, then the node j can choose the search sample point, when the ants selected the node j, $tabu(N)$ will be set 0, then the ants cannot choose the node.

Assume $X = \{X_i \mid X_j = (x_{i1}, x_{i2}, \ldots, x_{im})\}, i = 1, 2, \ldots, n\}$ is a collection of data to be clustered, $\tau_{ij}(t)$ is amount of information between X_i and X_j at time t. When all the ants have completed a path search, it is said the algorithm carried out a search cycle. In the t search period, the path selection probability can be expressed as:

$$p_{ij}(t) = \begin{cases} \dfrac{\tau_{ij}^{a}(t)\cdot\eta_{ij}^{b}}{\sum\limits_{(i,s)\in S,k\notin tabu(j)} \tau_{is}^{a}(t)\cdot\eta_{is}^{b}}, & j = 1,2,\ldots,m \\ 0, & \text{otherwise} \end{cases} \tag{10}$$

where $S = \{X_s \mid d_{sj} \leq r_j, s = 1, 2, \ldots, N\}$, and the other parameters are consistent with the above.

When the i value is determined, make j from 1 to m, search the maximum $p_{ij}(t)$, then X_i is merged into the X_j field. Make $C_j = \{X_i \mid d_{ij} < r_j, i = 1, 2, \ldots, k\}$, C_j represents all the data sets that are merged into X_j, and we find the cluster center:

$$\bar{c}_j = \frac{1}{k}\sum_{i=1}^{k} X_i \tag{11}$$

When the ant colony completes a search period, the probability of each parameter point attributed to a cluster is obtained according to $p_{ij}(t)$, and the general initial value of the fuzzy clustering membership matrix is obtained, and $\bar{c}_j$ is used as the initial of fuzzy clustering center.

As the ant colony algorithm itself has a certain computational complexity, each fuzzy clustering cycle using multiple ant colony algorithm will produce over optimization phenomenon. We adopt the following strategy: in the first initial cycles, we use ant colony algorithm to determine the initial value $p_{ij}(t)$ (in this paper we set 4 cycles) and $\bar{c}_j$, then iterate according to Equations (8) and (9). When the optimization process slows down, we use the ant colony algorithm once or twice to optimize until the accuracy requirements are reached.

2.5. Method Validation

In terms of monthly rainfall, the relevant factors include historical monthly mean temperature, mean air pressure, mean humidity, season etc. Because it is the region's total rainfall forecast, we do not take into account the effects of terrain and ocean currents.

In this way, a five-dimensional vector is formed, namely monthly rainfall, monthly average temperature, monthly mean humidity, monthly mean pressure, seasonal type. Through the fuzzy clustering, the historical data will be grouped to form a database. In this way, when we predict future weather data more precisely, we can search from the historical database to find useful data for rainfall forecast. For seasonal data, we need to map the value, as shown in Table 1.

Table 1. Seasonal mapping values.

Month	12, 1, 2	3, 4, 5	6, 7, 8	9, 10, 11
value	0.1	0.3	0.5	0.7

In order to test the performance of the algorithm, on a computer with the processor i7, memory 4g, the membership degree matrix is simulated. The first is to generate (0, 1) random number, the second is to use ant colony optimization (that is $p_{ij}(t)$). When calculating $p_{ij}(t)$, set $\rho = 0.7$, $a = 1$, $b = 1$, $\eta = 1$, $\tau_{ij}(0) = 0$. The results are shown in Figure 3.

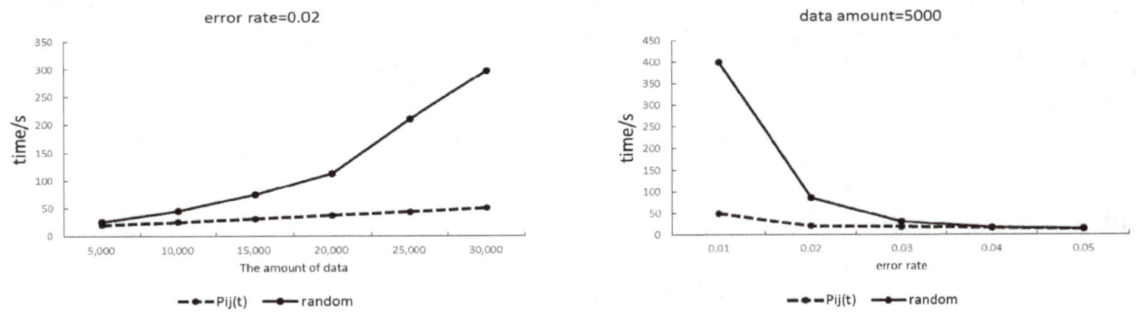

(**a**) (**b**)

Figure 3. Clustering calculation time. (**a**) calculation time when error rate is constant; (**b**) calculation time when data amount is constant.

Figure 3a shows that the fuzzy clustering with the random number as the initial value of the membership matrix increases rapidly with the increase of the data volume, while the fuzzy clustering algorithm with $p_{ij}(t)$ improves the number of samples with the increase of the number. Figure 3b shows that when the error rate is large, the random number and fuzzy clustering calculation using $p_{ij}(t)$ are not very different, but when the error rate becomes smaller, the $p_{ij}(t)$ fuzzy clustering is not changed, and the fuzzy clustering using random number calculation time is rising rapidly. Therefore, the clustering method adopted in this paper is more scientific and effective.

3. Rainfall Forecasting Model Based on CE

The historical data is huge and contains useful data. There are some abnormal data, so we have to choose specific methods according to actual situations in order to ensure the accuracy of the forecast. It is important to predict the future rainfall by grasping the key information of the rainfall in the historical year and using the data mining technology to classify the rainfall-related data reasonably; So Fuzzy Clustering by Ant Colony Algorithm is used for data clustering and the historical data are classified.

3.1. Combined Forecasting Model

The combined forecasting model comprises m single forecasting models and the relative effectiveness of a single forecasting model determined by the historical data. If the combined forecast value at time t is y_t, ω_{ij} is the weight of the i-th model at time t, and $\hat{y}_{it}$ is the predicted value of the i-th model at time t, then the problem of combined forecasting is described as follows:

$$y_t = \sum_{i=1}^{m} \omega_{it} \hat{y}_{it} \tag{12}$$

Here two factors influence the final results of combined forecasting: a single model and the weight of a single forecasting model. In this study, we focus on the latter.

There are no uniform rules for selecting a single method, but instead we must consider the actual problem and the needs of the model. The factors considered in this study include: independence, diversity, and the accuracy of the algorithm. We use a single forecasting method to include the ARIMA time series model, GM, and the RBF.

3.2. The CE Model

According to the definition of entropy, a method for calculating the difference in information between two random vectors is defined as the CE. The CE model determines the extent of the mutual support degree by assessing the degree of intersection between different information sources. Moreover, the mutual support degree can be used to determine the weights of the information sources, where a greater weight represents higher mutual support. This is also called the *Kullback-Leibler* (*K-L*) distance. The CE of two probability distributions is expressed as $D(f||g)$ [26,27].

For the discrete case:

$$D(g||f) = \sum_{1}^{n} g_i \ln \frac{g_i}{f_i} \tag{13}$$

and for the continuous case:

$$\begin{aligned} D(g||f) &= \int g(x) \ln \tfrac{g(x)}{f(x)} dx \\ &= \int g(x) \ln g(x) dx - \int g(x) \ln f(x) dx \end{aligned} \tag{14}$$

where $D(g||f)$ represents the f to g distance, and f and g denote the probability vector in the discrete case and the probability density function in the continuous case, respectively.

The CE model quantifies the "distance" between the amounts of information. However, the *K-L* distance is not the real length distance, but instead it is the difference between two probability distributions. In this paper, g is the combined forecast function, f is the single method. CE value should be smallest when two pdf are identical. For the combined forecasting model based on CE, the CE model represents the support for combined forecasting. Therefore, the objective function is to assign weights between different single methods, so that there is the most similar case between the total predictive function and the true value.

To use the CE model, two major problems should be solved: establishing the probability density function and generating the CE objective function and solving the weight coefficient by iteration.

The rainfall is treated as a sequence of discrete random variables in the forecast period. For a certain point in the sequence, the value of the rainfall at a certain prediction time is continuous, so it can be regarded as a continuous random variable. Therefore, rainfall prediction can be treated as a sequence of discrete times but continuous values.

The probability density function for predicting rainfall $f(x)$ can be regarded as the probability density function $f_i(x)$ of the single forecasting method multiplied by the corresponding weight. According to the central limit theorem, if a variable is the sum of many independent random factors, we can treat the variable as following a normal distribution, and thus the rainfall value at a certain time can be considered as satisfying a normal distribution. The minimum CE is used to determine the probability distribution of the different forecasting methods, so the combined probability distribution of the rainfall is obtained.

The probability density function for method i is ($i = 1, 2, \ldots, m$):

$$f_i(x) = \frac{1}{\sqrt{2\pi}\sigma_i} e^{-(x-\mu_i)^2/(2\sigma_i^2)} \tag{15}$$

where μ_i is the mean value and σ_i is the variance.

Thus, the combined probability density function of the predicted rainfall can be obtained based on the probability density function of the single prediction method:

$$f(x) = \sum_{i=1}^{m} \omega_{it} f_i(x) \tag{16}$$

where ω_{ij} is the weight of the i-th single method.

Therefore:

$$f(x) \sim N(\sum_{i}^{m} \omega_{it}\mu_{it}, \sum_{i}^{m} (\omega_{it}\sigma_{it})^2) \tag{17}$$

From (17), the objective function of the minimum CE optimization problem is set as:

$$\min F = \min D(f_i(x)||f(x)) \tag{18}$$

s.t.

$$\begin{cases} 0 \leq \omega_{it} \leq 1 \\ \sum_{i=1}^{m} \omega_{it} = 1 \end{cases}$$

Selecting the appropriate weight vector to obtain the minimum F involves determining the support for different algorithms.

The weight coefficient is derived based on the Lagrange function method. The *K-L* distance can be transformed into a sampling function $g^*(x)$ and $f(x;\omega_{ij})$ to ensure that $-\int g^*(x) \ln f(x;\omega_{it})dx$ reaches the minimum value, which is equivalent to the maximum value problem:

$$\max \int g^*(x) \ln f(x;\omega_{it})dx \tag{19}$$

where:

$$g^*(x) = \frac{I_{\{S(x)>\gamma\}} f(x,\omega_0)}{l} \tag{20}$$

and $I_{\{S(x)>\gamma\}}$ is called the indicator function:

$$I_{\{S(x)>\gamma\}} = \begin{cases} 1, S(x) > \gamma \\ 0, S(x) \leq \gamma \end{cases} \tag{21}$$

where $S(x)$ is also $f(x;\omega_{ij})$, ω_0 is the initial weight, γ is the target estimation parameter, and L represents the estimated target value of a low probability event.

Based on the idea of CE, a low probability sampling method is used to convert the optimization problem into the following CE problem:

$$\max \overline{D}(\omega_{it}) = \max \frac{1}{N} \sum_{k=1}^{N} I_{\{S(x)>\gamma\}} \ln f(x_k;\omega_{it}) \tag{22}$$

where N is a random number of samples.

Note that $\sum_{i=1}^{m} \omega_{it} = 1$, and thus we can construct a Lagrange function:

$$H(x) = \frac{1}{N} \sum_{k=1}^{N} I_{\{S(x)>\gamma\}} \ln f(x_k;\omega_{it}) + \lambda(\sum_{i=1}^{m} \omega_{it} - 1) \tag{23}$$

where λ is the Lagrange multiplier.

Note that:

$$\ln f(x_k;\omega_{it}) = -\sum_{i=1}^{m} \frac{(x_k - \mu_{it})^2}{2\sigma_{it}^2} \ln \frac{\omega_{it}}{\sqrt{2\pi}\sigma_{it}} \tag{24}$$

By taking the partial derivative to ω_{ij} and λ to zero, we can obtain:

$$\omega_{it} = \sqrt{\frac{\pi}{2}} \frac{\sum_{k=1}^{N} I_{\{S(x)>\gamma\}} (x_k - \mu_{it})^2}{N\sigma_{it}\lambda} \tag{25}$$

By substituting this into $\sum_{i=1}^{m} \omega_{it} = 1$, we can obtain:

$$\lambda = \frac{1}{N}\sqrt{\frac{\pi}{2}}\sum_{i=1}^{m} \frac{\sum_{k=1}^{N} I_{\{S(x)>\gamma\}}(x_k - \mu_{it})^2}{N\sigma_{it}} \tag{26}$$

The expression for the weight coefficient is obtained as follows:

$$\omega_{it} = \frac{\sum_{k=1}^{N} I_{\{S(x)>\gamma\}} \frac{(x_k-\mu_{it})^2}{\sigma_{it}}}{\sum_{i=1}^{m}\sum_{k=1}^{N} I_{\{S(x)>\gamma\}} \frac{(x_k-\mu_{it})^2}{\sigma_{it}}} \tag{27}$$

Iterative process:

A. Set $t = 1$;

B. Set $w_{it} = w_0$, set iteration number $z = 1$;

C. Generate sample sequence $X = \{X_1, X_2, \ldots, X_N\}$ by $f(x; \omega_{it})$, and sort it from small to large, calculate $S(x_k) = f(X_k, \omega_{it})$, and thus the estimated value γ is:

$$\gamma(z) = f(X_{1-\rho}; \omega_{it}) \tag{28}$$

D. Calculate (27) and obtain the z-th iteration result $\omega_{it}(z)$. Set $z = z + 1$;

E. Return to Step B to obtain $\gamma(z)$, and calculate $|\gamma(z) - \gamma(z-1)|$. If the results is less than a certain error ε, return to F; otherwise, return to C;

F. Stop the iterations, where $\omega_{it}(z)$ is the optimal weight and the rainfall prediction value is

$$f_t = \sum_{i=1}^{m} \omega_{it}(z) f_{it} \tag{29}$$

G. Set $t = t + 1$. Assess whether t is less than or equal to T. If yes, return to step 2 to calculate some combined forecast values at other times; if not, finish the computation.

The overall forecasting process is shown in Figure 4.

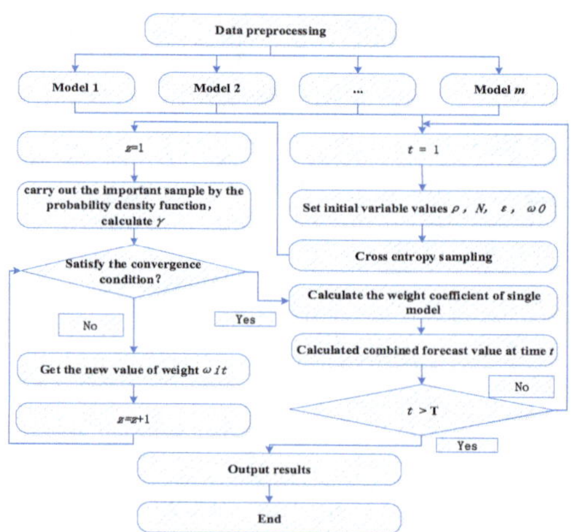

Figure 4. Flowchart illustrating the algorithm.

4. Results and Analysis

This study selects the monthly and annual rainfall data from 1960–2010 as training samples and chooses the rainfall data from 2011–2013 as test samples. Then a monthly rainfall forecast is carried on.

The selected single models include ARIMA, GM, RBF, CE. The performance of the single forecasting model and combined forecasting model is characterized by Root mean squared error (RMSE) and Maximum relative percentage error (MRPE).

4.1. Predictive Stability Comparison Results

After the single algorithm is determined, the rainfall from 2011 to 2013 is predicted and the absolute error is calculated. To compare the stability of various algorithms, true value, predictive value and the absolute error trend curve (a total of 36 prediction points) are shown as follows:

From Figure 5, it is clear that in a single algorithm, the absolute error of prediction is large or small, and stable results cannot be maintained over a long time scale. As a result, the reliability is not high; in combined prediction, the absolute error is relatively flat, which greatly improves the stability of prediction. The prediction results are more authentic. The error analysis of the results of the 2011–2013 is shown in Table 2.

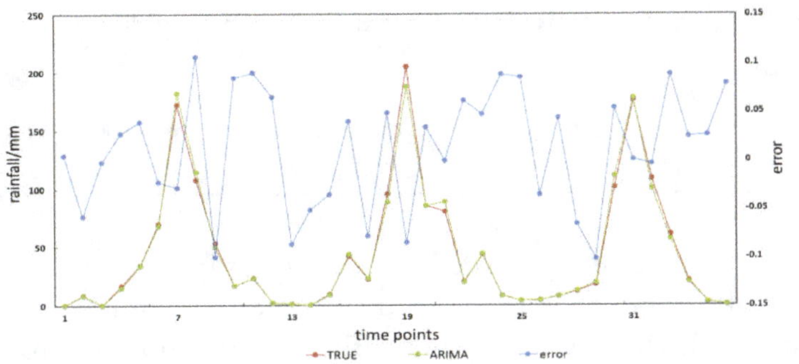

(**a**)

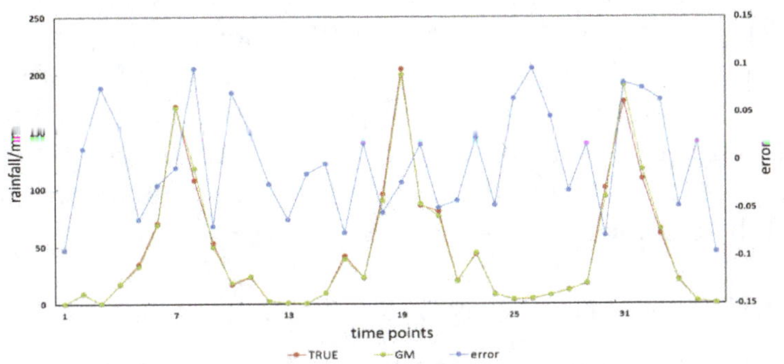

(**b**)

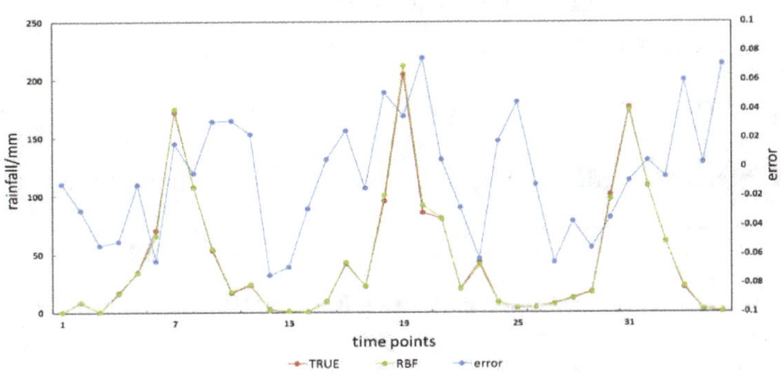

(**c**)

Figure 5. *Cont.*

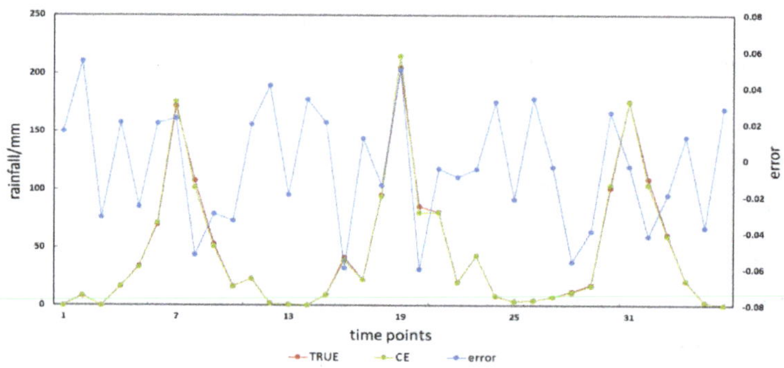

(d)

Figure 5. Prediction curve. (**a**) ARIMA; (**b**) GM; (**c**) RBF; (**d**) CE.

Table 2. Error analysis of the results.

	ARIMA		GM		RBF		CE	
	MRPE	RMSE	MRPE	RMSE	MRPE	RMSE	MRPE	RMSE
2011	12.13%	5.53%	11.15%	4.49%	10.99%	4.21%	10.02%	3.93%
2012	9.44%	4.99%	13.15%	4.97%	8.65%	5.01%	9.45%	4.32%
2013	11.10%	7.12%	7.22%	6.19%	7.01%	4.49%	6.88%	4.17%
Average	10.89%	5.88%	10.51%	5.22%	8.88%	4.57%	8.78%	4.14%

From Table 2 and Figure 5 it can be seen that: (1) compared to a single method of RMSE, the combination of forecasting methods in a certain prediction point may not be optimal, but the overall error is small compared with GM, ARIMA, RBF of RMSE, and the error reduces. The combined forecasting method has a higher accuracy: in the average MRPE index, compared to ARIMA, GM, RBF, CE reduces error of 2.11%, 1.73%, 0.1%; in the average RMSE index, compared to ARIMA, GM, RBF, CE were reduced by 1.74%, 1.08%, 0.43%; (2) the MRPE index in a single method can be very high (MRPE of GM in 2012), and there is a risk of failure of the model. The combined forecasting method greatly reduces the maximum error, and has better prediction stability (MRPE of CE in 2012); (3) in a single method, the prediction error sometimes is large, the error of the combined forecast will be also relatively large (for example in 2012), and the accuracy of the single forecasting model has a certain effect on the accuracy of the combined model.

4.2. The Influence of Clustering Method on Prediction Results

To illustrate the effect of clustering on the accuracy of prediction, two scenarios are created:

Scenario 1: Traditional c-means clustering method
Scenario 2: We do not cluster historical data.

The error results are shown in Table 3 and Figure 6.

Table 3. Influence of clustering method on prediction results.

	S1		S2	
	MRPE	RMSE	MRPE	RMSE
2011	10.04%	3.89%	13.01%	7.12%
2012	9.41%	4.34%	12.04%	8.14%
2013	6.92%	4.17%	10.29%	7.71%
Average	8.79%	4.19%	11.78%	7.66%

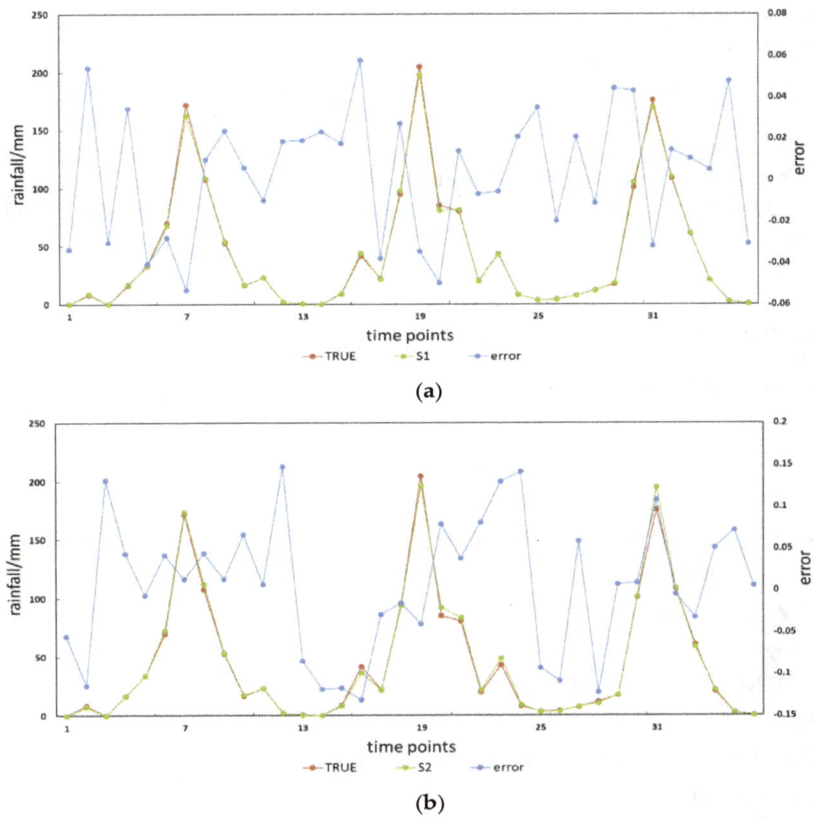

Figure 6. Prediction curve. (**a**) S1; (**b**) S2.

From the comparison between Tables 2 and 3, it is easy to know that the clustering method used in this paper is consistent with the accuracy of traditional clustering method to meet the forecasting requirements. In the average MRPE index, compared to S1, S2, the clustering method used in this paper reduces the error by 0.01%, 3%; in the average RMSE index, compared to ARIMA, GM and RBF, CE were reduced by 0.05%, 1.08%, 3.52%. However, from Figure 3, the method has great advantages in calculating the speed when the data volume is large. In addition, comparing with the case of S2, we can see that there is a large deviation in the result of prediction in the case of no clustering, which indicates that the selection of historical data has important influence on the precision of prediction and it is necessary to fully excavate and classify the historical information to reach reasonable effect.

5. Conclusions

Because the meteorological conditions are stochastic, the rainfall variation is a non-stationary time series, and the accuracy of prediction and reliability of traditional single forecasting methods cannot be ensured. The combined forecasting model based on CE is proposed to solve these problems. Besides, the forecasting results of several single methods are used as the input variables for training, and the output weights are predicted by analyzing the total rainfall. The simulation results show that the combined forecasting model enhances the accuracy and reliability of the rainfall forecasting model. Additionally, clustering method is modeled and analyzed so as to improve the accuracy of prediction. The results demonstrate that the accuracy of combined forecasting can be improved by using the fuzzy clustering. The chosen method characterizes the change laws of the rainfall well and to improve the stability of the prediction. The prediction results can help agriculture, water conservancy departments to improve the ability of drought and flood disaster prevention and control. In the future, more steps can be taken for improvement: (1) the single forecasting method with higher accuracy and more suitable single forecasting method should be chosen and the accuracy of combined model prediction should be further improved to explore the rule of selecting different single methods in combined forecasting method; (2) the data in similar year should be collected to predict and historical data should be utilized more effectively so that the result is more scientific and convincing.

Acknowledgments: This work was supported by the National Key R&D Program of China (Grant No. 2016YFC0401406), and the Famous Teachers Cultivation planning for Teaching of North China Electric Power University (the Fourth Period), the 2014 Education Reform Project of North China Electric Power University (Beijing Department) (Grant No. 2014JG57).

Author Contributions: The combined forecasting method based on SVM was developed to forecast the rainfall. The results demonstrate that this model can improve the accuracy of daily rainfall forecasting and it can be applied in practice. Baohui Men designed the study, Rishang Long and Yangsong Li performed the experiments and wrote the paper, Baohui Men reviewed and edited the manuscript. All authors read and approved the manuscript.

References

1. Serreze, M.C.; Etringer, A.J. Precipitation characteristics of the Eurasian Arctic drainage system. *Int. J. Climatol.* **2003**, *23*, 1267–1291. [CrossRef]
2. Bustamante, J. Evaluation of April 1999 Rainfall Fore-casts Over South American using the Eta Model Climanalise, Divulgacao Científica. *Cachoeira Paulista* **1999**, *6*, 3563–3569.
3. Black, T. The New NMC Mesoscale ETA Model: Descriptionand Forecast Examples. *Whether Forecast.* **1994**, *9*, 265–278. [CrossRef]
4. Mossad, A.; Alazba, A.A. Drought Forecasting Using Stochastic Models in a Hyper-Arid Climate. *Atmosphere* **2015**, *6*, 410–430. [CrossRef]
5. Bian, H.J.; Lei, H.J.; Wang, Y. Application of Grey Theory to Regional Rainfall Rorecast. *Anhui Agri. Sci.* **2009**, *37*, 6059–6060.
6. Yang, J.L.; Wu, Y.N.; Xie, M.; Li, H.Y. Application of Monte Carlo Method in Rainfall Forecast of Flood Season in Nenjiang Basin. *S. N. Water Transf. Water Sci. Technol.* **2011**, *9*, 28–32.
7. Ramirez, M.C.V.; de Campos Velho, H.F.; Ferreira, N.J. Artificial Neural Network Technique for Rainfall Forecasting Applied to the Sao Paulo Region. *J. Hydrol.* **2005**, *301*, 146–162. [CrossRef]
8. Manzato, A. Sounding-derived indices for neural network based short-term thunderstorm and rainfall forecasts. *Atmos. Res.* **2007**, *83*, 349–365. [CrossRef]
9. Yang, S.; Rui, J.; Feng, H. Application of Support Vector Machine (SVM) Method in Precipitation Classification Forecast. *J. Southwest. Agric. Univ.* **2006**, *28*, 252–257.
10. Cui, L.; Chi, D.; Qu, X. Based on Wavelet De-noising of Stationary Time Series Analysis Method in Rainfall Forecasting. *China Rural Water Hydropower* **2010**, *9*, 31–35.
11. Merlinde, K. The Application of TAPM for Site Specific Wind Energy Forecasting. *Atmosphere* **2016**, *7*, 23. [CrossRef]
12. Lauret, P.; Lorenz, E.; David, M. Solar Forecasting in a Challenging Insular Context. *Atmosphere* **2016**, *7*, 18. [CrossRef]
13. Men, B.; Liu, C.; Xia, J.; Liu, S.; Lin, Z. Application of R/S Analysis Method of Water Runoff Trend in West Route of South-to-North Water Transfer Project. *J. Glaciol. Geocryol.* **2005**, *27*, 568–573.
14. Soro, G.E.; Noufé, D.; Goula Bi, T.A.; Shorohou, B. Trend Analysis for Extreme Rainfall at Sub-Daily and Daily Timescales in Côte d'Ivoire. *Climate* **2016**, *4*, 37. [CrossRef]
15. Chen, C.-S.; Duan, S.; Cai, T. Short-term photovoltaic generation forecasting system based on fuzzy recognition. *Trans. China Electrotech. Soc.* **2011**, *26*, 83–88.
16. Liu, Y.; Fu, X.; Zhang, W. Short-term load forecasting method based on fuzzy pattern recognition and fuzzy cluster theory. *Trans. China Electrotech. Soc.* **2002**, *17*, 83–86.
17. Kolentini, E.; Sideratos, G.; Rikos, V. Developing a Matlab tool while exploiting neural networks for combined prediction of hour's ahead system load along with irradiation, to estimate the system loadcovered by PV integrated systems. In Proceedings of the IEEE Conferences on Clean Electrical Power, Capri, Italy, 9–11 June 2009; pp. 182–186.
18. Bates, J.; Granger, C. The combination of forecast. *Oper. Res. Q.* **1969**, *20*, 451–468. [CrossRef]

19. Wu, Q.; Peng, C. Wind Power Generation Forecasting Using Least Squares Support Vector Machine Combined with Ensemble Empirical Mode Decomposition, Principal Component Analysis and a Bat Algorithm. *Energies* **2016**, *9*, 261. [CrossRef]
20. Pedersen, J.W.; Lund, N.S.V.; Borup, M.; Löwe, R.; Poulsen, T.S.; Mikkelsen, P.S.; Grum, M. Evaluation of Maximum a Posteriori Estimation as Data Assimilation Method for Forecasting Infiltration-Inflow Affected Urban Runoff with Radar Rainfall Input. *Water* **2016**, *8*, 381. [CrossRef]
21. Box, G.E.P.; Jenkins, G.M. *Time Series Analysis, Forecasting and Control*; Holden-Day: San Francisco, CA, USA, 1976.
22. Men, B.; Long, R.; Zhang, J. Combined Forecasting of Stream flow Based on Cross Entropy. *Entropy* **2016,** *18* 336. [CrossRef]
23. Cui, D. Application of Combined Model in Rainfall Forecast. *Comput. Simul.* **2012**, *29*, 163–166.
24. Xiong, L.; O'Connor, K.M.; Kieran, M. Comparison of four updating models for real-time river flow forecasting. *Hydrol. Sci. J.* **2002**, *47*, 621–640. [CrossRef]
25. Lu, C.; Zhang, Y.; Zhou, J.; Vijay, P.; Guo, S.; Zhang, J. Real-time error correction method combined with combination flood forecasting technique for improving the accuracy of flood forecasting. *J. Hydrol.* **2015**, *521*, 157–169.
26. Singh, V.P. *Entropy Based Parameter Estimation in Hydrology*; Kluwer Academic Publishers: Dordrecht, The Netherlands, 1998.
27. Cui, H.; Singh, V.P. Entropy spectral analyses for groundwater forecasting. *J. Hydrol. Eng.* **2017**, 22, 06017002. [CrossRef]
28. Chen, L.; Singh, V.P.; Xiong, F. An Entropy-Based Generalized Gamma Distribution for Flood Frequency Analysis. *Entropy* **2017**, *19*, 239. [CrossRef]
29. Chen, L.; Singh, V.P. Generalized Beta Distribution of the Second Kind for Flood Frequency Analysis. *Entropy* **2017**, *19*, 254. [CrossRef]
30. Chen, L.; Singh, V.P.; Guo, S.L.; Zhou, J.; Ye, L. Copula entropy coupled with artificial neural network for rainfall-runoff simulation. *Stoch. Environ. Res. Risk Assess.* **2014**, *28*, 1755–1767. [CrossRef]
31. Li, R.; Liu, H.L.; Lu, Y.; Han, B. A combination method for distribution transformer life prediction based on cross entropy theory. *Power Syst. Prot. Control* **2014**, *42*, 97–101.
32. Chen, N.; Sha, Q.; Tang, Y.; Zhu, L. A Combination Method for Wind Power Predication Based on Cross Entropy Theory. *Proc. CSEE* **2012**, *32*, 29–34.
33. Mehdi, N.; Hossein, N. Image denoising in the wavelet domain using a new adaptive thresholding function. *Neurocomputing* **2009**, *72*, 1012–1025.
34. Asgari, M.S.; Abbasi, A. Comparison of ANFIS and FAHP-FGP methods for supplier selection. *Kybernetes* **2016**, *45*, 474–489. [CrossRef]

4

Modeling NDVI using Joint Entropy Method Considering Hydro-Meteorological Driving Factors in the Middle Reaches of Hei River Basin

Gengxi Zhang [1], Xiaoling Su [1,*], Vijay P. Singh [2] and Olusola O. Ayantobo [1,3]

1 College of Water Resources and Architectural Engineering, Northwest A & F University, Yangling 712100, China; gengxizhang@nwsuaf.edu.cn (G.Z.); ayantobo.olusola@nwsuaf.edu.cn or ayantobooo@funaab.edu.ng (O.O.A.)

2 Department of Biological & Agricultural Engineering and Zachry Department of Civil Engineering, Texas A&M University, 2117 TAMU, College Station, TX 77843, USA; vsingh@tamu.edu

3 Department of Water Resources Management and Agricultural-Meteorology, Federal University of Agriculture, PMB 2240, Abeokuta 110282, Nigeria

* Correspondence: xiaolingsu@nwsuaf.edu.cn

Abstract: Terrestrial vegetation dynamics are closely influenced by both hydrological process and climate change. This study investigated the relationships between vegetation pattern and hydro-meteorological elements. The joint entropy method was employed to evaluate the dependence between the normalized difference vegetation index (NDVI) and coupled variables in the middle reaches of the Hei River basin. Based on the spatial distribution of mutual information, the whole study area was divided into five sub-regions. In each sub-region, nested statistical models were applied to model the NDVI on the grid and regional scales, respectively. Results showed that the annual average NDVI increased at a rate of 0.005/a over the past 11 years. In the desert regions, the NDVI increased significantly with an increase in precipitation and temperature, and a high accuracy of retrieving NDVI model was obtained by coupling precipitation and temperature, especially in sub-region I. In the oasis regions, groundwater was also an important factor driving vegetation growth, and the rise of the groundwater level contributed to the growth of vegetation. However, the relationship was weaker in artificial oasis regions (sub-region III and sub-region V) due to the influence of human activities such as irrigation. The overall correlation coefficient between the observed NDVI and modeled NDVI was observed to be 0.97. The outcomes of this study are suitable for ecosystem monitoring, especially in the realm of climate change. Further studies are necessary and should consider more factors, such as runoff and irrigation.

Keywords: joint entropy; NDVI; temperature; precipitation; groundwater depth; Hei River basin

---

1. Introduction

Terrestrial vegetation plays a key role in energy, water, and biogeochemical cycles, while variation in vegetation can significantly influence atmospheric and hydrological processes. Vegetation is a sensitive indicator of global change and can also be a natural link between the atmosphere, land surface, soil, and water [1,2]. Changes in vegetation cover are influenced by climate change, human activities, and the atmospheric CO_2 fertilization effect [3–5]. It is necessary to explore the response mechanism of vegetation dynamics to hydrological processes and climate change.

Vegetation emergence and senescence are closely related to characteristics of the lower atmosphere, including the annual cycle of weather pattern shifts, temperature and precipitation [6]. In particular, the amount and duration of precipitation and temperature play a significant role in controlling vegetation development [2,7–9]. During the last decades, the global average land surface air temperature has been increasing systematically in the northern middle and high latitude [10]. Elevated air temperature increased plant growth by lengthening the growing season [11], enhancing photosynthesis, and altering nitrogen availability by accelerating decomposition or mineralization. Precipitation is the dominant controlling climatic factor in water-limited, semi-arid and arid regions such as the northern China [10]. The increase in precipitation contributes to the growth of vegetation. Several studies have reported relationships between vegetation indices and precipitation and temperature [3,12–17]. However, most of these studies focused solely on either precipitation or temperature for vegetation models, thereby resulting in low precision and low correlation coefficients between vegetation indices and climatic variables [17]. For instance, Ichii et al. (2002) found a weaker relationship between vegetation growth and precipitation [18], while Nicholson et al. (1990) reported a weaker relationship between NDVI and precipitation, indicating a nonlinear overall relationship [19]. However, apart from the influence of climatic change, vegetation is also strongly related to the groundwater table level (GTL) in arid and semi-arid regions [2,20–22]. Groundwater is a key impact factor in sustaining the ecological environment, by supplying soil moisture through capillarity for maintaining the root system. The shallower the groundwater depth the more available soil moisture, and vice versa [22]. Many investigators have also reported the relationship between vegetation and groundwater availability [2,21]. For instance, Jin et al. (2016) found a linear correlation between NDVI and groundwater depth in the Qaidam basin [2]. The distribution of GTL is becoming more heterogeneous in the middle reaches of the Hei River as a result of increasing cultivated land area and runoff from Yingluoxia station [21]. The distribution of vegetation is also highly heterogeneous and depends upon multiple factors. In this study, different statistical tools, as well as the joint entropy method, were employed in order to understand the relationships between NDVI and precipitation, temperature, and GTL. This will be necessary for investigating the relevance of coupling variables.

The joint entropy expresses the total amount of uncertainty contained in the union of events [23] either in a linear or nonlinear system [24] and has been employed for various purposes. For instance, Li et al. (2012) employed joint entropy to design appropriate hydrometric networks [25], while Mishra et al. (2012) employed the entropy theory to overcome the nonlinear relationship between precipitation and vegetation, and further identified the downscaling method yielding higher mutual information [23]. Entropy theory has also been applied in ecological studies [5,17,26,27]. For instance, Sohoulande, Djebou and Singh (2015) employed joint entropy to retrieve vegetation growth patterns from climatic variables. However, this study emphasizes the mutual information of NDVI and the influencing factors.

In this paper, the mutual information of NDVI was investigated by coupling cumulative precipitation, the average temperature of growing season, and GTL. To that end, the study area was divided into 5 sub-regions, based on 3 spatial distributions of mutual information. In each sub-region, NDVI was modeled using nested statistical models at the grid and regional scales, respectively.

2. Methodology

2.1. Trend Analysis

Trends for each pixel were represented to reflect the characteristics of vegetation cover over different periods of time [28]. The rate of change of NDVI was calculated as [29].

$$\theta_{slope} = \frac{n \times \sum_{i=1}^{n} i \times NDVI_i - \sum_{i=1}^{n} i \sum_{i=1}^{n} NDVI_i}{n \times \sum_{i=1}^{n} i^2 - (\sum_{i=1}^{n} i)^2} \quad (1)$$

where θ_{slope} is the slope of the trend line of annual NDVI; i represents the years; n is the time span; and $NDVI_i$ is the NDVI value for the i-th year.

θ_{slope} describes the trend in the annual NDVI within the study area. If $\theta_{slope} > 0$, the NDVI increases; or else, the NDVI decreases ($\theta_{slope} < 0$) or remains constant ($\theta_{slope} = 0$).

2.2. *Joint Entropy and Mutual Information*

Mutual information was employed to explore the relationship between NDVI and influencing factors. Considering a random variable X, for each value of X, x_i represents an event with a corresponding probability of occurrence, p_i. The entropy $H(X)$ can be expressed as:

$$H(X) = -\sum_{i=1}^{n} p_i \log_2(p_i) \tag{2}$$

Where i means the ith event. The logarithm is based on 2, because it is more convenient to use than logarithms based one or 10, however, the base can be taken as other than 2 without alteration [24]. The entropy is thereby measured in bits. The probability density function (PDF) of variable X is obtained using discrete intervals for the values of X. The discrete PDF is defined for n equal-width bins defined for the range of X [17,24,30]. The entropy $H(X)$ is a measurement of information or uncertainty [24]. Similarly, the joint entropy can be computed for a joint probability distribution of two or more variables [24,25]. Specifically, with three variables, the joint PDF is computed based on a three-dimensional contingency analysis as illustrated in Table 1. Considering the three variables precipitation, temperature and NDVI, the process generates a discrete PDF made of n^3 discrete probabilities such that:

$$\sum_{i=1}^{n}\sum_{j=1}^{n}\sum_{k=1}^{n} p(P = P_i,\ T = T_j,\ NDVI = NDVI_k) = 1 \tag{3}$$

The joint entropy of the three variables can be defined as:

$$H(P,\ T,\ NDVI) = -\sum_{i=1}^{n}\sum_{j=1}^{n}\sum_{k=1}^{n} p(P_i,\ T_j,\ NDVI_k)\log_2[p(P_i,\ T_j,\ NDVI_k)] \tag{4}$$

where n means the number of events.

Table 1. Illustration of a three-dimension contingency employed to compute the joint PDF. Consider 3 variables X, Y, and Z.

Variables			**Discrete Probabilities**
X = {X$_1$,X$_2$}	**Y = {Y$_1$,Y$_2$}**	**Z = {Z$_1$,Z$_2$}**	
X = X$_1$	Y = Y$_1$	Z = Z$_1$	$p(X = X_1, Y = Y_1, Z = Z_1)$
		Z = Z$_2$	$p(X = X_1, Y = Y_1, Z = Z_2)$
	Y = Y$_2$	Z = Z$_1$	$p(X = X_1, Y = Y_2, Z = Z_1)$
		Z = Z$_2$	$p(X = X_1, Y = Y_2, Z = Z_2)$
X = X$_2$	Y = Y$_1$	Z = Z$_1$	$p(X = X_2, Y = Y_1, Z = Z_1)$
		Z = Z$_2$	$p(X = X_2, Y = Y_1, Z = Z_2)$
	Y = Y$_2$	Z = Z$_1$	$p(X = X_2, Y = Y_2, Z = Z_1)$
		Z = Z$_2$	$p(X = X_2, Y = Y_2, Z = Z_2)$

Note: Each variable is categorized into 2 classes (n = 2).

The mutual information represents the amount of information common to both X and Y and provides a general measure of dependence between the random variables. It is superior to the Pearson correlation coefficient, because it captures both linear and nonlinear dependence while the Pearson correlation coefficient is only suitable for linear relationships [25]. It equals the difference between the sum of two marginal entropies and the total entropy:

$$M(X, Y) = H(X) + H(Y) - H(X, Y) \tag{5}$$

The two-dimensional cases of the amount of information transmitted can be extended to three or more dimensional cases [31]. Considering the three variables of precipitation, temperature, and NDVI, precipitation and temperature are considered inputs while NDVI is considered an output. Therefore:

$$M(P, T; NDVI) = H(P, T) + H(NDVI) - H(P, T, NDVI) \tag{6}$$

Moreover, applying the same methodology, the coupling of P and GTL; T and GTL along with NDVI were further explored. The ensemble E of candidate equations can be summarized as:

$$E = \{NDVI_{GTL,\,P} = f(GTL, P);\; NDVI_{GTL,\,T} = f(GTL, T);\; NDVI_{T,\,P} = f(T, P)\} \tag{7}$$

3. Study Area and Data

The study area occupied the middle reaches of the Hei River basin (38°30′–39°55′N, 98°55′–100°55′E; Figure 1), which is located in the middle of Hexi corridor of Gansu Province, northwest of China, and belongs to an arid climate zone. The total area is about 10,000 km^2. The mean annual temperature is around 8 °C, the mean annual precipitation is between 69 and 216 mm and is concentrated between June and September, while the mean annual potential evaporation is between 1453 and 2351 mm [32].

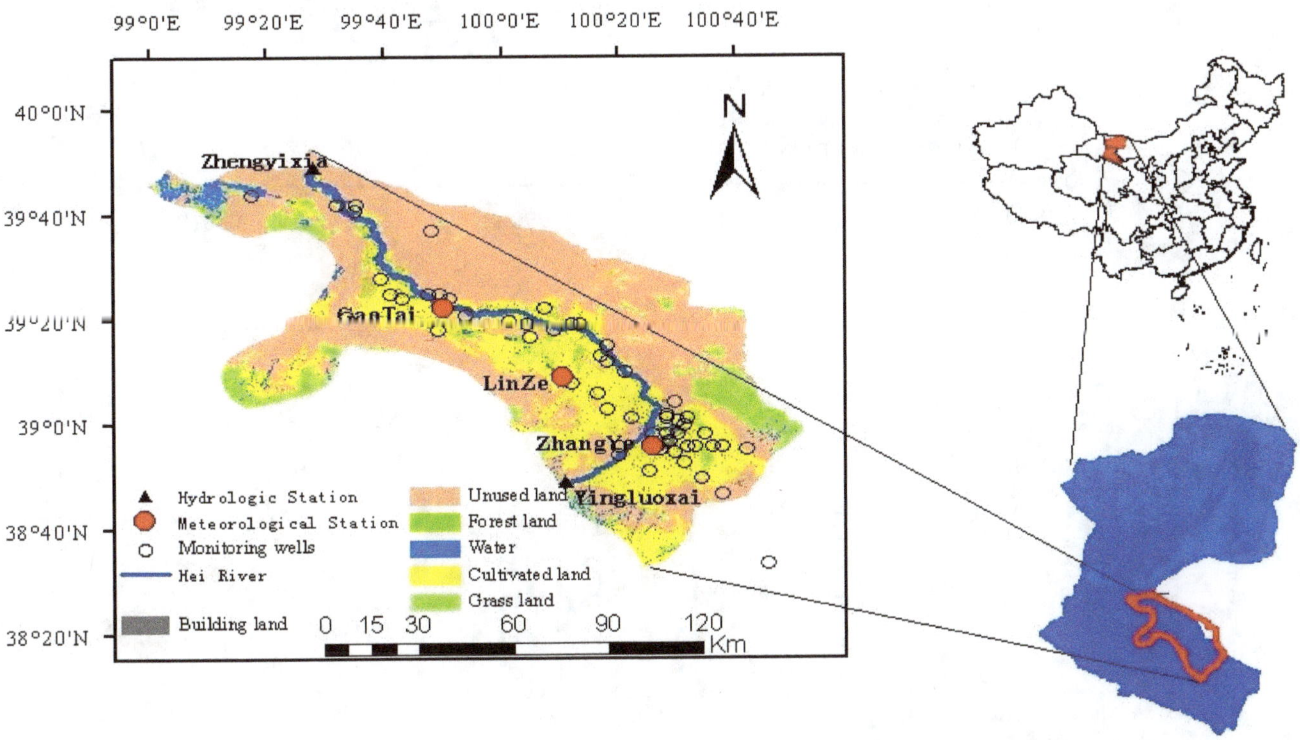

Figure 1. Map of the middle reaches of Hei River basin and the monitoring wells.

The Hei River is the second largest inland river in China with a total length of 821 km and a runoff of 16.0 × 10^8 m^3·year^{-1}. It flows through Yingluoxia and Zhengyixia stations, entering and exiting middle reaches. Runoff has shown an increasing/decreasing trend in the past decade. The annual runoff at Yingluoxai station increased to 16.0 × 10^8 m^3 in the 1990s from around 14.4 × 10^8 m^3 in the 1960s (Figure 2), while the runoff at Zhengyixai station decreased by 3 × 10^8 m^3 during the same time period (Figure 2). This situation accelerated desertification within the northern parts of the basin [33,34]. In order to mitigate deteriorating ecosystems in the Hei River basin, an Ecological Water Diversion Project (EWDP) was initiated by the Chinese government in 2000 to ensure the delivery of water to the lower reaches for ecological water needs [35]. After that, runoff at Zhengyixia station has increased to levels not recorded since the 1960s [33,36]. The implementation of the EWDP alleviated the ecological deterioration in the lower reaches of the Hei River but aggravated the water resources shortage in the middle reaches. In order to meet the increasing demand for water resources, groundwater exploration provided an option. After 2004, the recharge of groundwater also increased due to the increase in precipitation, irrigation and runoff replenishment. The temporal variation in groundwater levels of the observation wells over the period from2001 to 2011 is shown in Figure 3c which shows a rising trend in groundwater level along the Hei River [37]. However, the water level in the region away from the river showed a declining trend.

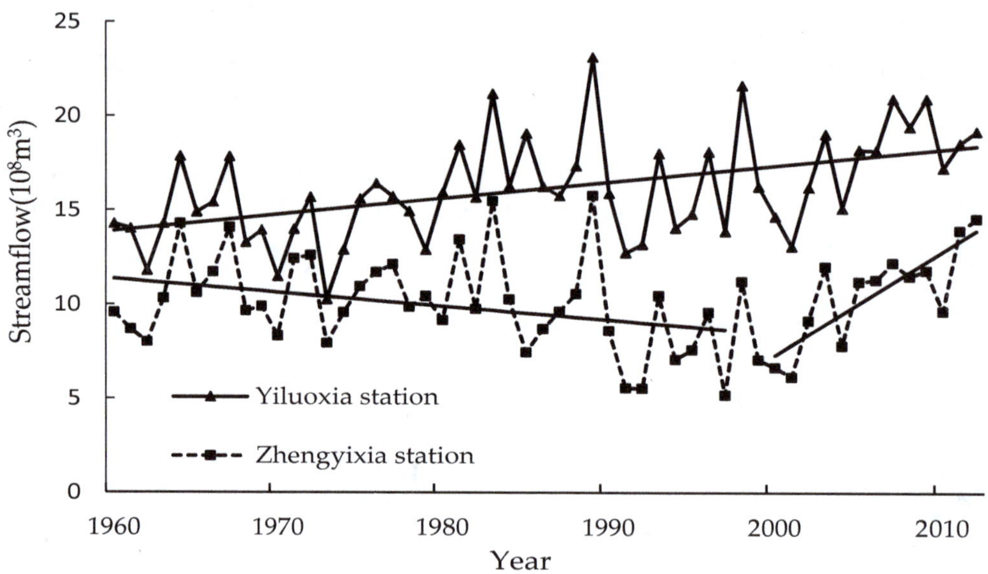

Figure 2. Annual discharge of Hei River at Yinluoxia and Zhengyixia stations.

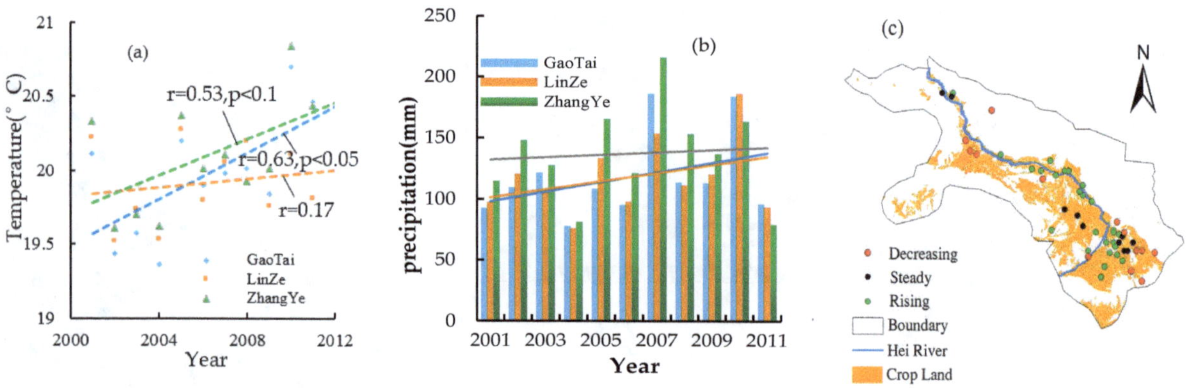

Figure 3. The temporal variation of average temperature of growing season (**a**); cumulative precipitation (**b**) and average ground depths (**c**).

The annual NDVI of each pixel was calculated by the annual maximum value of NDVI in order to eliminate the noise from the cloud and solar altitude. The Moderate-resolution imaging spectroradiometer (MODIS)-based NDVI products(MYD13A2 and MOD13A2)used for this study were acquired from the National Natural Science Foundation of China "Environmental and Ecological

Science Data Center in the West of China (http://westdc.westgis.ac.cn/)" with a spatial resolution of 1 km and a temporal resolution of 1 day, groundwater depths of 53 wells were obtained from the National Natural Science Foundation of China "Hei River Project Data Management Center (http://westdc.westgis.ac.cn/)" while monthly precipitation and temperature datasets were acquired from the China Meteorological Network (http://data.cma.cn/), with a 0.5° spatial resolution, interpolated according to the observed data of 2747 stations in China.

The ordinary kriging method was employed to interpolate the observed values of ground water depths. For consistency regarding the NDVI spatial resolution, the interpolated groundwater table levels, precipitation, and temperature datasets were rescaled to a 1 km spatial resolution.

4. Results and Discussion

4.1. Variation of Hydrometeorological Variables and NDVI

During the period from2001 to 2011, both the average temperature and cumulative precipitation increased at the three hydrologic stations (Figure 3a,b). The temperature increases significantly at Gaotai station ($p < 0.05$). However, the trend of temperature increase is not obvious at another two stations. The distributions of cumulative precipitation and average temperature are shown in Figure 4a,b. Temperature gradually rises from south to north, while precipitation decreases were also observed. This is mainly determined by topography.

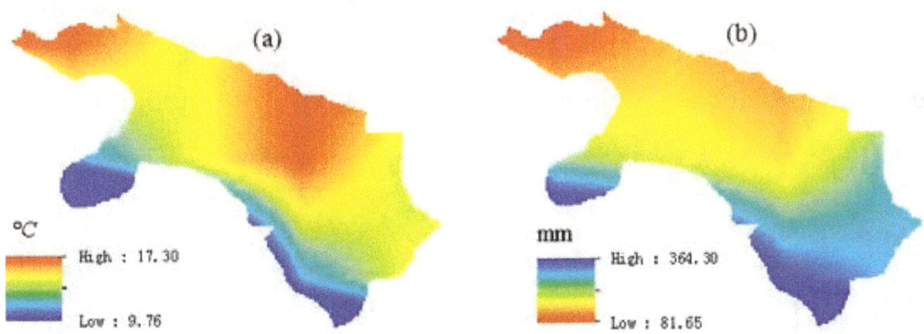

Figure 4. The spatial distribution of (**a**) average temperature of growing season; (**b**) cumulative precipitation.

Figure 5a illustrates the annual maximum NDVI in the middle reaches of the Hei River basin from 2000 to 2011. The slope of the linear regression trend-line was 0.005. The NDVI time series corresponded to the average of all the pixels in the study area and reflected the overall trend. Figure 5b shows the spatial distribution of the linear regression slope (θ_{slope}). In most cases, the area in which NDVI increased occupied about 94% of the total area. It can be seen that NDVI changes obviously. In order to explore the influencing factors on NDVI variation, the whole region was divided into 5 sub-regions based on the joint entropy method, and different regression models were constructed considering hydro-meteorological variables.

4.2. Mutual Information of NDVI with Coupling of Variables

To understand the relationship between NDVI and hydro-meteorological variables, mutual information was employed to measure the dependence between the NDVI time series and each coupled variable. The greater the values of mutual information the higher the correlation. The spatial distribution of mutual information between NDVI and coupled variables are shown in Figure 6a–c, respectively. It can be observed from the figures that the mutual information of NDVI and coupled variables of precipitation and GTL generally decreased from southeast to northwest. The mutual information values of NDVI and the coupled variables of precipitation and temperature were dispersedly distributed, increasing from the center to the edge of regions, while the mutual information of NDVI and the coupled variables of temperature and groundwater table levels were greater in the northwestern and southeastern regions. In order to appropriately model the NDVI using dependent variables, the study region was artificially divided into 5 sub-regions (Figure 6d), based on three spatial distributions of mutual information, topographic and land use maps. In order to simply construct models, as far as possible each sub-region was made to have the same characteristics. For instance, sub-region I and sub-region IV are mainly mountainous and M (P, T; NDVI) is higher than M (T, GTL; NDVI) and M (P, GTL; NDVI) in most areas of these sub-regions. As the same time, the land use types are desert. Likewise, in most areas of sub-region II and sub-region V, M (GTL, P; NDVI) is higher than M (P, T; NDVI) and M (T, GTL; NDVI), and these areas are distributed as artificial oases.

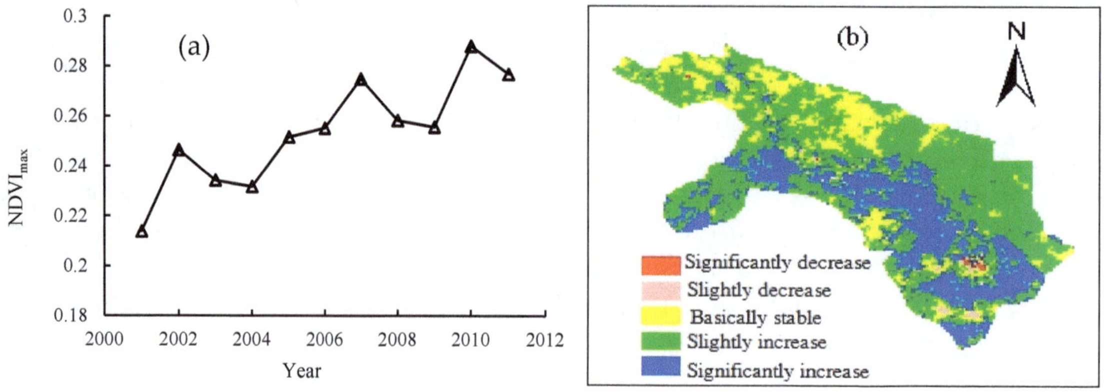

Figure 5. Inter-annual variability of annual NDVI (**a**) and the trends of NDVI (**b**).

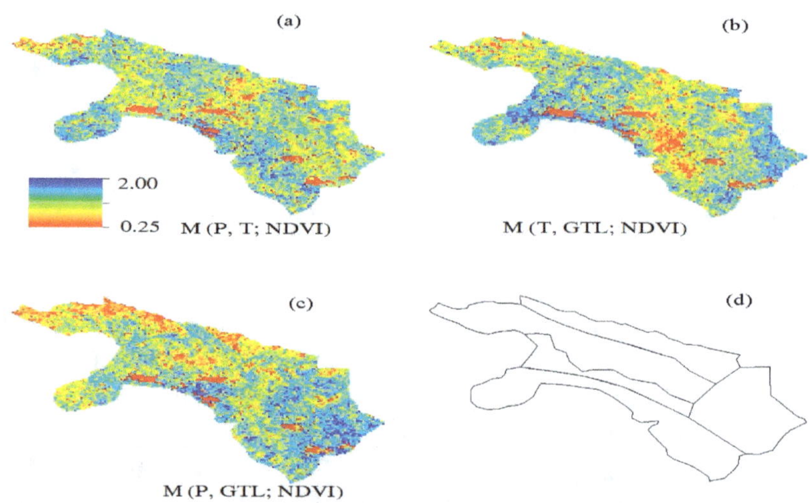

Figure 6. The distribution of mutual information of NDVI and three coupling hydro-meteorological variables (**a**–**c**) and the divided sub-regions (**d**).

4.3. Modelling NDVI

In sub-regions I and IV, precipitation and temperature were used to model the NDVI, since the mutual information of NDVI and the variables of precipitation and temperature were relatively higher. GTL and precipitation were used to model the NDVI in sub-regions II and V, while temperature and GTL were selected to model the NDVI in sub-region III.

Nested statistical models were developed for modeling NDVI in each sub-region. In most cases, the linear fitting of the NDVI time series to precipitation, temperature, and GTL resulted in poor correlation coefficients and high root means square error [17]. The vegetation growth may not be monotonic in relation to the atmospheric variables. For example, the temperature sensitivity of photosynthesis increases up to an optimum and later decreases as temperature gets higher [17]. A two-dimensional fitting was used to model NDVI using each of the hydro-meteorological variables as a factor. In order to select appropriate functional forms for modeling NDVI, four simple functional forms including linear function, quadratic function, exponent function and logarithm function were compared using Akaike information criterion (AIC) criteria. AIC is a measure of the relative quality of statistical models for a given set of data. The comparison of AIC values between different functional forms are as shown in Table 2. The smaller the AIC value, the more appropriate the form of the regression function. It was discovered that a logarithmic transformation of precipitation produced a better correlation with NDVI in some sub-regions (I, II, IV). Considering GTL and temperature, a negative exponential and a quadratic model showed a better relationship with NDVI. However, the logarithm regression model exhibited a better relationship between GTL and NDVI in sub-region III.

Table 2. The comparison of AIC values between different functional forms on 5 sub-regions; AIC in bold indicates most appropriate functional form.

NDVI and Temperature					
Function \ Regin	**Region I**	**Region II**	**Region III**	**Region IV**	**Region V**
liner	−3.6153		−4.7407	−6.1732	
quadratic	**−3.8733**		**−5.4122**	**−6.5508**	
exponent	−0.6537		−1.3973	−2.9187	
logarithm	−3.7049		−5.2447	−6.5427	
NDVI and Groundwater					
Function \ Regin	**Region I**	**Region II**	**Region III**	**Region IV**	**Region V**
liner		−1.8812	−3.8162		−2.7076
quadratic		−3.8876	−5.1042		−4.3212
exponent		**−3.9708**	−2.2641		**−4.3329**
logarithm		0.3744	**−5.1401**		0.4571
NDVI and Precipitation					
Function \ Regin	**Region I**	**Region II**	**Region III**	**Region IV**	**Region V**
liner	−3.5737	−1.7001		1.9864	−2.8516
quadratic	−4.5808	−4.714		−7.1646	**−4.4654**
exponent	−4.5530	−4.5093		−7.1620	−4.4443
logarithm	**−4.7402**	**−4.7974**		**−7.2847**	−2.5677

Different candidate equations were used to fit NDVI by coupling pair-wise precipitation, temperature, and GTL. The aim of the fitting procedures was to maximize the correlation coefficient and minimize the sum of squared residuals. However, the main importance of the fitting procedure was to maximize the proportion of the area in which the correlation coefficient met with significance testing.

In each sub-region, analysis was performed at the grid scale. In sub-regions I and IV, the coupling of precipitation and temperature resulted in a model of NDVI time series with an average correlation coefficient of 0.68 and 0.58, respectively. Li et al. (2016) also reported that in the desert regions in middle reaches of the Hei River, precipitation is a dominant influencing factor on annual NDVI variation [38]. Zhou et al. (2013) also demonstrated that precipitation and temperature are important factors driving the increase of NDVI in the study area [39]. However, the percentage of the area in which the correlation coefficient met with significance testing was only 42% ($p < 0.05$) and 15% ($p < 0.01$) in sub-region IV. This was mainly because the temperature was relatively high, thereby increasing evapotranspiration and water demand for vegetation growth that precipitation was unable to satisfy. In sub-regions II and V, the best fitting models were obtained by combining pair-wise negative exponentially transformed GTL respectively with the logarithmically and linearly transformed precipitation. In these regions, groundwater depths became an important influencing factor. Zhao et al. (2014) have also reported that in an oasis ecosystem, groundwater plays an important role in the formation of vegetation productivity [40]. In sub-region V, the percentage area in which the correlation coefficient met with the significance testing was only 50% ($p < 0.05$) and 16% ($p < 0.01$). There was a large area of arable land and artificial grassland in sub-region V (Figure 1). In sub-regions III and V, the growth of vegetation was strongly influenced by human activities, e.g., irrigation. Therefore, irrigation was the main influencing factor for vegetation growth. Beyond that, there was a considerable amount of water in sub-region III, thereby influencing prediction accuracy. After various fittings in different sub-regions, the correlation coefficients were combined for the entire study area (Figure 7). Results showed that there was about 53% of the area in which relationships were significant.

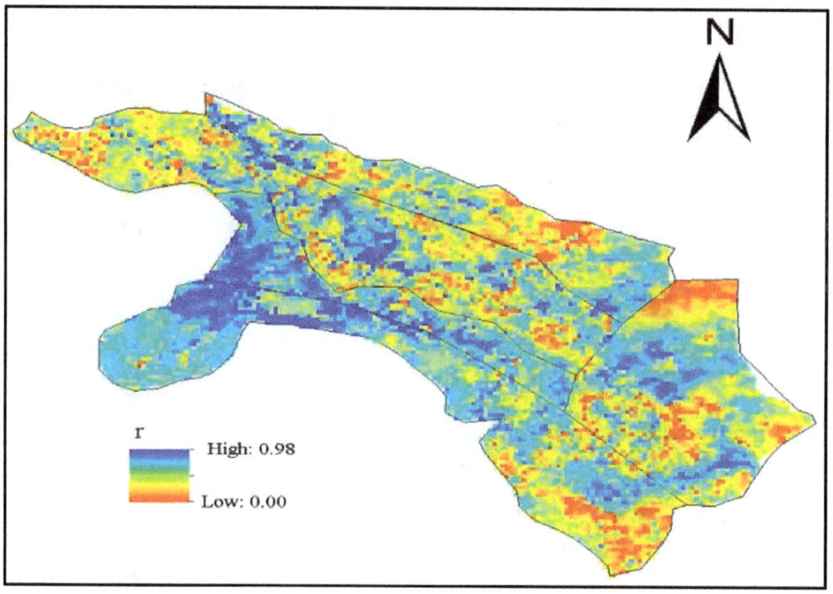

Figure 7. The distribution of correlation coefficients of estimated NDVI by coupling hydro-meteorological variables.

Analysis was also performed on a regional scale. Using the average NDVI time series of different sub-regions as predicted and each of the average hydro-meteorological variables as a predictor, a two-dimensional fitting was employed in the five sub-regions, respectively. This prediction also employed the same formulas used for the prediction on a grid scale (Table 3). Figure 8a–e showed the results of the three-dimensional scatter-plot couplings. Comparisons of these figures demonstrated that sub-regions I and II had a better accuracy of fitting. Figure 8f presents the overall relationship between observed NDVI and modeled NDVI for the whole region, where the correlation coefficient was observed to be 0.97, which indicated a high accuracy. This showed that this method was suitable for NDVI modeling.

Table 3. Fitting models, correlation coefficients and area proportion meeting significance testing in five sub-regions.

Sub-Region No.	Fitting Formula	Average Correlation Coefficient	Area Proportion ($p < 0.05$)	Area Proportion ($p < 0.01$)
I	$NDVI_{T,P} = a \cdot \ln(b \cdot P) + c \cdot T^2 + d \cdot T + e$	0.68	0.75	0.40
II	$NDVI_{GTL,P} = u \cdot e^{-(b \cdot GTL)} + c \cdot \ln(d \cdot P) + e$	0.66	0.72	0.42
III	$NDVI_{GTL,T} = a \cdot \ln(b \cdot GTL) + c \cdot T^2 + d \cdot T + e$	0.47	0.36	0.16
IV	$NDVI_{T,P} = a \cdot \ln(b \cdot P) + c \cdot T^2 + d \cdot T + e$	0.58	0.42	0.15
V	$NDVI_{GTL,P} = a \cdot GTL^2 + b \cdot GTL + c \cdot P^2 + d \cdot P + e$	0.57	0.50	0.16

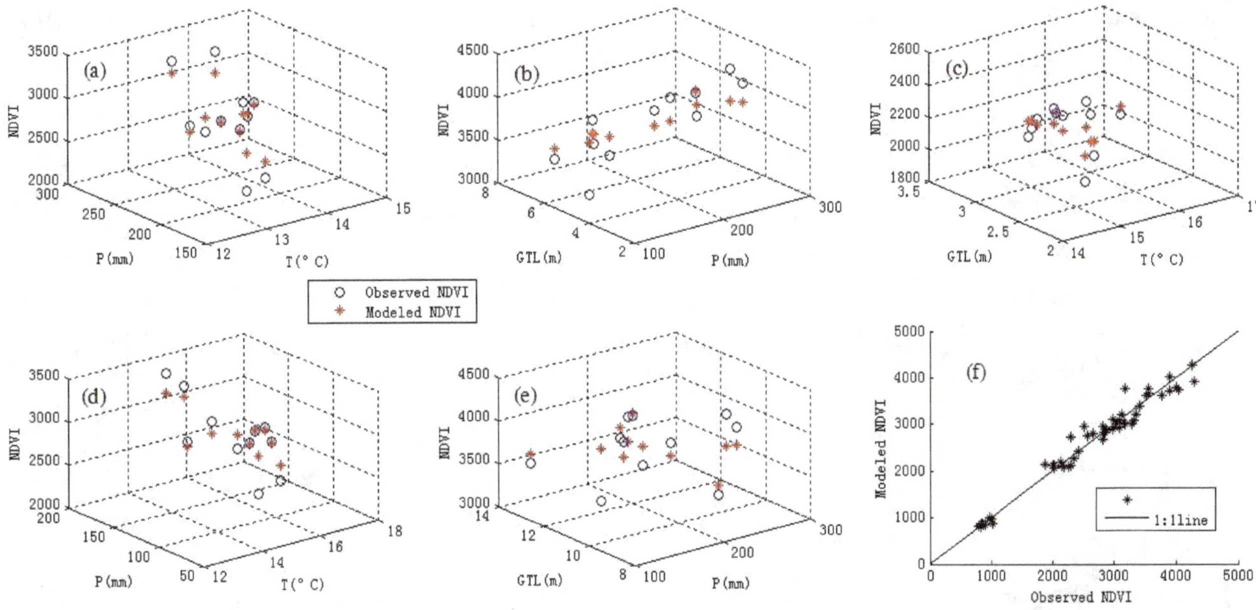

Figure 8. Three-dimensional scatter plots of NDVI (multiplied by 10^4) in five sub-regions (**a–e**) and the relationship between observed NDVI modeled NDVI in the whole region (**f**).

5. Conclusions

In this study, the spatio-temporal variation of vegetation cover in the middle reaches of theHei River basin and the main driving factors during the period from 2001 to 2011 were analyzed using MODIS NDVI data and hydro-meteorological data. Considering the nonlinear relationship between hydro-meteorological variables, the entropy theory was employed to calculate the mutual information between the NDVI time series and coupled hydro meteorological variables. Based on the spatial distribution of mutual information, the whole region was further divided into 5 sub-regions, and nested statistical models were employed to simulate NDVI in each sub-region. The main conclusions are as follows:

(1) The average annual NDVI increased at a rate of 0.005/a over the past 11 years in the middle reaches of the Hei River. The percentage area in which NDVI increased occupied 94% of the total area.

(2) In the desert sub-regions (I and IV), temperature and precipitation are the main driving factors for vegetation growth. In sub-region I, NDVI is consistent with the trend of temperature and precipitation (Figure 9a). However, in sub-region IV, the trend of temperature change is not obvious and the change in NDVI is mainly due to the increase in precipitation (Figure 9b). In the oasis regions (sub-region II and sub-region III), groundwater was an important factor for vegetation growth.

(3) In coupling hydro-meteorological variables, a nested statistical model was proposed for modeling NDVI on a regional scale. The overall correlation coefficient between observed NDVI and modeled NDVI was observed to be 0.97. This high simulation accuracy further proves the suitability of this method.

(4) Due to the influence of human activities, the modeling accuracy was not effective within the artificial oasis (sub-region III and sub-region V). For instance, in irrigation areas, vegetation can absorb water from irrigation to sustain growth but in non-irrigation areas the over-exploitation of groundwater caused by the increase in the amount of irrigation breaks the natural ecological balance, affecting the growth of natural vegetation. Additionally, the shortage of the dataset may also be a factor influencing the modeling accuracy. For instance, the scarcity of temperature and precipitation data may cause nearby cells to have similar values of T and P. Therefore, further

studies are necessary for modeling NDVI that consider additional factors such as runoff and irrigation using long-term and higher resolution datasets.

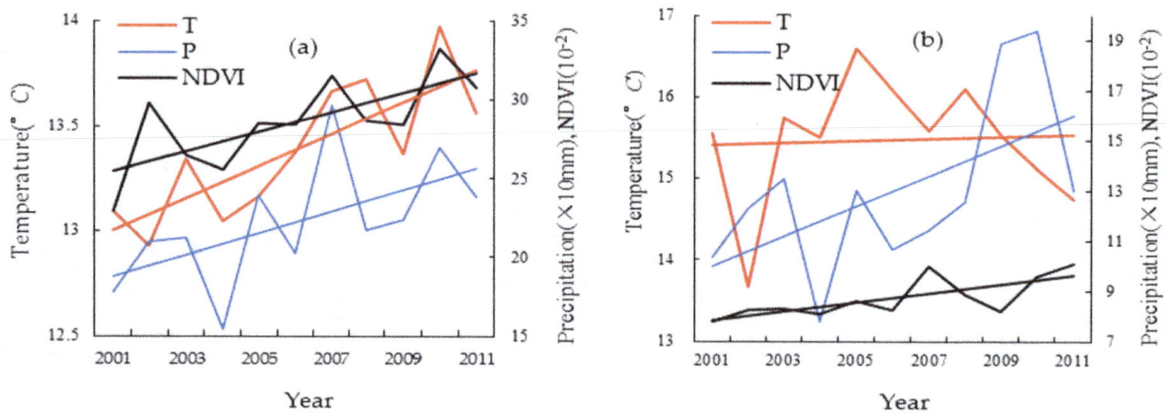

Figure 9. The variation of NDVI, temperature and precipitation on region I (**a**) and region IV (**b**). Note: The black, red and blue lines express the trend of variation in NDVI, temperature and precipitation respectively.

The results can provide some scientific guidance for water resources management and the reasonable exploitation of groundwater, especially in the artificial oasis areas. The outcomes demonstrated the ability to successfully employ the joint entropy method in vegetation dynamics analysis in relation to hydro-meteorological variables.

Acknowledgments: We are grateful for the grant support from the National Natural Science Fund in China (91425302 and 51279166).

Author Contributions: Gengxi Zhang and Xiaoling Su designed the computations and wrote the paper; Vijay P. Singh and Olusola O Ayantobo revised the paper and gave some suggestions; All authors have read and approved the final manuscript.

References

1. Kutiel, P.; Cohen, O.; Shoshany, M.; Shub, M. Vegetation establishment on the southern Israeli coastal sand dunes between the years 1965 and 1999. *Landsc. Urban Plann.* **2004**, *67*, 141–156. [CrossRef]
2. Jin, X.M.; Liu, J.T.; Wang, S.T.; Xia, W. Vegetation dynamics and their response to groundwater and climate variables in Qaidam Basin, China. *Int. J. Remote Sens.* **2016**, *37*, 710–728. [CrossRef]
3. Piao, S.L.; Mohammat, A.; Fang, J.Y.; Cai, Q.; Feng, J.M. NDVI-based increase in growth of temperate grasslands and its responses to climate changes in China. *Glob. Environ. Chang.* **2006**, *16*, 340–348. [CrossRef]
4. Dai, S.P.; Zhang, B.; HaiJun, W.J.; Wang, Y.M. Vegetation cover change and the driving factors over northwest China. *J. Arid Land* **2011**, *3*, 25–33. [CrossRef]
5. Liu, J.H.; Wu, J.J.; Wu, Z.T.; Liu, M. Response of NDVI dynamics to precipitation in the Beijing–Tianjin sandstorm source region. *Int. J. Remote Sens.* **2013**, *34*, 5331–5350. [CrossRef]
6. Reed, B.C.; Brown, J.F.; Darrel, V.Z.; Loveland, T.R.; Merchant, J.W.; Ohlen, D.O. Measuring phenological variability from satellite imagery. *J. Veg. Sci.* **1994**, *5*, 703–714. [CrossRef]
7. Clerici, N.; Weissteiner, C.J.; Gerard, F.G. Exploring the Use of MODIS NDVI-Based Phenology Indicators for Classifying Forest General Habitat Categories. *Remote Sens.* **2012**, *4*, 1781–1803. [CrossRef]
8. Feilhauer, H.; He, K.S.; Rocchini, D. Modeling Species Distribution Using Niche-Based Proxies Derived from Composite Bioclimatic Variables and MODISNDVI. *Remote Sens.* **2012**, *4*, 2057–2075. [CrossRef]
9. Rousvel, S.; Armand, N.; Andre, L.; Tengeleng, S. Comparison between vegetation and rainfall of Bioclimatic Ecoregions in Central Africa. *Atmosphere* **2013**, *4*, 411–427. [CrossRef]

10. Xiao, J.F.; Zhou, Y.; Zhang, L. Contribution of natural and human factors to increases in vegetation productivity in China. *Ecosphere* **2015**, *6*, 1–20. [CrossRef]
11. Nemani, R.; White, M.; Thornton, P.; Nishid, K.; Reddy, S.; Jenkins, J.; Running, S. Recent trends in hydrologic balance have enhanced the terrestrial carbon sink in the United States. *Geophys. Res. Lett.* **2002**, *29*, 1–4. [CrossRef]
12. Brunsell, N.A.; Young, C.B. Land surface response to precipitation events using MODIS and NEXRAD data. *Int. J. Remote Sens.* **2008**, *29*, 1965–1982. [CrossRef]
13. Kawabata, A.; Ichii, K.; Yamaguchi, Y. Global monitoring of interannual changes in vegetation activities using NDVI and its relationships to temperature and precipitation. *Int. J. Remote Sens.* **2001**, *22*, 1377–1382. [CrossRef]
14. Pettorelli, N.; Pelletier, F.; Hardenberg, A.; Festa-Bianchet, M.; Cote, S.D. Early onset of vegetation growth vs. rapid green-up, Impacts on juvenile mountain ungulates. *Ecology* **2007**, *88*, 381–390. [CrossRef] [PubMed]
15. Piao, S.L; Nan, H.J.; Huntingford, C.; Ciais, P.; Friedlingstein, P. Evidence for a weakening relationship between interannual temperature variability and northern vegetation activity. *Nat. Commun.* **2014**, *5*, 5018. [CrossRef] [PubMed]
16. Pielke, R.A.; Avissar, R. Interactions between the atmosphere and terrestrial ecosystems, influence on weather and climate. *Glob. Chang. Biol.* **1998**, *4*, 461–475. [CrossRef]
17. Sohoulande, D.C.; Singh, V.P. Retrieving vegetation growth patterns from soil moisture, precipitation and temperature using maximum entropy. *Ecol. Model.* **2015**, *309*, 10–21. [CrossRef]
18. Ichii, K.; Kawabata, A.; Yamaguchi, Y. Global correlation analysis for NDVI and climatic variables and NDVI trends, 1982–1990. *Int. J. Remote Sens.* **2002**, *23*, 3873–3878. [CrossRef]
19. Nicholson, S; Davenport, M. A comparison of the vegetation response to rainfall in the Sahel and East-Africa, using normalized difference vegetation index from NOAA AVHRR. *Clim. Chang.* **1990**, *17*, 209–241. [CrossRef]
20. Jin, X.M.; Guo, R.H.; Zhang, Q.; Zhou, Y.X.; Zhang, D.R.; Yang, Z. Response of vegetation pattern to different landform and water-table depth in Hailiutu River basin, Northwestern China. *Environ. Earth Sci.* **2014**, *71*, 4889–4898. [CrossRef]
21. Lv, J.J.; Wang, X.S.; Zhou, Y.X.; Qian, K.Z.; Wan, L.; Eamus, D.; Tao, Z.P. Groundwater-dependent distribution of vegetation in Hailiutu River catchment, a semi-arid region in China. *Ecohydrology* **2013**, *6*, 142–149. [CrossRef]
22. Jin, X.M.; Schaepman, M.E.; Clevers, J.; Su, Z.B.; Hu, G.C. Groundwater Depth and Vegetation in the Ejina Area, China. *Arid Land Res. Manag.* **2011**, *25*, 194–199. [CrossRef]
23. Mishra, A.K.; Ines, A.V.M.; Singh, V.P.; Asen, J.W. Extraction of information content from stochastic disaggregation and bias corrected downscaled precipitation variables for crop simulation. *Stoch. Environ. Res. Risk Assess.* **2012**, *27*, 449–457. [CrossRef]
24. Singh, V.P. *Entropy Theory and Its Application in Environmental and Water Engineering*; Wiley: New York, NY, USA, 2013.
25. Li, C.; Singh, V.P.; Mishra, A.K. Entropy theory-based criterion for hydrometric network evaluation and design, Maximum information minimum redundancy. *Water Resour. Res.* **2012**, *48*, 1–15. [CrossRef]
26. Phillips, S.J.; Anderson, R.P.; Schapire, R.E. Maximum entropy modeling of species geographic distributions. *Ecol. Model.* **2006**, *190*, 231–259. [CrossRef]
27. Pueyo, S.; He, F.; Zillio, T. The maximum entropy formalism and the idiosyncratic theory of biodiversity. *Ecol. Lett.* **2007**, *10*, 1017–1028. [CrossRef] [PubMed]
28. Urbani, F.; Alessandro, P.; Frasca, R.; Biondi, M. Maximum entropy modeling of geographic distributions of the flea beetle species endemic in Italy (Coleoptera, Chrysomelidae, Galerucinae, Alticini). *Zoologischer Anzeiger* **2015**, *258*, 99–109. [CrossRef]
29. Song, Y.; Ma, M. Varition of AVHRR NDVI and its relationship with climate in Chinese arid and cold regions. *J. Remote Sens.* **2008**, *12*, 499–505.
30. Sohoulande Djebou, D.C.; Singh, V.P.; Frauenfeld, O.W. Analysis of watershed topography effects on summer precipitation variability in the southwestern United States. *J. Hydrol.* **2014**, *511*, 838–849. [CrossRef]
31. McGill, W.J. Mutivariate information transmisson. *Psychometrica* **1954**, *19*, 97–116. [CrossRef]
32. Nian, Y.; Li, X.; Zhou, J.; Hu, X. Impact of land use change on water resource allocation in the middle reaches of the Heihe River Basin in northwestern China. *J. Arid Land* **2013**, *6*, 273–286. [CrossRef]

33. Wang, Y.; Roderick, M.L.; Shen, Y.; Sun, F. Attribution of satellite-observed vegetation trends in a hyper-arid region of the Heihe River basin, Western China. *Hydrol. Earth Syst. Sci.* **2014**, *18*, 3499–3509. [CrossRef]
34. Guo, Q.; Feng, Q.; Li, J. Environmental changes after ecological water conveyance in the lower reaches of Heihe River, northwest China. *Environ. Geol.* **2009**, *58*, 1387–1396. [CrossRef]
35. Zhang, A.; Zheng, C.; Wang, S.; Yao, Y. Analysis of streamflow variations in the Heihe River Basin, northwest China, Trends, abrupt changes, driving factors and ecological influences. *J. Hydrol. Reg. Stud.* **2015**, *3*, 106–124. [CrossRef]
36. Qin, D.; Zhao, Z.; Han, L.; Qian, Y. Determinnation of groundwater recharge regim and flowpath in the Lower Heihe River basin in an arid area of Northwest China by using environmental tracers, Implications for vegetation degradation in the Ejina Oasis. *Appl. Geochem.* **2012**, *27*, 1133–1145. [CrossRef]
37. Mi, L.; Xiao, H.; Zhang, J.; Yin, Z.; Shen, Y. Evolution of the groundwater system under the impacts of human activities in middle reaches of Heihe River Basin (Northwest China) from 1985 to 2013. *Hydrogeol. J.* **2016**, *24*, 971–986. [CrossRef]
38. Li, F.; Zhao, W. Changes in normalized difference vegetation index of deserts and dunes with precipitation in the middle Heihe River Basin. *Chin. J. Plant Ecol.* **2016**, *40*, 1245–1256. [CrossRef]
39. Zhou, W.; Wang, Q.; Zhang, C.B.; Li, J. Spatiotemporal variation of grassland vegetation NDVI in the middle and upper reaches of the Hei River and its response to climatic factors. *Acta Prataculturae Sin.* **2013**, *22*, 138–147.
40. Zhao, W.; Chang, X. The effect of hydrologic process changes on NDVI in the desert-oasis ecotone of the Hexi Corridor. *Sci. China Earth Sci.* **2014**, *57*, 3107–3117. [CrossRef]

5

Bayesian Technique for the Selection of Probability Distributions for Frequency Analyses of Hydrometeorological Extremes

Lu Chen [1,*], Vijay P. Singh [2] and Kangdi Huang [1]

1 College of Hydropower & Information Engineering, Huazhong University of Science & Technology, Wuhan 430074, China; hkd921110@163.com
2 Department of Biological and Agricultural Engineering & Zachry Department of Civil Engineering, Texas A&M University, College Station, TX 77843-2117, USA; vsingh@tamu.edu
* Correspondence: chen_lu@hust.edu.cn

Abstract: Frequency analysis of hydrometeorological extremes plays an important role in the design of hydraulic structures. A multitude of distributions have been employed for hydrological frequency analysis, and more than one distribution is often found to be adequate for frequency analysis. The current method for selecting the best fitted distributions are not so objective. Using different kinds of constraints, entropy theory was employed in this study to derive five generalized distributions for frequency analysis. These distributions are the generalized gamma (GG) distribution, generalized beta distribution of the second kind (GB2), Halphen type A distribution (Hal-A), Halphen type B distribution (Hal-B), and Halphen type inverse B (Hal-IB) distribution. The Bayesian technique was employed to objectively select the optimal distribution. The method of selection was tested using simulation as well as using extreme daily and hourly rainfall data from the Mississippi. The results showed that the Bayesian technique was able to select the best fitted distribution, thus providing a new way for model selection for frequency analysis of hydrometeorological extremes.

Keywords: entropy theory; frequency analysis; hydrometeorological extremes; Bayesian technique; rainfall

1. Introduction

Frequency analysis of hydrometeorological extremes plays an important role in the design of structures, such as dams, bridges, culverts, levees, highways, sewage disposal plants, waterworks, and industrial buildings [1–5]. From a frequency analysis, the probability of an extreme event can be estimated, and the value of a T-year design event (e.g., rainfall or flood) can be calculated. One of the objectives of frequency analysis of hydrometeorological extremes therefore is to establish a relationship between a flood or rainfall magnitude and its recurrence interval or return period.

A multitude of distributions have been employed for frequency analysis of hydrometeorological extremes. For example, the Pearson Type three (P-III) distribution is recommended in China; the Log-Pearson type three (LPT 3) is used in the U.S and Australia; and generalized extreme value (GEV) distribution is usually employed in Europe. Frequency analysis of hydrometeorological extremes at a given site or location is usually performed based on an appropriate probability distribution, which is selected on the basis of statistical tests for extreme hydrometeorological data [6]. However, no single distribution has gained global acceptance [7,8]. The traditional method is to try a variety of distributions and choose the best fitted distribution based on a particular mathematical norm, such as a least square error or a likelihood norm [9]. The disadvantages of this method of choosing are that it is laborious because too many different distributions need to be tried and empirical choices of candidate distributions make the results subjective [9–11]. In order to overcome these disadvantages, the generalized distributions have recently gained a lot of attention because they have been shown to be an effective tool for frequency analysis of hydrometeorological extremes. The greatest advantage of these generalized distributions is that they provide sufficient flexibility to fit a large variety of data sets, which facilitates the selection and comparison of different distributions. For example, Papalexiou and Koutsoyiannis [9] concluded that the generalized beta distribution of the second kind (GB2), which includes commonly used exponential, Weibull, and gamma distributions as special cases, was a suitable model for rainfall frequency analysis because of its ability to describe both J-shaped and bell-shaped data. Chen et al. [10] and Chen and Singh [11] also used the generalized gamma (GG) and GB2 distributions for hydrological frequency analysis, respectively. The results demonstrated that these two distributions could fit hydrometeorological data well. The generalized distributions can be derived using entropy theory by specifying appropriate constraints. The theory also provides a way for efficient parameter estimation [12].

Selection of the most appropriate distribution is of fundamental importance in hydrometeorological frequency analysis, since a wrong choice could lead to significant error and bias in design flood or rainfall estimates, particularly for higher return periods, leading to either under- or over-estimation, which may have serious implications in practice [13].

A distribution is often selected on the basis of statistical tests or by graphical methods [14]. Selection criteria based on the Akaike Information Criterion (AIC), the Bayesian Information Criterion (BIC), and the Anderson–Darling Criterion (ADC) are widely used in hydrology [4,15]. Laio et al. [16] presented an objective model selection criterion based on the AIC, the Bayesian Information Criterion (BIC), and the Anderson–Darling Criterion (ADC). Using a rigorous numerical framework, they found that the ability of these criteria to recognize the correct parent distribution from the available data varied from case to case, and these were more effective for two parameter distributions [13]. In this study, a more objective method based on a Bayesian technique is introduced to select the distribution with more parameters for frequency analysis of hydrometeorological extremes.

Bayesian method has been widely used for hydrological analysis, such as model selection and hydrological uncertainty analysis. Duan et al. [17] used Bayesian model averaging for multi-model ensemble hydrologic prediction. Hsu et al. [18] used a sequential Bayesian approach for hydrologic model selection and prediction. Najafi et al. [19] used Bayesian Model Averaging method to assess the uncertainties of hydrologic model selection. Robertson and Wang [20] introduced a predictor selection method for the Bayesian joint probability modeling approach to seasonal streamflow forecasting at multiple sites. In addition, Bayesian model method was also used for model uncertainty analysis [21,22].

The objective of this study is therefore to present a more objective method based on a Bayesian technique to select the most appropriate generalized distribution for frequency analysis of hydrometeorological extremes. The entropy theory was employed to derive generalized distributions for hydrometeorological extremes and estimate their parameters based on the principle of maximum entropy. A simulation test was carried out to evaluate the performance of the proposed Bayesian model selection technique. The proposed method was then tested using annual maximum hourly and daily precipitation data from Mississippi.

2. Entropy Theory

Since the entropy theory was used for the derivation of these generalized distributions and estimation of their parameters, in this section, the entropy theory combined with the principle of maximum entropy (POME) method is introduced.

The entropy, defined by Shannon in 1848, can be expressed by

$$H(X) = -\int_{0}^{\infty} f(x) \log f(x) \mathrm{d}x \tag{1}$$

where $f(x)$ is the probability density function (PDF) of X. $f(x)$ can be derived by maximizing the entropy subject to given constraints, which can be expressed by

$$\max H(X) \tag{2a}$$

$$\text{s.t.} \int_0^\infty f(x)\mathrm{d}x = 1; \tag{2b}$$

$$\int_0^\infty g_i(x)f(x)\mathrm{d}x = C_i \qquad (i = 1, \ldots, m) \tag{2c}$$

Employing the method of Lagrange multipliers, the PDF of X from Equations (1) and (2) can be derived as

$$f(x) = \exp(-\lambda_0 - \lambda_1 g_1(x) - \lambda_2 g_2(x) - \ldots - \lambda_m g_m(x)) \tag{3}$$

where m is the number of constraints; and $\lambda_0, \ldots, \lambda_m$ are the Lagrange multipliers. According to (2b), λ_0 can be defined as

$$\exp(\lambda_0) = \int_0^\infty \exp(-\lambda_1 g_1(x) - \lambda_2 g_2(x) - \ldots - \lambda_m g_m(x))\mathrm{d}x. \tag{4}$$

When different constraints are used, different PDFs can be obtained. According to the POME theory, all of the generalized distributions discussed in the following can be written in the form of Equation (3).

3. Generalized Distributions

Five generalized distributions, namely the GG distribution, the GB2 distribution, and three Halphen family distributions, were used in this study. The principle of maximum entropy (POME) method was used for parameter estimation, and it involves the following steps: (1) specification of constraints and maximization of entropy using the method of Lagrange multipliers; (2) derivation of the relation between Lagrange multipliers and constraints; (3) derivation of the relation between Lagrange multipliers and distribution parameters; and (4) derivation of the relation between distribution parameters and constraints. Detailed information on obtaining the equations for parameter estimation of those generalized distributions is given in [10,11,23]. In this paper, we mainly focus on model selection based on the Bayesian method.

3.1. Generalized Gamma Distribution

The probability density function of the GG distribution is given by

$$f(x) = \frac{r_2}{\beta\Gamma(\frac{r_1}{r_2})}\left(\frac{x}{\beta}\right)^{(r_1-1)} \exp\left(-\left(\frac{x}{\beta}\right)^{r_2}\right) \tag{5a}$$

where $\Gamma(\cdot)$ is the gamma function; r_1 and r_2 are the shape parameters, $r_1 > 0$, $r_2 > 0$; and beta is the scale parameter, $\beta > 0$.

For deriving Equation (5a) from the entropy theory, the following constraints are specified:

$$\int_0^\infty f(x)\mathrm{d}x = 1 \tag{5b}$$

$$\int_0^\infty f(x)\ln x \mathrm{d}x = E(\ln X) \tag{5c}$$

$$\int_0^\infty f(x)x^q \mathrm{d}x = E(X^q). \tag{5d}$$

The probability density function (PDF) of the GG distribution can then be expressed as [10]:

$$f(x) = \exp(-\lambda_0 - \lambda_1 \ln(x) - \lambda_2 x^q) \tag{6}$$

where λ_0, λ_1, and λ_2 are the Lagrange multipliers, and q is the parameter $q = r_2$ [10].

The relations between Lagrange multipliers and parameters can be summarized as

$$\begin{cases} q = r_2 \\ \lambda_1 = 1 - r_1 \\ \lambda_2 = \beta^{-r_2} \end{cases}. \tag{7}$$

The equations for parameter estimation based on the POME method can be given as [10]

$$\begin{cases} \frac{1}{r_2}\ln(\beta^{-r_2}) - \frac{1}{r_2}\varphi(\frac{r_1}{r_2}) = -E(\ln X) \\ \beta^{r_2}\frac{r_1}{r_2} = -E(\ln(X^{r_2})) \\ \frac{1}{r_2{}^2}\varphi'(\frac{r_1}{r_2}) = \mathrm{var}(\ln X) \end{cases} \tag{8}$$

where φ $(\cdot)$ is the digamma function; and $\varphi'(\cdot)$ is the tri-gamma function.

As seen in Equation (8), there are three unknown parameters, r_1, r_2, and β, in the three equations, and the variable X represents the observed hydrometeorological extreme series, which have been known before. By solving this equation set, the parameter of the GG distribution can be determined. The estimation procedures for other distributions are the same as those for the GG distribution.

3.2. Generalized Beta Distribution of the Second Kind

The PDF of the GB2 distribution is given by

$$f(x) = \frac{r_3}{bB(r_1, r_2)}\left(\frac{x}{b}\right)^{r_1 r_3 - 1}\left(1 + \left(\frac{x}{b}\right)^{r_3}\right)^{-(r_2 + r_1)} \tag{9a}$$

where $B(\cdot)$ is the beta function; and r_1, r_2, and r_3 are the shape parameters, $r_1 > 0$, $r_2 > 0$ and $r_3 > 0$; and b is the scale parameter, $b > 0$.

For deriving Equation (9a) from the entropy theory, the following constraints are specified:

$$\int_0^\infty f(x)\mathrm{d}x = 1 \tag{9b}$$

$$\int_0^\infty f(x)\ln x \mathrm{d}x = E(\ln X) \tag{9c}$$

$$\int_0^\infty f(x)\ln{(1 + px^q)}^{1/p}\mathrm{d}x = E(\ln{(1 + pX^q)}^{1/p}). \tag{9d}$$

According the maximum entropy theory, the PDF of the GB2 distribution can be expressed as [11]

$$f(x) = \exp(-\lambda_0 - \lambda_1 \ln(x) - \lambda_2 \ln{(1 + px^q)}^{1/p}) \tag{10}$$

where p and q are two parameters, which are also related to the parameters of the GB2 distribution, $p = (\frac{1}{\beta})^{r_3}$, and $q = r_3$.

The relations between Lagrange multipliers and parameters can be summarized as

$$\begin{cases} \lambda_1 = 1 - r_1 q \\ \lambda_2 = p(r_2 + \frac{1-\lambda_1}{q}) \\ p = (\frac{1}{\beta})^{r_3} \\ q = r_3 \end{cases} . \tag{11}$$

The equations for parameter estimation based on the POME method can be given as [11]

$$\begin{cases} -\ln\beta - \frac{1}{r_3}\varphi(r_1) + \frac{1}{r_3}\varphi(r_2) = -E(\ln X) \\ \varphi(r_2) - \varphi(r_1 + r_2) = -E(\ln(1 + (\frac{X}{\beta})^{r_3})) \\ \frac{1}{r_3{}^2}\varphi'(r_1) + \frac{1}{r_3{}^2}\varphi'(r_2) = \mathrm{var}(\ln X) \\ \varphi'(r_2) - \varphi'(r_1 + r_2) = \mathrm{var}(\ln(1 + (\frac{X}{\beta})^{r_3})) \end{cases} . \tag{12}$$

3.3. *Halphen Type A (Hal-A) Distribution*

The PDF of the Hal-A distribution is given as

$$f(x) = \frac{1}{2m^v K_v(2\alpha)} x^{v-1} \exp[-\alpha(\frac{x}{m} + \frac{m}{x})] \qquad x > 0 \tag{13a}$$

where $K_0(\cdot)$ is the modified Bessel function of the second kind of order v, $v \in R$; and m and α are parameters, $m > 0$ and $\alpha > 0$.

For deriving Equation (13a) from the entropy theory, the following constraints are specified:

$$\int_0^\infty f(x)\mathrm{d}x = 1 \tag{13b}$$

$$\int_0^\infty f(x)\ln x\mathrm{d}x = E(\ln X) \tag{13c}$$

$$\int_0^\infty xf(x)\mathrm{d}x = E(X) \tag{13d}$$

$$\int_0^\infty \frac{1}{x}f(x)\mathrm{d}x = E(\frac{1}{X}). \tag{13e}$$

From the entropy theory, the PDF of the Halphen type A distribution can be expressed as [23]

$$f(x) = \exp(-\lambda_0 - \lambda_1\ln x - \lambda_2 x - \lambda_3\frac{1}{x}) \quad x > 0 \tag{14}$$

where λ_3 is also the Lagrange multiplier.

The relations between Lagrange multipliers and parameters can be summarized as

$$\begin{cases} \lambda_1 = 1 - v \\ \lambda_2 = \frac{\alpha}{m} \\ \lambda_3 = m\alpha \end{cases} . \tag{15}$$

The equations for parameter estimation based on the POME method can be given as

$$\begin{cases} \ln m + \frac{1}{K_v(2\alpha)} \frac{\partial K_v(2\alpha)}{\partial v} = E(\ln X) \\ \frac{mK_{v+1}(2\alpha)}{K_v(2\alpha)} = E(X) \\ \frac{K_{v-1}(2\alpha)}{mK_v(2\alpha)} = E(\frac{1}{X}) \end{cases} . \tag{16}$$

3.4. Halphen Type B (Hal-B) Distribution

The PDF of the Hal-B distribution can be given as

$$f(x) = \frac{2}{m^{-2v} ef_v(\alpha)} x^{2v-1} \exp[-(\frac{x}{m})^2 + \alpha(\frac{m}{x})] \qquad x > 0 \tag{17a}$$

where $ef_v(\cdot)$ is the exponential factorial function, defined as $ef_v(\alpha) = 2\int_0^\infty x^{2v-1} \exp[-x^2 + \alpha x] dx$ $(x > 0)$, $m > 0$ are scale parameters, and $v > 0$ and $\alpha \in \Re$ are shape parameters.

For deriving Equation (17a) from the entropy theory, the following constraints are specified:

$$\int_0^\infty f(x) dx = 1 \tag{17b}$$

$$\int_0^\infty f(x) \ln x dx = E(\ln X) \tag{17c}$$

$$\int_0^\infty x^2 f(x) dx = E(X^2) \tag{17d}$$

$$\int_0^\infty x f(x) dx = E(X). \tag{17e}$$

From the entropy theory, the PDF of the Halphen type B distribution can be expressed as [23]

$$f(x) = \exp(-\lambda_0 - \lambda_1 \ln x - \lambda_2 x^2 - \lambda_3 x) \qquad x > 0. \tag{18}$$

The relations between Lagrange multipliers and parameters can be summarized as

$$\begin{cases} \lambda_1 = 1 - 2v \\ \lambda_2 = \frac{1}{m^2} \\ \lambda_3 = -\frac{\alpha}{m} \end{cases} . \tag{19}$$

The equations for parameter estimation based on the POME method can be given as

$$\begin{cases} \ln m + \frac{1}{2ef_v(\alpha)} \frac{\partial ef_v(\alpha)}{\partial v} = E(\ln X) \\ \frac{m^2 ef_{v+1}(\alpha)}{ef_v(\alpha)} = E(X^2) \\ \frac{m \cdot ef_{v+\frac{1}{2}}(\alpha)}{ef_v(\alpha)} = E(X) \end{cases} . \tag{20}$$

3.5. *Halphen Type Inverse B (Hal-IB) Distribution*

The PDF of the Hal-IB distribution can be given as

$$f(x) = \frac{2}{m^{-2v} ef_v(\alpha)} x^{-2v-1} \exp[-(\frac{m}{x})^2 + \alpha(\frac{m}{x})] \quad x > 0 \tag{21a}$$

where $m > 0$ is a scale parameter, and $\alpha \in \Re$ and $v > 0$ are shape parameters.

For deriving Equation (21a) from the entropy theory, the following constraints are specified:

$$\int_0^\infty f(x)\mathrm{d}x = 1 \tag{21b}$$

$$\int_0^\infty f(x) \ln x \mathrm{d}x = E(\ln X) \tag{21c}$$

$$\int_0^\infty \frac{1}{x^2} f(x)\mathrm{d}x = E(\frac{1}{X^2}) \tag{21d}$$

$$\int_0^\infty \frac{1}{x} f(x)\mathrm{d}x = E(\frac{1}{X}). \tag{21e}$$

From the entropy theory, the PDF of the Halphen type inverse B can be expressed as [23]

$$f(x) = \exp(-\lambda_0 - \lambda_1 \ln x - \lambda_2 \frac{1}{x^2} - \lambda_3 \frac{1}{x}) \qquad x > 0. \tag{22}$$

The relations between Lagrange multipliers and parameters can be summarized as

$$\begin{cases} \lambda_1 = 2v + 1 \\ \lambda_2 = m^2 \\ \lambda_3 = -m\alpha \end{cases}. \tag{23}$$

The equations for parameter estimation based on the POME method can be given as

$$\begin{cases} \ln m - \frac{1}{2ef_v(\alpha)} \frac{\partial ef_v(\alpha)}{\partial v} = E(\ln X) \\ \frac{ef_{v+1}(\alpha)}{m^2 ef_v(\alpha)} = E(\frac{1}{X^2}) \\ \frac{ef_{v+\frac{1}{2}}(\alpha)}{mef_v(\alpha)} = E(\frac{1}{X}) \end{cases}. \tag{24}$$

4. Model Selection Based on the Bayesian Technique

First, the five generalized distributions given above were used to fit a given data set **D**, and the equation sets derived by the POME method were applied for estimating their parameters. Second, the Bayesian technique introduced as follows was used to select the most appropriate distribution from the set of distributions for the data set **D**. In this study, the data **D** can be simulated data and observed data.

Let I be the background information. The posterior probabilities over a set of distributions can be expressed as

$$P(M_i|\mathbf{D}, I) = \frac{P(M_i|I) \cdot P(\mathbf{D}|M_i, I)}{P(\mathbf{D}|I)} \tag{25}$$

where $P(M_i|\mathbf{D}, I)$ is the posterior probability of distribution or model M_i and indicates the probability of this distribution to be true given the data series **D** and background information I. The largest approximate posterior probability among all of the distributions should be chosen as the most appropriate distribution. $P(M_i|I)$ is the prior model probability of distribution M_i; $P(\mathbf{D}|M_i, I)$ is the probabilistic evidence or integrated likelihood of data $\mathbf{D}$ conditional on model M_i. $P(\mathbf{D}|I)$ is a normalization constant and is calculated using the sum and product rules of probability theory as

$$P(\mathbf{D}|I) = \sum_{i=1}^{N} P(M_i|I)P(\mathbf{D}|M_i, I) \tag{26}$$

where N is the number of distributions that are used for the frequency analysis.

To obtain the posterior probability, one needs to calculate the probabilistic evidence $P(\mathbf{D}|M_i, I)$, which can be obtained by integrating a joint distribution $P(\boldsymbol{\lambda}, \mathbf{D}|M_i, I)$ with respect to vector $\boldsymbol{\lambda}$, and can be expressed as

$$P(\mathbf{D}|M_i, I) = \int_{-\infty}^{+\infty} P(\boldsymbol{\lambda}, \mathbf{D}|M_i, I)\mathrm{d}\boldsymbol{\lambda} \tag{27}$$

since

$$P(\boldsymbol{\lambda}, \mathbf{D}|M_i, I) = P(\boldsymbol{\lambda}|M_i, I)P(\mathbf{D}|\boldsymbol{\lambda}, M_i, I) \tag{28}$$

where $P(\boldsymbol{\lambda}|M_i, I)$ is the prior PDF for the Lagrangian multipliers given distribution M_i and background information I. Equation (27) can be obtained as

$$P(\mathbf{D}|M_i, I) = \int_{-\infty}^{+\infty} P(\boldsymbol{\lambda}|M_i, I)P(\mathbf{D}|\boldsymbol{\lambda}, M_i, I)\mathrm{d}\boldsymbol{\lambda} = E[P(\mathbf{D}|\boldsymbol{\lambda}, M_i, I)] \tag{29}$$

where $P(\mathbf{D}|\boldsymbol{\lambda}, M_i, I)$ is the likelihood function of the data in terms of the set of Lagrangian multipliers, and can be expressed by

$$P(\mathbf{D}|\boldsymbol{\lambda}, M_i, I) = \prod_{k=1}^{n} f(D_k|\boldsymbol{\lambda}, M_i, I) \tag{30}$$

where n is the sample size, and D_k denotes a specific value in data set $\mathbf{D}$. For a given sample size $\mathbf{D}$, model M_i and background information I, $P(\mathbf{D}|\boldsymbol{\lambda}, M_i, I)$ can be calculated by the multiplication of all PDF values of D_k.

The multivariate Gaussian distribution was selected as the prior distribution for the Lagrangian multiplier vector $\boldsymbol{\lambda}$. The mean value of Lagrangian multipliers was the estimated $\boldsymbol{\lambda}$. The covariance matrix Σ was calculated based on the Hessian matrix H, $\Sigma = \mathrm{H}^{-1}$. The equation for calculating the Hessian matrix can be expressed as

$$H = \begin{bmatrix} \frac{\partial^2 \lambda_0}{\partial \lambda_1^2} & \frac{\partial^2 \lambda_0}{\partial \lambda_1 \partial \lambda_2} & \cdot & \cdot & \cdot & \frac{\partial^2 \lambda_0}{\partial \lambda_1 \partial \lambda_m} \\ \frac{\partial^2 \lambda_0}{\partial \lambda_2 \partial \lambda_1} & \frac{\partial^2 \lambda_0}{\partial \lambda_2^2} & \cdot & \cdot & \cdot & \frac{\partial^2 \lambda_0}{\partial \lambda_2 \partial \lambda_m} \\ \cdot & \cdot & \cdot & \cdot & \cdot & \cdot \\ \cdot & \cdot & \cdot & \cdot & \cdot & \cdot \\ \frac{\partial^2 \lambda_0}{\partial \lambda_m \partial \lambda_1} & \frac{\partial^2 \lambda_0}{\partial \lambda_m \partial \lambda_2} & & \cdot & \cdot & \frac{\partial^2 \lambda_0}{\partial \lambda_m^2} \end{bmatrix}. \tag{31}$$

From Equation (29), $P(D|M_i, I)$ can be obtained by integration. Since the integration in Equation (29) is often a complex and high-dimensional function in Bayesian statistics, the quantity $P(\mathbf{D}|M_i, I)$ was calculated based on the calculation of $E[P(\mathbf{D}|\boldsymbol{\lambda}, M_i, I)]$.

A Markov Chain Monte Carlo (MCMC) method was used in this study to calculate $P(\mathbf{D}|M_i, I)$ and the posterior distribution of each distribution. The idea of MCMC sampling was first introduced by [24]. Since the target distribution is very complex, we cannot sample from it directly. The indirect method for obtaining samples from the target distribution is to construct an Markov chain with state space E, and whose stationary (or invariant) distribution is $\pi(\cdot)$, as discussed in [25]. Then, if we run the chain for sufficiently long, simulated values from the chain can be treated as a dependent sample from the target distribution. Using the MCMC simulation, pairs of Lagrangian multipliers $\boldsymbol{\lambda}$ were drawn from the joint distribution $P(\boldsymbol{\lambda}, \mathbf{D}|M_i, I)$. The quantity $P(D|M_i, I)$ was finally calculated based on the calculation of $E[P(D|\boldsymbol{\lambda}, M_i, I)]$.

In the following, simulated data and real-world data were used for testing the proposed method. The flow chart can be found in Figure 1.

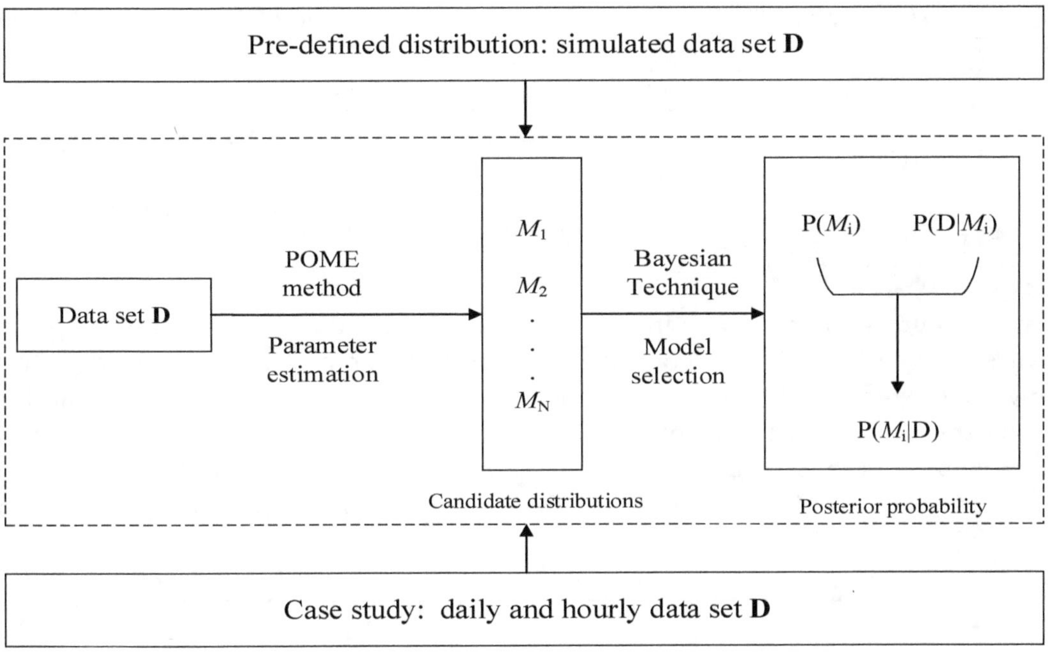

Figure 1. Flowchart of the whole paper. POME: principle of maximum entropy.

5. Performance Evaluation

Before using the proposed method in a practical application, a simulation test was carried out to evaluate the performance of the proposed Bayesian technique for model selection. The simulation test involves the following steps.

First, a distribution with given parameters was pre-defined.

Second, simulated datasets D were randomly drawn from the pre-defined distributions.

Third, the Gaussian, lognormal, Gamma, and Weibull distributions were used to fit the data set **D**, and the POME method was applied for parameter estimation.

Fourth, the proposed Bayesian technique was applied for model selection, and the best fitted distributions with the highest posterior probabilities were determined. The results were compared with the pre-defined distributions.

Fifth, the Bayesian model selection technique was compared with commonly used methods in hydrology, such as the root mean square error of the empirical and theoretical probabilities and the AIC criterion.

According to the steps mentioned above, this test focuses on the evaluation of the reliability of the Bayesian model selection for different distributions and data sample sizes. In order to show the performance of the proposed method, some simple and widely used distributions were considered, including the Gaussian, lognormal, Gamma, and Weibull distributions, which involve the Gaussian and non-Gaussian cases. The parameters used for the simulation are given in Table 1. Simulated datasets were randomly drawn from the pre-defined distributions given in Table 1 with sample sizes of 40, 80, 120, 160, 200, and 240. The proposed Bayesian technique was then applied to determine the best fitted distributions for each dataset. The multivariate Gaussian distribution was used for the prior distribution, in which the mean values are the estimated Lagrangian multiplier, and the covariance matrix Σ was calculated based on the Hessian matrix H, $\Sigma = H^{-1}$. Usually, the estimated parameters were around the true values, so the Gaussian distribution was used. Additionally, the Hessian matrix was calculated to represent the covariance matrix. It is not straightforward to try other distributions, since it is a multivariate problem for which the multivariate Gaussian distribution is widely used.

Table 1. Parameters of different distributions for simulation test.

Number	Distribution	Probability Density Function (PDF)	Parameters
1	Gaussian	$f(x) = \frac{1}{\sqrt{2\pi}\sigma}\exp(-\frac{(x-\mu)^2}{2\sigma^2})$	$\mu = 10$, $\sigma = 3.162$
2	Lognormal	$f(x;\mu,\sigma) = \begin{cases} \frac{1}{\sqrt{2\pi}\sigma x}\exp[-\frac{1}{2\sigma^2}(\ln x - \mu)^2],\ x > 0 \\ 0,\ x \leq 0 \end{cases}$	$\mu = 2$, $\sigma = 0.6$
3	Gamma	$f(x;\ \beta,\alpha) = \frac{\beta^\alpha}{\Gamma(\alpha)}x^{\alpha-1}e^{-\beta x},\ x > 0$	$\alpha = 10$, $\beta = 1$

The simulation results are shown in Figure 2, which indicate that when the data was sampled from the Gaussian distribution, for all of the sample size, the posterior probabilities of the Gaussian distribution were the highest. For the other tests, namely the lognormal distribution and the gamma distribution as the pre-defined distributions, respectively, the highest posterior probabilities for all of the sample size were the lognormal distribution and gamma distribution as well. Therefore, the proposed Bayesian technique can select the best fitted distribution even for a small sample size (sample size = 40).

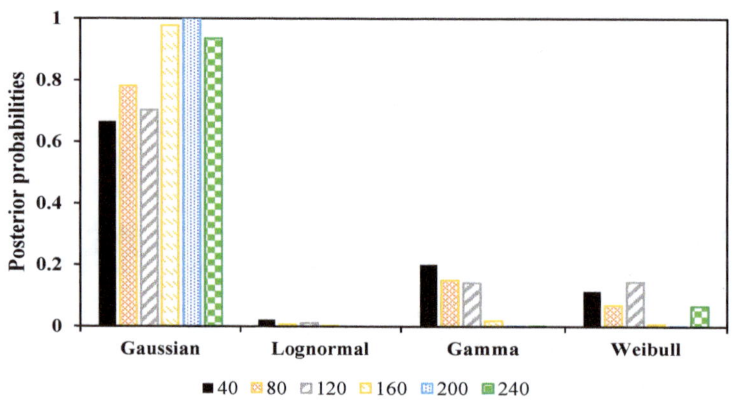

(**a**) Gaussian distribution

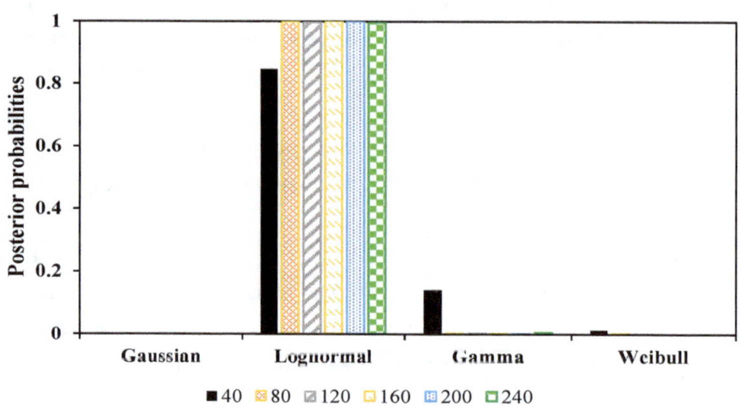

(**b**) Lognormal distribution

Figure 2. *Cont.*

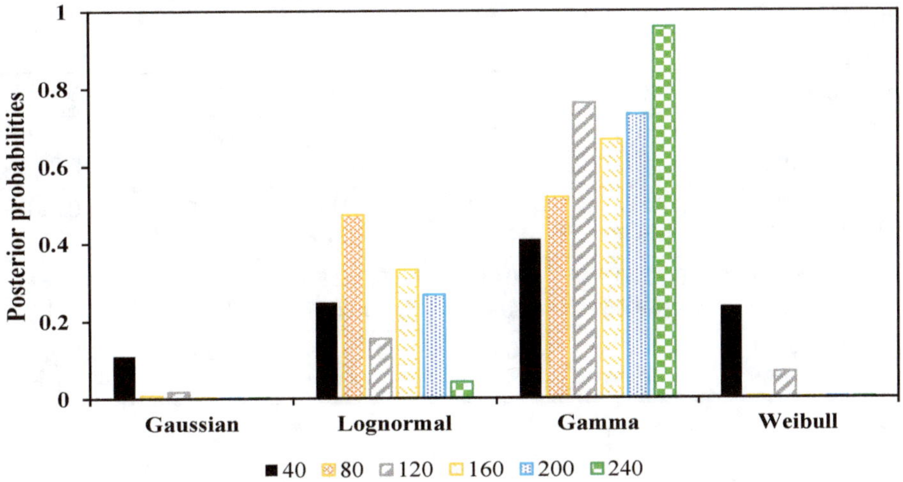

(c) Gamma distribution

Figure 2. The posterior probabilities of the simulation tests with the Gaussian distribution, Lognormal distribution, and Gamma distribution as the pre-defined distributions, respectively.

The proposed method was compared with the traditional root mean square error (RMSE) and AIC values, which are also used to select the most appropriate distribution. The results are given in Tables 2 and 3, in which the best fitted distributions with the smallest RMSE and AIC values are in bold. According to the smallest RMSE and AIC values, the correct distribution cannot always be selected. Take the Gaussian distribution as an example. When the sample size was 40, 80, 120, and 160, the best fitted distribution was, respectively, gamma, Weibull, Weibull, and Weibull. When the sample size became larger, greater than 160, the Gaussian distribution was detected as the correct distribution. The RMSE and AIC values of different distributions did not show significantly different results. In other words, the differences in the RMSE and AIC values among those distributions were not large. In Table 3, generally the AIC and RMSE values can show the best fitted distribution. However, in some cases the RMSE and AIC values of different distributions were nearly the same, such as the sample size equaling 160 and 200 in Table 3.

According to the performance test, the Bayesian technique can obtain the correct distribution at any time no matter what the sample size is. On the contrary, the traditional RMSE and AIC do not always work effectively. The RMSE and AIC for the data fitted using different distributions do not shown large differences. Therefore, the proposed method can provide an effective way for model selection in hydrological frequency analysis.

Table 2. The root mean square error (RMSE) and Akaike Information Criterion (AIC) values for the simulation test, in which the pre-defined distribution is the Gaussian distribution.

Distributions	Criteria	40	80	120	160	200	240
Gaussian	RMSE	0.03	0.0175	0.0247	0.0243	**0.0118**	**0.0115**
	AIC	−169.07	−443.78	−577.56	−792.76	**−1244.68**	**−1482.07**
lognormal	RMSE	0.0385	0.044	0.025	0.05	0.057	0.0437
	AIC	−155.12	−279.77	−558.39	−525.84	−683.75	−913.47
Gamma	RMSE	**0.0239**	0.0293	0.0161	0.0377	0.0367	0.026
	AIC	**−177.05**	−332.5793	−642.25	−607.66	−801.37	−1116.59
Weibull	RMSE	0.03	**0.0168**	**0.016**	**0.0168**	0.017	0.0118
	AIC	−167.34	**−441.3**	**−661.23**	**−858.42**	−1082.21	−1457.45

Table 3. The RMSE and AIC values for the simulation test, in which the pre-defined distribution is the Gamma distribution.

Distributions	Criteria	40	80	120	160	200	240
Gaussian	RMSE	0.0409	0.0317	0.33	0.0333	0.0318	0.033
	AIC	−152.98	−372.77	−509.91	−693.5	−879.39	−1030.29
lognormal	RMSE	0.0217	0.0161	0.0211	**0.0098**	0.0158	0.0179
	AIC	−185.92	−437.1	−582.86	−1018.23	**−1094.31**	−1311.04
Gamma	RMSE	**0.0229**	**0.0141**	**0.0169**	0.0108	**0.0124**	0.0126
	AIC	**−189.06**	**−486.4**	**−632.77**	**−1026.04**	−1191.77	**−1421.8**
Weibull	RMSE	0.042	0.0398	0.0314	0.036	0.0344	0.0332
	AIC	−150.55	−373.95	−523.74	−672.99	−853.67	−1053.41

6. Case Study

Rainfall data for many different timescales were investigated. The timescales of these rainfall dates in the Mississippi River basin ranged from hourly to yearly. The annual maximum daily and hourly series were extracted for frequency analysis, and detailed information of daily and hourly data is shown in Table 4, in which the length of data, the mean value, standard deviation, and the minimum and maximum values are shown. The daily and hourly rainfall histograms for each gauging station are given in Figure 3.

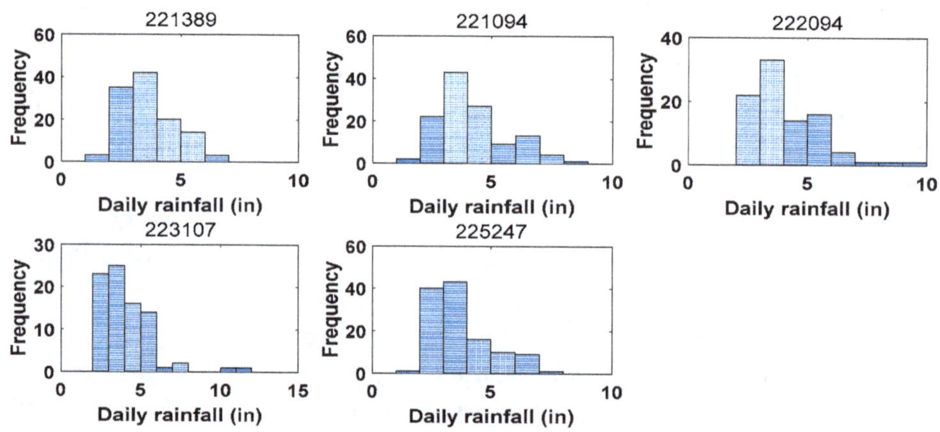

(**a**) Daily rainfall histograms

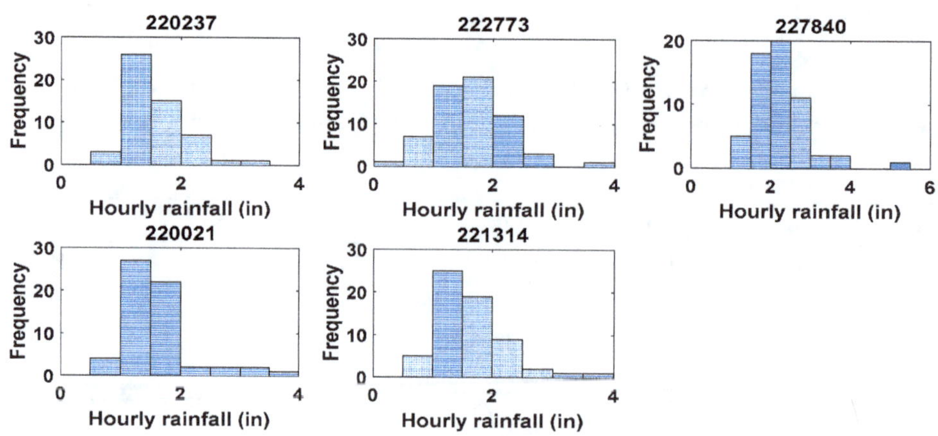

(**b**) Daily rainfall histograms

Figure 3. Daily and Hourly rainfall histograms for each gauging station.

Table 4. Detailed information of daily and hourly annual maximum rainfall series.

Times	Gauging Station	Number	Length of Data	Mean Values	SD	Max	Min
Daily	Canton gauging	221389	1893–2012	3.6	1.1	6.8	1.65
	Brookhaven City, MS	221094	1894–2014	4.14	1.44	8.08	1.85
	Crvstal Spgs Exp Stn, MS	222094	1893–1954, 1985–2014	4	1.4	9.04	2.02
	Forest, MS	223107	1930–2012	4.02	1.66	11.75	2
	Louisville, MS	225247	1895–2014	3.65	1.28	7.47	1.7
Hourly	Arkabutla dam	220237	1949–2001	1.57	0.46	3.12	0.88
	Enid dam MS	222773	1949–2012	1.62	0.59	4	0.2
	Saucier experimental forest MS	227840	1955–2013	2.23	0.7	5.13	1.2
	Aberdeen MS	220021	1952–2011	1.55	0.62	3.8	0.7
	Calhoun city MS	221314	1948–2009	1.6	0.57	3.88	0.8

The five generalized distributions were used to fit the data set, and the entropy method was used to estimate the parameters of these distributions, as given in Table 5 (for daily data) and Table 6 (for hourly data). A full Newton method was used to find the solution of the non-linear equation sets derived before. The "nleqslv" package in R language was used to solve the equation set. The initial value was set as 1 for all potential parameters. The proposed Bayesian technique was used to select the most appropriate distribution for rainfall frequency analysis. The multivariate Gaussian distribution was used for the prior distribution, in which the mean values are the estimated Lagrangian multiplier, and the covariance matrix Σ was calculated based on the Hessian matrix H, $\Sigma = H^{-1}$. The posterior probabilities are also in Table 5 (for daily data) and Table 6 (for hourly data). The RMSE, AIC, and BIC were also calculated as given in Tables 5 and 6. Both the AIC and BIC indexes are based on the likelihood values, and a penalty term was introduced for the number of parameters in the model. However, the differences between them are that the penalty term is larger in BIC than in AIC. In this study, it is seen from Tables 5 and 6 that the selected model by the two methods is the same. Therefore, only the results given by AIC are discussed hereafter. The results indicate that for some of the cases, the selected model based on the three criteria are the same, e.g., gauging stations 225247, 220237, 227840, 220021, and 221314. For some of the stations, the results given by the three methods were not coincident. However, for these cases, the distribution with the lowest AIC value usually had the second-highest posterior probability. Take the gauging station 221094 in Table 5 for example. The AIC and RMSE criteria suggested that the GB2 distribution was the best, for which the posterior probability was 0.34, smaller than the highest one 0.58 (Hal-A). According to the simulation test in Section 4, the performance of the proposed method was better than the traditional AIC and RMSE values. The Bayesian method amplified the differences among the generalized distributions. In order to further compare the performance of these models, the theoretical and empirical exceedance probabilities of the daily rainfall data for the gauging station 223107 are shown in Figure 4a.

According to the results given in Table 5, the best fitted distribution for the gauging station 223107 recommended by the RMSE, AIC, and Bayesian methods, was GB2, Hal-A, and Hal-IB, respectively. As shown in Figure 4a, if the Hal-A distribution was used, the design values for large return periods would be underestimated. The fitting curves of the GB2 and Hal-IB distributions were nearly the same. Thus, the distribution Hal-A recommended by the AIC is not appropriate, and compared with GB2, the Hal-IB with less parameters and higher posterior probability was chosen finally.

The theoretical and empirical exceedance probabilities of the hourly rainfall data for the gauging station 222773 are shown in Figure 4b. According to the results given in Table 6, the best fitted distribution for the gauging station 222773 recommended by the RMSE, AIC and Bayesian methods was GG, Hal-B, and GB2, respectively. As shown in Figure 4b, if the GG and Hal-B distributions were used, the design values for large return periods would be underestimated.

Table 5. Parameters, RMSE, AIC, and posterior probabilities for daily data calculated by five generalized distributions.

Number	Distribution	Para1	Para2	Para3	Para4	RMSE	AIC	BIC	Posterior Probabilities
221389	GG	13.251	0.8397	0.1335		0.0205	−613.05	−604.76	0.02
	Gb2	0.9955	1.5808	34.95	16.84	0.0195	−605.68	−597.39	0.08
	Hal-A	5.235	3.431	0.0327		**0.0183**	**−624.56**	**−616.27**	0.14
	Hal-B	−20.927	7.363	5.381		0.0237	−593.21	−584.93	0.02
	Hal-IB	−9.103	2.855	4.776		0.0233	−544.21	−535.92	**0.74**
221094	GG	13.308	0.684	0.0527		0.0362	−482.72	−477.32	0.08
	Gb2	1.678	2.203	11.994	4.894	**0.0203**	**−607.37**	**−601.98**	0.34
	Hal-A	3.507	7.812	−5.534		0.0229	−584.34	−578.95	**0.58**
	Hal-B	−20.918	8.0322	5.6835		0.0557	−409.26	−410.15	0.00
	Hal-IB	−10.755	2.319	3.699		0.0525	−539.94	−531.55	0.00
222094	GG	14.855	0.628	0.0255		0.0307	−412.5	−404.94	0.04
	Gb2	1.3803	1.38	24.977	6.573	0.0213	−462.61	−455.04	0.20
	Hal-A	2.5669	12.3193	−7.9616		**0.021**	**−466.21**	**−458.64**	0.32
	Hal-B	−12.2589	7.581	3.5971		0.04	−385.6	−378.04	0.01
	Hal-IB	−9.546	2.2708	4.2153		0.0218	−464.18	−456.61	**0.44**
223107	GG	13.5999	0.5606	0.0131		0.0294	−409.36	−402.11	0.00
	Gb2	2.199	1.837	9.931	2.4389	**0.0164**	−466.07	−458.81	0.08
	Hal-A	0.977	30.104	−8.194		0.0168	**−479.18**	**−471.93**	0.21
	Hal-B	−27.5605	14.5778	3.8889		0.0406	−362.58	−355.33	0.00
	Hal-IB	−3.8918	3.7076	3.2608		0.0165	−457.91	−450.65	**0.71**
225247	GG	13.596	0.6657	0.0384		0.0437	−433.23	−424.86	0.00
	Gb2	1.822	1.156	23.62	3.641	0.02456	−565.73	−557.36	0.00
	Hal-A	2.0554	14.775	−8.756		0.0287	−533.57	−525.21	0.04
	Hal-B	−46.515	14.128	6.0834		0.0653	−363.04	−354.68	0.00
	Hal-IB	−3.361	3.938	3.602		**0.0232**	**−575.37**	**−567.01**	**0.96**

Table 6. Parameters, RMSE, AIC, and posterior probabilities for hourly data calculated by five generalized distributions.

Number	Distribution	Para1	Para2	Para3	Para4	RMSE	AIC	BIC	Posterior Probabilities
220237	GG	−19.179	0.6794	0.0284		0.048	−187.74	−181.83	0.20
	GB2	2.963	2.5391	6.8892	2.4497	**0.0308**	**−233.74**	**−227.83**	**0.57**
	Hal-A	3.0259	15.972	−12.521		0.0344	−215.46	−209.55	0.08
	Hal-B	−27.458	8.384	6.6655		0.0523	−178.6	−172.69	0.10
	Hal-IB	−7.185	4.2898	5.7116		0.0312	−222.62	−216.72	0.06
222773	GG	4.169	1.858	1.656		**0.0318**	−291.31	−284.83	0.13
	GB2	4.5	3.221	0.775	2.153	0.0337	−291.92	−285.44	**0.70**
	Hal-A	5.36×10^{-2}	1.90×10^{-2}	6.8204		0.0333	−274.18	−267.70	0.05
	Hal-B	1.3105	1.6132	1.5702		0.0338	**−292.69**	**−286.21**	0.12
	Hal-IB	−50.5719	0.1565	3.0044		0.0419	−231.43	−224.95	0.00
227840	GG	17.0415	0.6977	0.0225		0.0473	−266.6	−260.37	0.05
	GB2	2.013	1.2966	11.7277	4.6042	**0.0266**	**−275.55**	−266.32	**0.84**
	Hal-A	2.7995	8.6585	−11.0919		0.0271	−274.68	**−268.45**	0.00
	Hal-B	−23.6777	4.7218	5.8388		0.0422	−249.55	−243.32	0.11
	Hal-IB	−10.601	1.865	5.7014		0.0316	−264.11	−257.88	0.00
220021	GG	11.875	0.65	0.0173		0.0582	−189.32	−183.04	0.10
	GB2	4.1081	0.9324	3.7284	0.9621	0.0468	−221.54	−212.26	0.33
	Hal-A	1.1704	10.6076	−8.8785		0.0532	−217.75	−211.47	0.15
	Hal-B	−30.3037	5.6947	4.203		0.0761	−181.74	−175.42	0.00
	Hal-IB	0.7828	2.248	2.3074		**0.0455**	**−228.4**	**−222.12**	**0.42**
221314	GG	14.1465	0.6738	0.0171		0.0405	−281.77	−275.39	0.00
	GB2	1.5734	0.3406	44.9999	4.516	0.0318	−293.19	−286.81	0.05
	Hal-A	1.7786	7.979	−9.475		0.0266	−316.06	**−309.68**	0.11
	Hal-B	−27.792	4.9387	4.622		0.0432	−263.62	−257.24	0.00
	Hal-IB	−6.217	1.438	4.185		**0.0249**	**−309.73**	−303.35	**0.84**

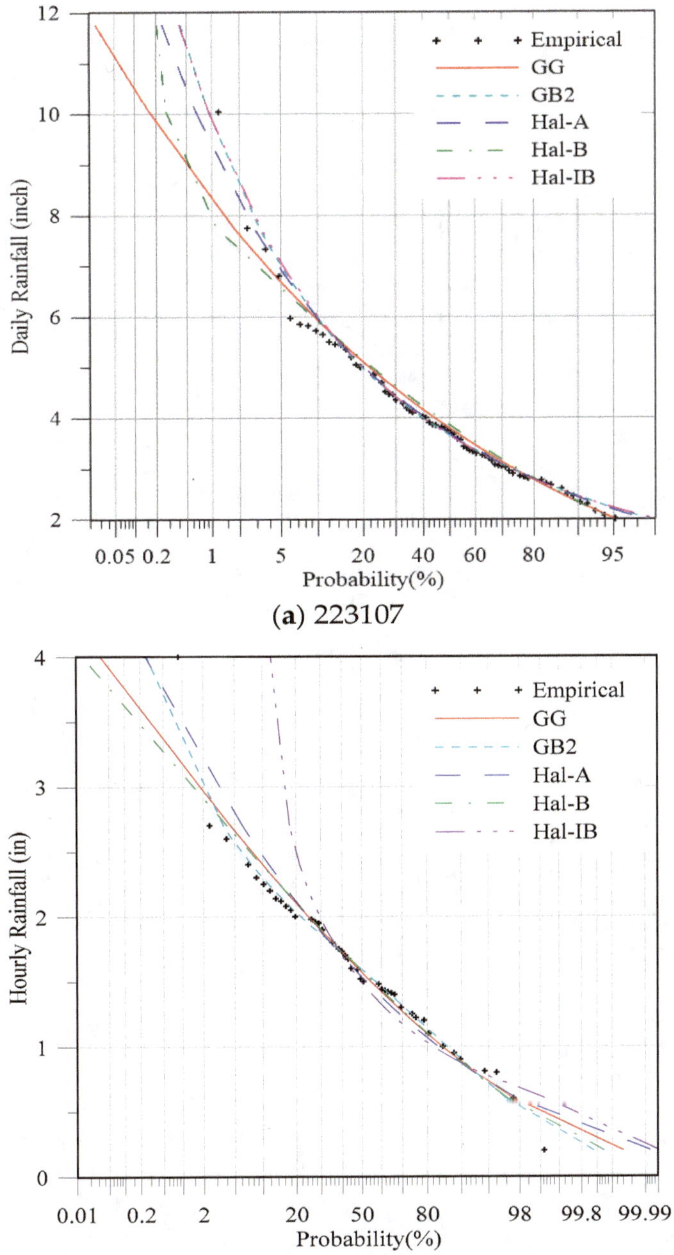

(a) 223107

(b) 222773

Figure 4. Theoretical and empirical exceedance probabilities of the annual maximum rainfall data at the stations 223107 and 222773.

In order to compare the fitting results more comprehensively, the Q-Q plot, P-P plot, and S-P plot were represented for the daily rainfall data from the gauging station 223107 as shown in Figure 5. It can be seen from Figure 5a that the fitting results of GB2 and Hal-IB are nearly the same. When the GG, Hal-A, and Hal-B distributions were used, the design rainfall for a large quantile would be underestimated, since the theoretical rainfall values calculated by the GG, Hal-A, and Hal-B distributions are significantly lower than the observed ones. For the P-P and S-P plots, the differences for large probability are not so obvious, and the plots in Figure 5b,c are well-distributed compared with the Q-Q plot. In Figure 5b, it is easily observed that the Hal-B distribution fits the worst, and the empirical probabilities in the middle part are significantly larger than the theoretical ones. S-P plots remove the impact of variance on the plot, and it is seen that the plots in the S-P figure are much more concentrated than those in the P-P figure.

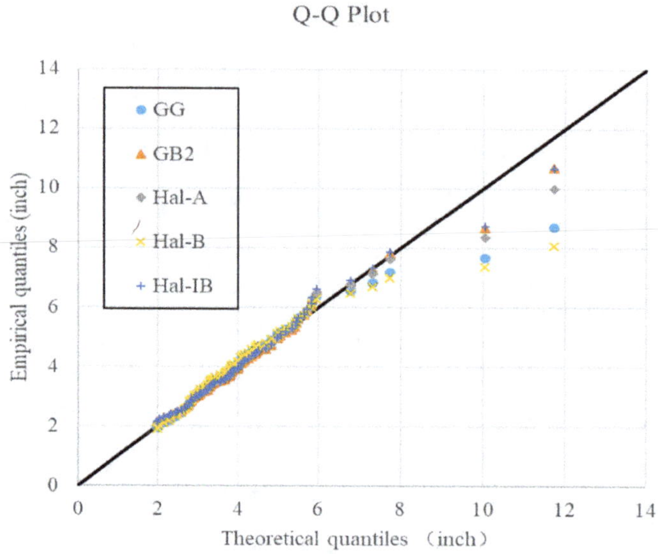

(a) Q-Q plot

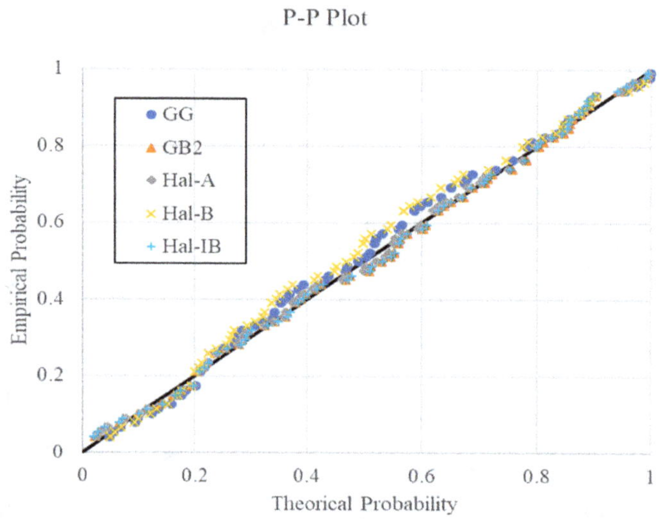

(b) P-P plot

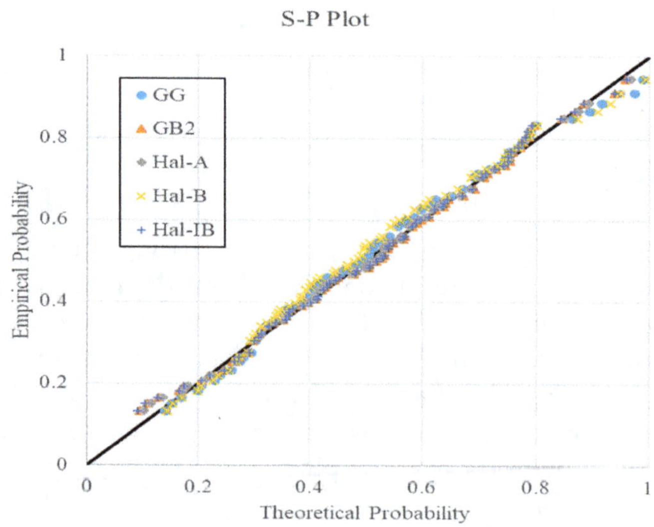

(c) S-P plot

Figure 5. Q-Q, P-P, and S-P plots for the daily rainfall data from the gauging station 223107.

Furthermore, in the U.S., the Log-Pearson three (LP3) distribution has been recommended for hydrological frequency analysis [26,27]. In order to compare the five generalized distributions with the commonly used LP3 distribution, the six distributions were considered and the proposed Bayesian method was used to select the best fitted one. The results are given in Table 7.

Table 7. Parameters, RMSE, AIC, and poster probabilities for 223107 daily data calculated by five generalized distributions and the Log-Pearson three (LP3) distribution.

Distributions	Para1	Para2	Para3	Para4	RMSE	AIC	BIC	Posterior Probabilities
GG	13.60	0.56	0.013		0.0294	−406.36	−402.11	0.002
GB2	2.20	1.84	9.93	2.44	**0.0164**	−463.07	−460.81	0.071
Hal-A	0.98	30.10	−8.19		0.0168	**−476.18**	**−471.93**	0.197
Hal-B	−27.56	14.58	3.89		0.0406	−359.58	−355.33	0.0003
Hal-IB	−3.89	3.70	3.26		0.0165	−454.91	−450.77	**0.674**
LP3	14.29	0.09	−0.03		0.0167	−448.65	−444.39	0.056

7. Conclusions and Discussion

The paper proposes a model selection approach based on a Bayesian technique to choose the best fitted distribution for hydrological frequency analysis. Five generalized distributions, including GG, GB2, Hal-A, Hal-B, and Hal-IB, which are also widely used in hydrology, were considered. The entropy-based method was used to express these distributions and the POME method was applied for parameter estimation. A simulation test was carried out to evaluate the performance of the proposed Bayesian method. Daily rainfall data from five stations and hourly rainfall data from another five stations from the Mississippi basin were selected as case studies. The main conclusions are summarized as follows.

(1) The entropy-based five generalized distributions are given, and their corresponding equation sets for parameter estimation are introduced. The results of simulation test and case study show that the POME method can provide an effective way for parameter estimation.

(2) Results of the simulation test demonstrate that the Bayesian technique can choose the most suitable distribution. Compared with the commonly used RMSE and AIC values, the proposed method gives a better performance.

(3) Results of the case study indicate that when using different criteria for model selection, the results are not always the same. For some of the cases, the three criteria choose the same distribution. For others, the results are slightly different. Since choosing the probable distribution for hydraulic design is very significant, especially for extreme magnitudes, the distribution should be selected carefully. According to the posterior probabilities calculated by the proposed method for daily and hourly data from 10 gauging stations, generally the Hal-IB distributions give better fits for daily data and GB2 distributions give better fits for hourly data.

(4) According to the results of the simulation test and case studies, the Bayesian model selection technique can give a more reliable result than the traditional RMSE and AIC values. Thus, the proposed method provides an effective way for model selection for hydrological frequency analysis.

(5) The significant contribution of this paper is that compared with the traditional method, the proposed method is based on entropy theory, and the posterior probabilities were calculated based on the generation of Lagrange multipliers. In addition, the five generalized distributions were involved in this paper, since previous research mainly focus on the commonly used distribution or standard distributions.

This contribution of this paper mainly concentrates on univariate hydrometeorological frequency analysis. Recently, multivariate hydrological analysis has also surged up, such as [2,4,28–31]. However, univariate frequency analysis is the basis of multivariate frequency analysis, which can provide the marginal distributions for joint distribution. Thus, before establishing the multivariate distributions, the univariate distribution should be built rationally and appropriately first.

In addition, in the common hydrological frequency analysis, the hydrological data set is assumed to be independent and identically distributed [1]. Since there are influences of climate change and human activities on streamflow, it is possible that the mean value or the variation of the whole series would be changed. In other words, the data set is non-stationary. Non-stationary hydrological frequency analysis is also another hot and difficult topic in hydrology recently. In this paper, we mainly focus on the stationary frequency analyses of hydrometeorological extremes. Non-stationary hydrological frequency analysis will be discussed in future research.

Although this paper discussed the model selection method based on the five generalized distributions, the traditional commonly used distribution, the LP3 distribution, is still an effective tool for frequency analysis and can be used for design rainfall or flood calculation.

Acknowledgments: This study was supported by the National Natural Science Foundation of China (51679094), the National Key R&D Program of China (2017YFC0405900), National Natural Science Foundation of China (51509273; 91547208), and Fundamental Research Funds for the Central Universities (2017KFYXJJ194, 2016YXZD048).

Author Contributions: Vijay P. Singh conceived and designed the experiments; Lu Chen and Kangdi Huang performed the experiments and analyzed the data; Lu Chen and Vijay P. Singh wrote the paper.

References

1. Rosbjerg, D. *Partial Duration Series in Water Resources*; Technical University of Denmark: Kongens Lyngby, Denmark, 1993.
2. Lin-Ye, J.; García-León, M.; Gracia, V.; Sánchez-Arcilla, A. A multivariate statistical model of extreme events: An application to the Catalan coast. *Coast. Eng.* **2016**, *117*, 138–156. [CrossRef]
3. Chen, L.; Zhang, Y.; Zhou, J.; Singh, V.P.; Guo, S.L.; Zhang, J. Real-time error correction method combined with combination flood forecasting technique for improving the accuracy of flood forecasting. *J. Hydrol.* **2015**, *521*, 157–169. [CrossRef]
4. Chen, L.; Singh, V.P.; Guo, S.; Zhou, J. Copula-based method for Multisite Monthly and Daily Streamflow Simulation. *J. Hydrol.* **2015**, *528*, 369–384. [CrossRef]
5. Chen, L.; Singh, V.P.; Guo, S.L.; Zhou, J.Z.; Zhang, J.H.; Liu, P. An objective method for partitioning the entire flood season into multiple sub-seasons. *J. Hydrol.* **2015**, *528*, 621–630. [CrossRef]
6. Yoon, P.; Kim, T.M.; Yoo, C. Rainfall frequency analysis using a mixed GEV distribution: A case study for annual maximum rainfalls in South Korea. *Stoch. Environ. Res. Risk Assess.* **2013**, *27*, 1143–1153. [CrossRef]
7. Perreault, L.; Bobée, B.; Rasmussen, P. Halphen distribution system. I: Mathematical and statistical properties. *J. Hydrol. Eng.* **1999**, *4*, 189–199. [CrossRef]
8. Zhang, J.; Chen, L.; Singh, V.P.; Cao, W.; Wang, D. Determination of the distribution of flood forecasting error. *Nat. Hazards* **2015**, *1*, 1389–1402. [CrossRef]
9. Papalexiou, S.M.; Koutsoyiannis, D. Entropy based derivation of probability distribution: A case study to daily rainfall. *Adv. Water Resour.* **2012**, *45*, 51–57. [CrossRef]
10. Chen, L.; Singh, V.P.; Xiong, F. An Entropy-Based Generalized Gamma Distribution for Flood Frequency Analysis. *Entropy* **2017**, *19*, 239. [CrossRef]
11. Chen, L.; Singh, V.P. Generalized Beta Distribution of the Second Kind for Flood Frequency Analysis. *Entropy* **2017**, *19*, 254. [CrossRef]
12. Singh, V.P. *Entropy Based Parameter Estimation in Hydrology*; Kluwer Academic Publishers: Dordrecht, The Netherlands, 1998.
13. Rahman, A.S.; Rahman, A.; Zaman, M.A.; Haddad, K.; Ahsan, A.; Imteaz, M. A study on selection of probability distributions for at-site flood frequency analysis in Australia. *Nat. Hazards* **2013**, *69*, 1803–1813. [CrossRef]
14. Bobee, B.; Perreault, L.; Ashkar, F. Two kinds of moment ratio diagrams and their applications in Hydrology. *Stoch. Hydrol. Hydraul.* **1993**, *7*, 41–65. [CrossRef]

15. Chen, L.; Singh, V.P.; Guo, S.L.; Zhou, J.; Ye, L. Copula entropy coupled with artificial neural network for rainfall-runoff simulation. *Stoch. Environ. Res. Risk Assess.* **2014**, *28*, 1755–1767. [CrossRef]
16. Laio, F.; Baldassarre, G.D.; Montanari, A. Model selection techniques for the frequency analysis of hydrological extremes. *Water Resour. Res.* **2009**, *45*, 162–174. [CrossRef]
17. Duan, Q.; Ajami, N.K.; Gao, X.; Sorooshian, S. Multi-model ensemble hydrologic prediction using Bayesian model averaging. *Adv. Water Resour.* **2007**, *30*, 1371–1386. [CrossRef]
18. Hsu, K.L.; Moradkhani, H.; Sorooshian, S. A sequential Bayesian approach for hydrologic model selection and prediction. *Water Resour. Res.* **2009**, *45*, 1079. [CrossRef]
19. Najafi, M.R.; Moradkhani, H.; Jung, I.W. Assessing the uncertainties of hydrologic model selection in climate change impact studies. *Hydrol. Process.* **2011**, *25*, 2814–2826. [CrossRef]
20. Robertson, D.E.; Wang, Q.J. A Bayesian Approach to Predictor Selection for Seasonal Streamflow Forecasting. *J. Hydrometeorol.* **2012**, *13*, 155–171. [CrossRef]
21. Vrugt, J.A.; Robinson, B.A. Treatment of uncertainty using ensemble methods: Comparison of sequential data assimilation and Bayesian model averaging. *Water Resour. Res.* **2007**, *43*, 223–228. [CrossRef]
22. Parrish, M.A.; Moradkhani, H.; Dechant, C.M. Toward reduction of model uncertainty: Integration of Bayesian model averaging and data assimilation. *Water Resour. Res.* **2012**, *48*, 3519. [CrossRef]
23. Chen, L.; Singh, V.P. Entropy-based derivation of generalized distributions for hydrometeorological frequency analysis. *J. Hydrol.* **2018**, *557*, 699–712. [CrossRef]
24. Metropolis, N.; Rosenbluth, A.W.; Rosenbluth, M.N.; Teller, A.H.; Teller, E. Equations of state calculations by fast computing machines. *J. Chem. Phys.* **1953**, *21*, 1087–1091. [CrossRef]
25. Smith, A.E.M.; Roberts, G. Bayesian computation via the Gibbs sampler and related Markov chain Monte Carlo methods. *J. R. Stat. Soc. B* **1993**, *55*, 3–23.
26. Griffis, V.W.; Stedinger, J.R. Log-Pearson Type 3 Distribution and Its Application in Flood Frequency Analysis. I: Distribution Characteristics. *J. Hydrol. Eng.* **2007**, *12*, 482–491. [CrossRef]
27. Lamontagne, J.R.; Stedinger, J.R.; Yu, X.; Whealton, C.A.; Xu, Z. Robust flood frequency analysis: Performance of EMA with multiple Grubbs-Beck outlier tests. *Water Resour. Res.* **2016**, *52*, 3068–3084. [CrossRef]
28. Lin-Ye, J.; Garcia-Leon, M.; Gracia, V.; Ortego, M.I.; Lionello, P.; Sanchez-Arcilla, A. Multivariate statistical modelling of future marine storms. *Appl. Ocean Res.* **2017**, *65*, 192–205. [CrossRef]
29. Chen, L.; Singh, V.P.; Guo, S.L. Measure of correlations for rivers flows based on copula entropy method. *J. Hydrol. Eng.* **2013**, *18*, 1591–1606. [CrossRef]
30. Chen, L.; Ye, L.; Singh, V.P.; Zhou, J.; Guo, S.L. Determination of input for artificial neural networks for flood forecasting using the copula entropy method. *J. Hydrol. Eng.* **2014**, *19*, 217–226. [CrossRef]
31. Chen, L.; Singh, V.P.; Lu, W.; Zhang, J.; Zhou, J.; Guo, S. Streamflow forecast uncertainty evolution and its effect on real-time reservoir operation. *J. Hydrol.* **2016**, *540*, 712–726. [CrossRef]

Entropy-Based Investigation on the Precipitation Variability over the Hexi Corridor in China

Liang Cheng, Jun Niu * and Dehai Liao

Center for Agricultural Water Research in China, China Agricultural University, Beijing 100083, China; cl182234@163.com (L.C.); liaodehai@cau.edu.cn (D.L.)

* Correspondence: junniu.cau@yahoo.com

Abstract: The spatial and temporal variability of precipitation time series were investigated for the Hexi Corridor, in Northwest China, by analyzing the entropy information. The examinations were performed on monthly, seasonal, and annual timescales based on 29 meteorological stations for the period of 1961–2015. The apportionment entropy and intensity entropy were used to analyze the regional precipitation characteristics, including the intra-annual and decadal distribution of monthly and annual precipitation amounts, as well as the number of precipitation days within a year and a decade. The regions with high precipitation variability are found in the western part of the Hexi corridor and with less precipitation, and may have a high possibility of drought occurrence. The variability of the number of precipitation days decreased from the west to the east of the corridor. Higher variability, in terms of both of precipitation amount and intensity during crop-growing season, has been found in the recent decade. In addition, the correlation between entropy-based precipitation variability and the crop yield is also compared, and the crop yield in historical periods is found to be correlated with the precipitation intensity disorder index in the middle reaches of the Hexi corridor.

Keywords: precipitation; variability; marginal entropy; crop yield; Hexi corridor

---

1. Introduction

The Hexi corridor in northwest China (Figure 1), located in 93°21′–104°05′ E and 37°15′–42°48′ N, plays an important role in grain production and regional economic development. The corridor is a long-narrow region in Northwest China with a length of about 1000 km, an average width of about 270 km, and a total area of about 270,000 km^2. The Qiliang Mountains are located in the southern part of the corridor and the Mazong and Longshou Mountains in the north, with the Wushaoling to the east and Dunhuang to the west. The corridor includes three inland water systems, namely the Shule River, the Hei River, and the Shiyang River, from west to east, respectively. The long-term mean annual precipitation is about 100–150 mm. Although it receives relatively limited precipitation, the Hexi corridor has sufficient light and heat resources, with a sunshine duration of about 3000 h/year. Therefore, it is an important crop-producing base, serving both the local food requirements and seeding production for other parts of China. Due to the shortage of irrigation water, the fluctuation of precipitation at different timescales may exert considerable effects on the farmland and oases in this area. The water resources is the most significant factor to affect the regional prosperity and the quality of life.

The crop growth relies heavily on water at different months or seasons for different grow phases. To rationally allocate regional water resources and mitigate the possible effects of drought, we need to analyze the long-term hydro-meteorological variables [1]. Together with the warming effects of climate change, the amount, timing, and distribution of precipitation may be affected, leading to changes in the amount of local available water. Many obvious/abrupt changes have been witnessed, such as the precipitation decline in Sahel [2], greater incidence of extreme precipitation events [3], and an increase in fall precipitation [4]. These observed changes could be partly attributed to the evolutions of large climate signals (e.g., [5]) in the context of climate change.

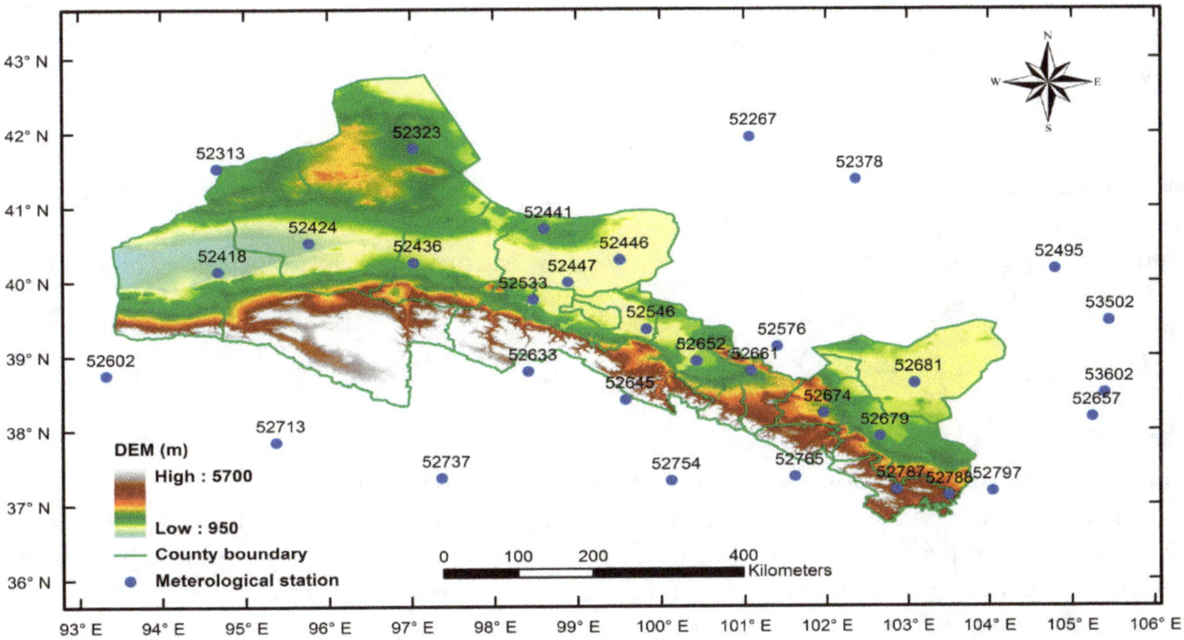

Figure 1. The Hexi corridor and the meteorological stations used in this study.

In order to reveal the underlying disorder features of precipitation variability during different months and different seasons, we use the entropy-based approach to study spatial and temporal precipitation variability over the Hexi corridor. The entropy theory presented by Shannon in the late 1940s [6] and the principle of maximum entropy presented by Jaynes in the late 1950s [7,8] have been applied in various research fields. Among them, there are many valuable hydrological applications of entropy theory (e.g., [9–21]). The aspects we want to address in this study include: (1) the detection of the spatial region and temporal year for high disorder features; (2) the identification of a monthly time series, which dominates the seasonal precipitation in Hexi corridor, and the identification of a seasonal time series, which dominates the annual precipitation in the corridor; (3) the correlations between drought-induced crop reduction and precipitation variability. The methods and data adopted in the present study are first presented in Sections 2 and 3, respectively. The entropy-based precipitation features are analyzed and discussed, and the correlation between crop production and precipitation variability is examined in Section 4. The conclusions are given in Section 5.

2. Materials and Methods

The entropy-based investigation is helpful to determine the least-biased probability distribution of a random variable. A discrete form of entropy $H(x)$ is written as [6,17]:

$$H(x) = -\sum_{k=1}^{K} p(x_k) \log_2[p(x_k)] \tag{1}$$

where k is time interval of the K events, x_k is an event corresponding to the interval k, and $p(x_k)$ is the probability of x_k. The application in this study is to measure the spatial and temporal variability/disorder features of precipitation in Hexi corridor in Northwest China. By following the methods in Mishra et al. [1], the several entropy measures, namely, intensity entropy, apportionment entropy, and marginal entropy, are employed for the research purpose. More details about the entropy and its applications can be found in Singh [9].

2.1. Marginal Entropy

The marginal entropy (ME) is the average information of a random variable x with the probability distribution $p(x)$, which is used to quantify uncertainty. The calculation processes are expressed in Equation (1). ME is applicable to different timescales, such as the daily, seasonal, monthly, and annual timescales in the present study.

2.2. Intensity Entropy

We consider the number of precipitation days (n_i) within a month i ($i = 1, 2, 3, ..., 12$) of a year and the total amount of precipitation days N ($N = \sum n_i$) in the year [1]. The probabilities of precipitation days in each month are expressed as $p_i = n_i/N$. The intensity entropy (IE) for an individual meteorological station for a month is given as:

$$\mathrm{IE} = -\sum p_i(\log_2 p_i) = -\sum (n_i/N)\log_2(n_i/N) \tag{2}$$

2.3. Apportionment Entropy

The apportionment entropy (AE) is used to study the distribution features of precipitation for different months over a year. Assume that the precipitation amount in the ith month is r_i ($i = 1, 2, 3, \ldots, 12$), and the total precipitation amounts for twelve months is R ($R = \sum r_i$). The probabilities of precipitation days in each month to the total days in a year are expressed as $p_i = r_i/R$. The AE for an individual meteorological station for a year is calculated as:

$$\mathrm{AE} = -\sum (r_i/R)\log_2(r_i/R) \tag{3}$$

The AE value is in the range of 0 and $\log_2(12)$ [1], in which the precipitation occurs only one out of twelve months and the annual precipitation amount is evenly distributed for twelve month respectively.

2.4. Entropy-Based Variability

The entropy-based variability is characterized by disorder index (DI). The DI is the difference between the maximum possible entropy and the calculated entropy from the individual series. In case the probability of each event is even, the DI reaches the maximum value. Therefore, the DI carries the information of disorder features of the analyzed time series. Accordingly, there is a marginal disorder index (MDI), an apportionment disorder index (ADI), and an intensity disorder index (IDI). In temporal domain, the DI is computed for annual, seasonal, and monthly time series. In spatial and temporal domain, the mean DI (MDI) is given by:

$$\mathrm{MDI} = \frac{1}{N}\sum_{i=1}^{N}\mathrm{DI} \tag{4}$$

where N is the length of entropy time scales. The higher the DI value is, the higher will be the variability [1].

3. Data

The precipitation data used in this study are derived from weather records of national standard meteorological stations for the Hexi corridor in Northwest China, obtained from China Meteorological Data Sharing Service System (http://data.cma.cn). A total of 29 stations, which have continuous precipitation records for the period 1961–2015, were selected for investigating precipitation variability. For the gap filling, the missing data were generally filled using the means of neighboring data, but if the missing period is longer than 1–2 days, the data of the same period from other years were considered occasionally. The geographical information of the 29 meteorological stations in Hexi corridor is listed in Table 1, and the basic statistical properties about annual total precipitation amount (ATP), monthly total precipitation amount (MTP), annual total precipitation days (ATD), and monthly total precipitation days (MTD) for the stations are listed in Table 2. The associated crop yield data were obtained from statistical data of the China Economic and Social Development Statistics Database (http://tongji.cnki.net/kns55/index.aspx), Gansu Water Statistical Yearbook.

Table 1. Geophysical information of 29 meteorological stations.

Station No.	Station ID	Latitude (Decimal Degrees)	Longitude (Decimal Degrees)
1	52267	41.95	101.07
2	52313	41.53	94.67
3	52323	41.80	97.03
4	52378	41.37	102.37
5	52418	40.15	94.68
6	52424	40.53	95.77
7	52436	40.27	97.03
8	52446	40.30	99.52
9	52495	40.17	104.80
10	52533	39.77	98.48
11	52546	39.37	99.83
12	52576	39.13	101.41
13	52602	38.75	93.33
14	52633	38.80	98.42
15	52645	38.42	99.58
16	52652	38.93	100.43
17	52657	38.18	105.25
18	52661	38.80	101.08
19	52674	38.23	101.97
20	52679	37.92	102.67
21	52681	38.63	103.08
22	52737	37.37	97.37
23	52754	37.33	100.13
24	52765	37.38	101.62
25	52787	37.20	102.87
26	52797	37.18	104.05
27	53502	39.47	105.45
28	53513	40.45	107.25
29	53602	38.50	105.40

Table 2. Statistical properties for precipitation time series (1981–2010) of the analyzed 29 stations.

Station No.	Station ID	ATP/MTP (mm)				ATD/MTD (Day)			
		Max.	Min.	Mean	STD	Max.	Min.	Mean	STD
1	52267	77.3/45.5	7.0/0	34.2/2.8	18.2/5.7	32/15	10/0	19.9/1.7	5.4/2.1
2	52313	95.3/40.9	16.9/0	50.5/4.2	19.5/6.9	42/10	22/0	30.0/2.5	5.7/2.1
3	52323	114.9/56.7	30.1/0	63.8/5.3	26.5/9.2	60/13	23/0	33.8/2.8	7.9/2.5
4	52378	105.2/54.6	13.9/0	42.1/3.5	23.9/7.2	26/10	12/0	18.4/1.5	3.6/1.8
5	52418	105.3/81.7	11.6/0	42.6/3.5	21.4/7.2	34/10	14/0	22.4/1.9	4.8/2.0
6	52424	83.6/53.0	19.0/0	50.8/4.2	15.7/6.8	40/10	19/0	25.6/2.1	5.2/2.2
7	52436	123.0/53.3	33.6/0	66.5/5.5	19.6/7.8	49/11	25/0	35.2/2.9	5.9/2.4
8	52446	108.5/58.6	28.4/0	60.4/5.0	21.6/8.6	42/14	19/0	30.8/2.6	6.2/2.7
9	52495	187.5/76.9	42.8/0	104.5/8.7	38.9/13.9	53/13	21/0	36.6/3.0	6.9/3.0
10	52533	157.3/67.8	41.4/0	88.4/7.4	30.3/10.9	60/13	28/0	39.4/3.3	7.5/2.8
11	52546	191.0/69.7	54.9/0	119.0/9.9	30.4/12.6	64/16	35/0	49.1/4.1	7.3/3.3
12	52576	197.6/98.0	76.6/0	134.8/11.2	30.8/16.7	65/16	34/0	48.8/4.1	7.9/3.6
13	52602	35.8/19.5	5.2/0	15.4/1.3	7.2/2.6	22/8	6/0	12.9/1.1	4.3/1.5
14	52633	404.4/152.4	180.7/0	309.7/25.8	57.7/33.5	109/23	74/0	88.3/7.4	10.3/6.3

Table 2. *Cont.*

Station No.	Station ID	ATP/MTP (mm)				ATD/MTD (Day)			
		Max.	Min.	Mean	STD	Max.	Min.	Mean	STD
15	52645	602.3/151.0	274.9/0	426.4/35.5	75.5/41.5	159/28	100/0	125.5/10.5	13.9/7.4
16	52652	216.3/90.3	71.6/0	132.6/11.0	36.6/14.4	73/15	37/0	53.9/4.5	9.4/3.4
17	52657	573.1/162.4	331.0/0	415.1/34.6	61.5/40.1	139/26	87/0	112.0/9.3	12.7/7.1
18	52661	301.2/107.8	97.8/0	202.6/16.9	49.8/20.8	93/17	51/0	68.4/5.7	10.1/3.8
19	52674	294.6/128.0	110.5/0	211.7/17.6	40.4/22.3	95/20	52/0	73.4/6.1	11.9/4.5
20	52679	251.3/111.9	101.6/0	171.0/14.3	40.8/17.8	83/14	46/0	60.9/5.1	7.9/3.6
21	52681	202.0/73.2	52.0/0	113.2/9.4	29.9/13.1	53/14	24/0	39.7/3.3	7.9/3.0
22	52737	354.9/145.1	128.4/0	214.3/17.9	58.1/24.2	87/20	36/0	59.8/5.0	12.2/4.5
23	52754	535.7/162.3	263.7/0	411.9/34.3	61.2/40.6	141/26	89/0	110.7/9.2	11.9/7.5
24	52765	730.7/201.2	404.0/0	531.1/44.3	72.3/44.2	153/26	107/0	130.8/10.9	13.7/6.9
25	52787	543.3/172.2	274.0/0	407.0/33.9	58.6/36.4	183/27	112/0	140.5/11.7	13.7/5.8
26	52797	298.4/95.4	94.8/0	179.8/15.0	54.3/19.1	77/16	42/0	56.1/4.7	9.7/3.9
27	53502	189.1/94.5	56.8/0	101.8/8.5	34.1/13.7	46/13	22/0	33.0/2.8	6.2/2.8
28	53513	267.9/193.6	55.9/0	148.9/12.4	48.7/21.5	52/15	22/0	36.6/3.0	7.4/3.0
29	53602	347.3/103.8	126.2/0	213.7/17.8	55.3/22.3	65/15	37/0	53.2/4.4	7.1/3.3

4. Results and Discussion

4.1. Variability of Annual, Seasonal, and Monthly Precipitation

Figure 2 shows the calculated marginal disorder index (MDI) for annual and seasonal (spring, summer, fall, and winter) precipitation time series over different meteorological stations. Overall, the MDI at annual timescale is lower than those of four seasons. The higher MDI values appear for winter time series, which indicates the higher variability associated with the analyzed stations in winter. Among them, the precipitation recorded by Station 52602 near the western part of the Hexi corridor shows maximum marginal entropy in both annual and four seasonal timescales, followed by Station 52378 (in the northeastern part) and 52446 (in the middle region of the Hexi corridor) in terms of the MDI value in Winter. The spatial distribution of the computed MDI at annual timescale is shown in Figure 3, in comparison with the corresponding mean annual precipitation for the period 1961–2015. It is observed that the annual precipitation in Hexi corridor is increasing from the northwest region to the southeast region, with a range of 50–400 mm/year. Regarding the MDI value, the highest value exits in the west corner of the Hexi corridor, which is in the area of Shule River basin. The MDI gradually decreases from west to the junction area between the Hei and Shiyang River basins, and it further slightly increases in the northeastern part of Shiyang River basin. Therefore, the disorder characteristics associated with the mean annual precipitation vary spatially for the Hexi corridor.

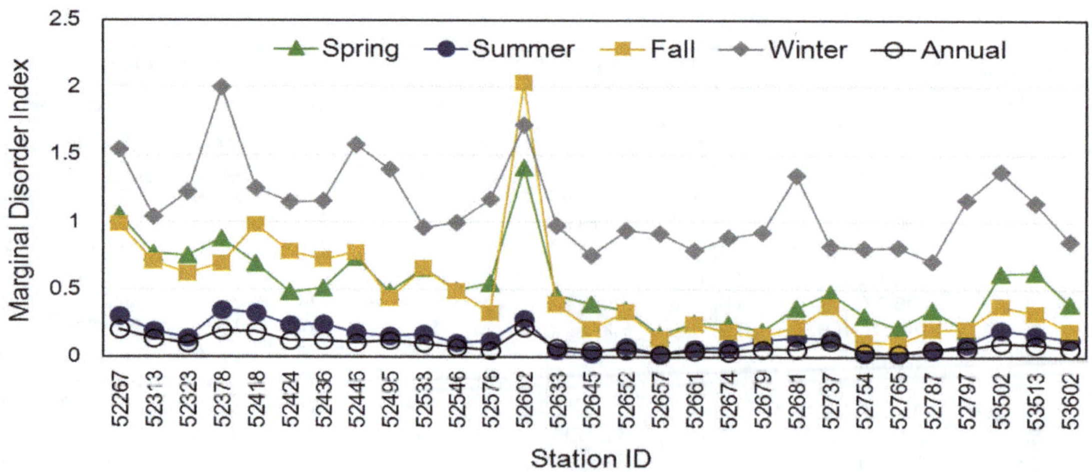

Figure 2. Marginal disorder indices of annual and seasonal precipitation time series over the Hexi region, Northwest China.

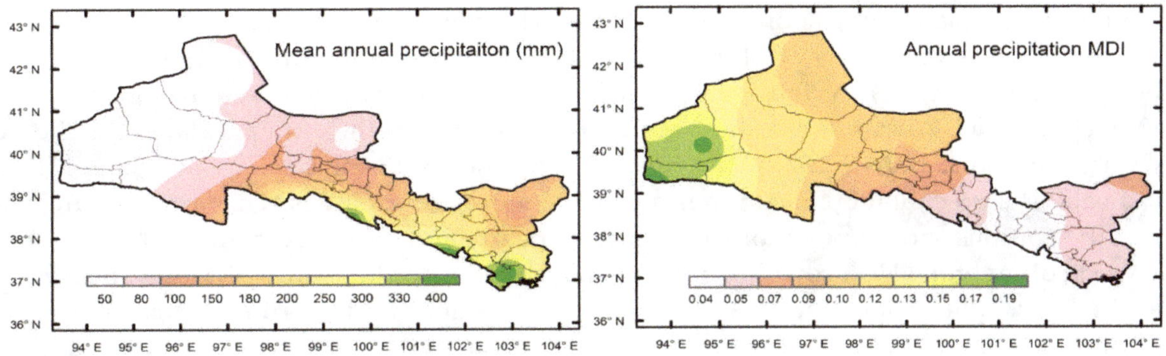

Figure 3. Spatial distribution of mean annual precipitation (**left panel**) and marginal disorder index (MDI) (**right panel**) for annual time series over the Hexi corridor.

The spatial distribution of MDI of seasonal precipitation time series is shown in Figure 4. It is found that the stations with high variability for the four seasons are: Stations 52424, 52436, and 52546 in spring; Station 52418 in summer; Stations 52313, 52418, 52424, and 52446 in fall; and Stations 52602, 52446, 52418, and 52681 in winter. It is interesting that the station 52418 is the only one with relatively higher disorder features in summer over the Hexi corridor. The stations with the higher winter MDI are not concentrated in one region, but distributed in the western, middle, and eastern parts. It is also observed that the winter MDI value is higher than that of other seasons, as shown in Figure 5. The distribution pattern of spring precipitation MDI is similar to that in fall MDI. The MDI value in summer is quite low, which indicates the disorder features are less obvious when the precipitation is relatively high in summer season over the Hexi corridor.

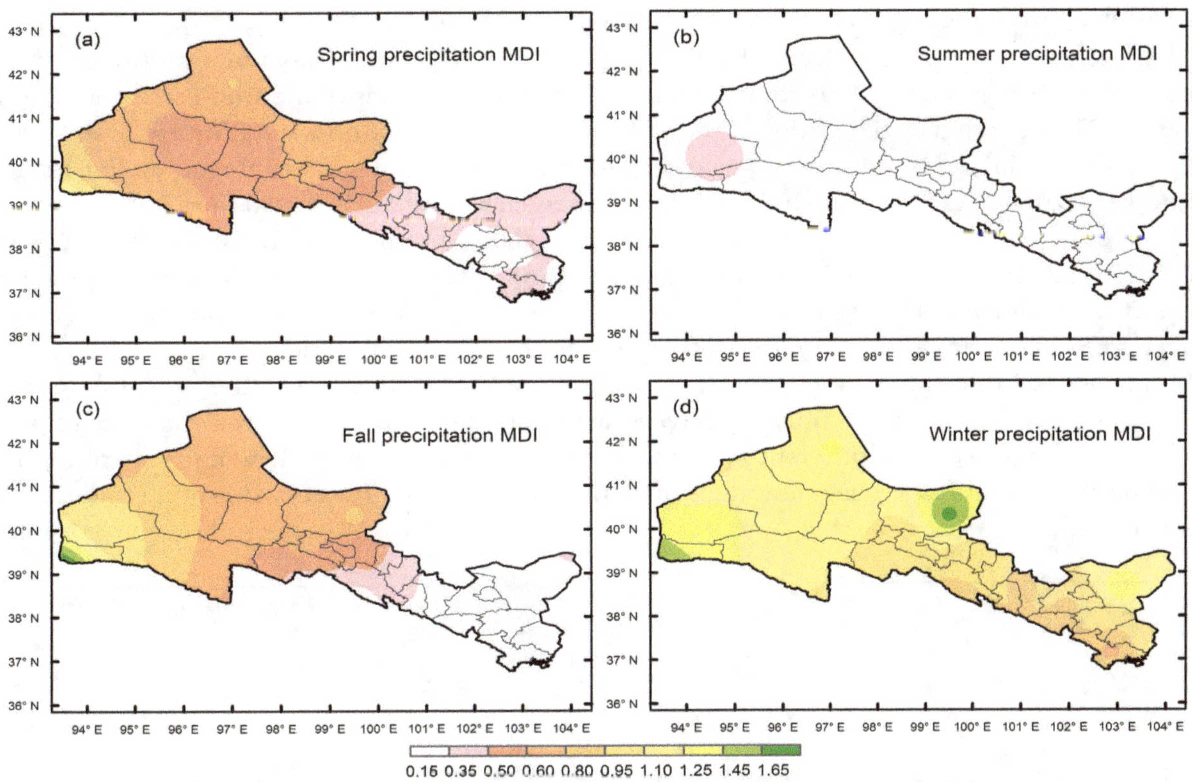

Figure 4. Spatial distribution of marginal disorder index (MDI) of precipitation time series for (**a**) spring; (**b**) summer; (**c**) fall; and (**d**) winter seasons over the Hexi corridor.

To understand which month is responsible for seasonal variability, it is useful to examine the intra-variability of months within a season [1]. Basically, the entropy-based variability of individual month is higher than that of corresponding season, which shows the disordered nature of precipitation. It is observed that the variability of March precipitation time series contributes the most in spring, while the variability in April contributes less. The summer variability is low in terms of the MDI value, and the MDI in July and August exerts the steadiest processes across the different stations in the studied region. Among it, the Stations 52267, 52378, 52418, 52436, and 53513 show relatively higher MDI value, which indicates the precipitation in summer may have more uncertainties. The farmland area located on or near the stations should pay more attention to the precipitation forecasting. The Station 52602, near the west corner of the Hexi corridor, has a distinguished disorder feature through all four seasons, with the most variability appearing in April, August, November, and February, respectively.

4.2. Variability of Precipitation Apportionment and Intensity

The apportionment features of precipitation can be reflected by the variability of precipitation amount of different months within a year. The apportionment disorder index (ADI) serves to quantify the precipitation characteristics, with the higher value indicating the higher unbalanced feature. Figure 6a summarizes the features for the whole studied region for the past 55 years (1961–2015). It is observed that the variability of precipitation amounts within a year are higher during the first 20-year period (1961–1980). These calculation results could be explained in two ways: (1) there are more zero values in precipitation records during 1961–1980 period, which probably demonstrates dire environment in this region compared to that of recent decades; and (2) the observed records may not reliable during the quite earlier years, and those zero values may not be effective. For the period 1980–2015, we can observe that the high variability of precipitation apportionment appears during the last decade in 20th century and after 2010. Meanwhile, Figure 7a reveals the station-based spatial features of ADI for the whole studied period. The stations with high apportionment variability include Stations 52267, 52378, 52418, 52446, and 52602. The variability generally decreases from the northwestern part of the region to the southeastern part area, as shown in Figure 8a.

The intensity entropy is reflected by the number of precipitation days in a month or in a year. The intensity disorder index (IDI) is computed for the involved stations of the studied region. Figure 6b shows the IDI over the past 5 decades. The relatively higher IDI are observed for the first 2 decades, which may be explained by the similar two reasons presented above for ADI results. For the recent 35 years (1980–2015), the years with high precipitation intensity are during 1980s and 2010s. The spatial features over all stations are summarized in Figure 7b. The stations with distinct intensity peak are 52267, 52378, 52418, 52446, and 52602. Overall, the spatial features of IDI generally are in line with the apportionment results in Figure 7a. This can be more easily observed in Figure 8a,b. In addition, Figure 8b reflects precipitation amount contribution within the Hexi corridor.

It is observed that the most contributions are along the south edge of the corridor, and those that are relatively higher are in Shiyang River basin. However, the contributions of precipitation days, shown in Figure 8d, are different from that of precipitation amount. The high contribution area includes both Shiyang River basin and the northern part of the Hei River basin in the middle part of the Hexi corridor.

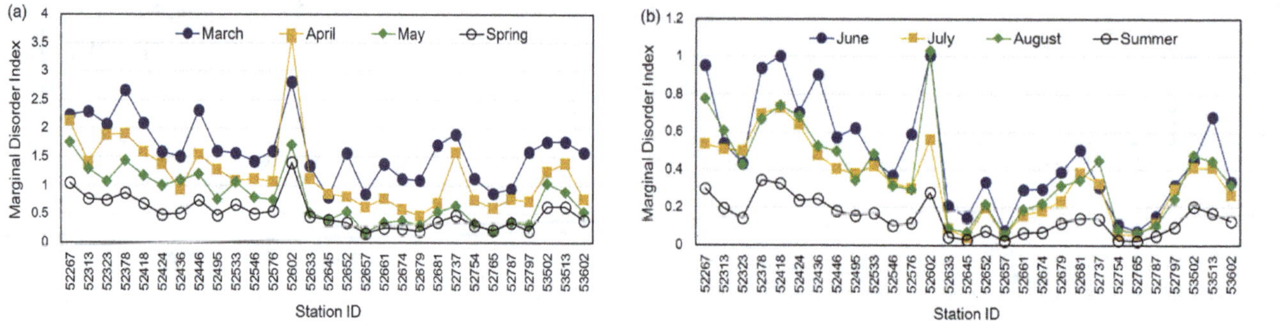

Figure 5. *Cont.*

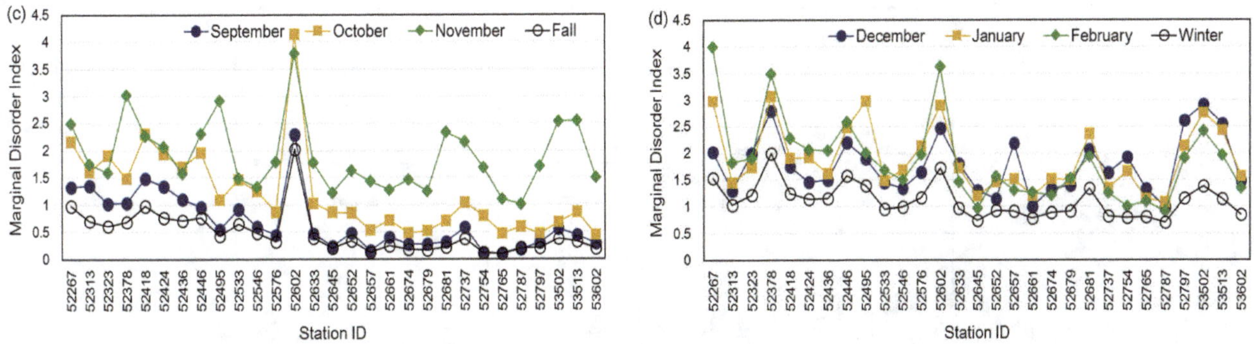

Figure 5. Marginal disorder index (MDI) of precipitation time series over different stations for (**a**) spring; (**b**) summer; (**c**) fall; and (**d**) winter seasons and its respective months.

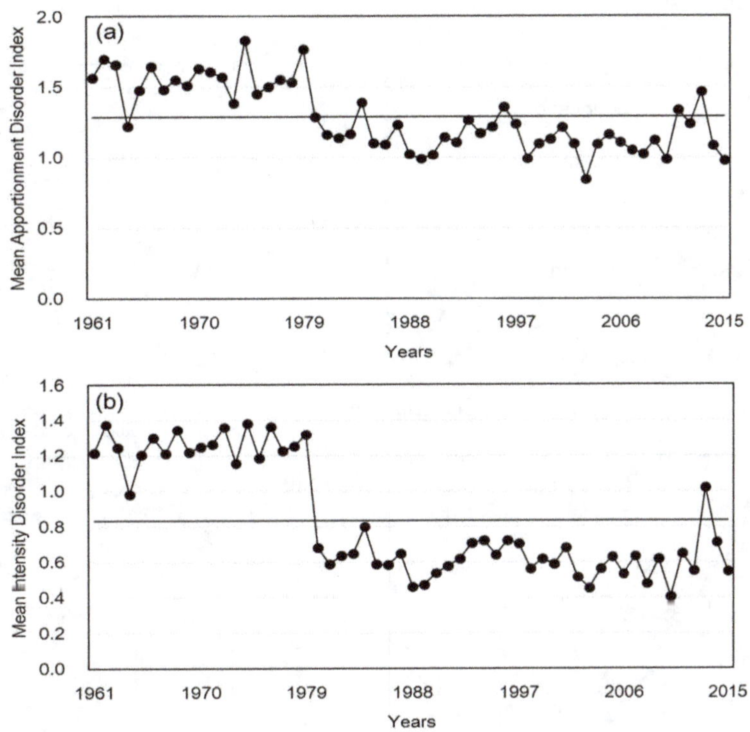

Figure 6. (**a**) Mean apportionment disorder index (MDI) of precipitation time series of all stations for the period of 1961–2015; the central line is the long-term mean value; (**b**) same as (**a**), but for mean intensity disorder index (IDI).

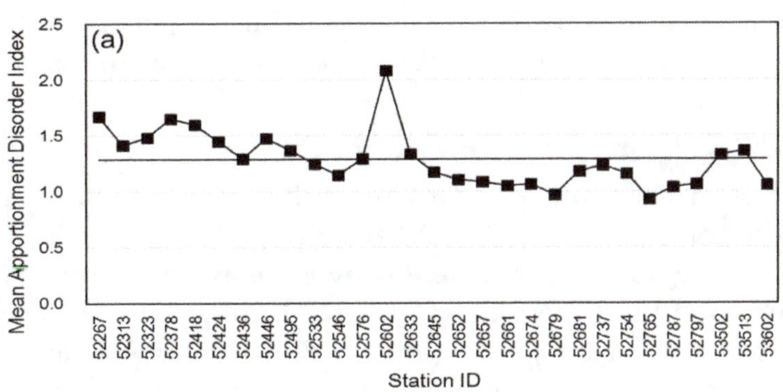

Figure 7. *Cont.*

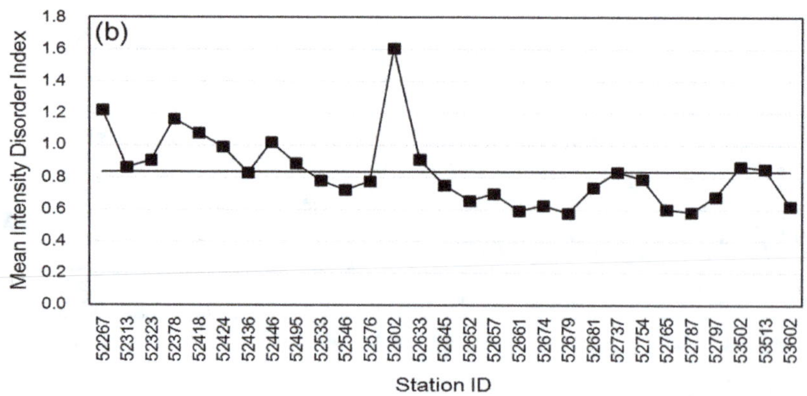

Figure 7. (**a**) Mean apportionment disorder index (MDI) of precipitation time series for different stations for the period of 1961–2015. The central line is the long-term mean value; (**b**) same as (**a**), but for mean intensity disorder index (IDI).

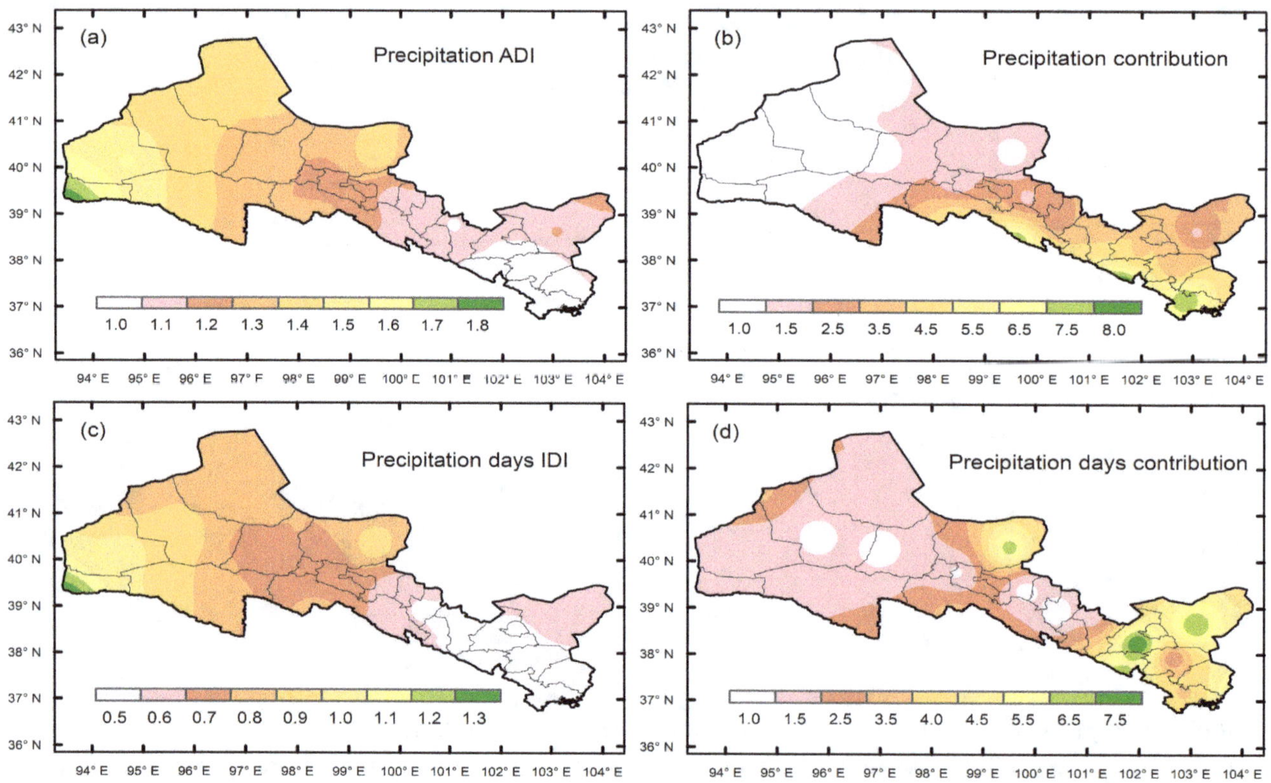

Figure 8. (**a**) The distribution of apportionment disorder index of precipitation time series over the Hexi corridor; (**b**) the contribution of individual station to overall precipitation in the region; (**c**) same as (**a**), but for intensity disorder index; (**d**) same as (**b**), but for a number of precipitation days in the area.

4.3. Decadal Variability of Precipitation Amount and Precipitation Days

To reveal the long-term changing trend of precipitation variability over the Hexi corridor, the decadal apportionment disorder index (DADI) and decadal intensity disorder index (DIDI) were computed for the recent three decades. Figure 9 shows the average entropy value for all stations at annual, seasonal, and monthly time scales for three individual decades. It is observed from Figure 9a that the stations in February, June, and July in the last decade (2001–2010) appear to have more variability in precipitation amount as compared with the two decades (1981–1990 and 1991–2010). In contrast to the variability of precipitation amount, the variability on precipitation days in February in the last decade is less than that during 1981–1990 but slightly higher than the decade of 1991–2010, as shown in Figure 9b. The variability reduction of precipitation days, reflected by DIDI, also occurred in August and October. The variability of precipitation amount and intensity changes for the crop-growing season over the Hexi corridor are very important factors for regional agricultural activities.

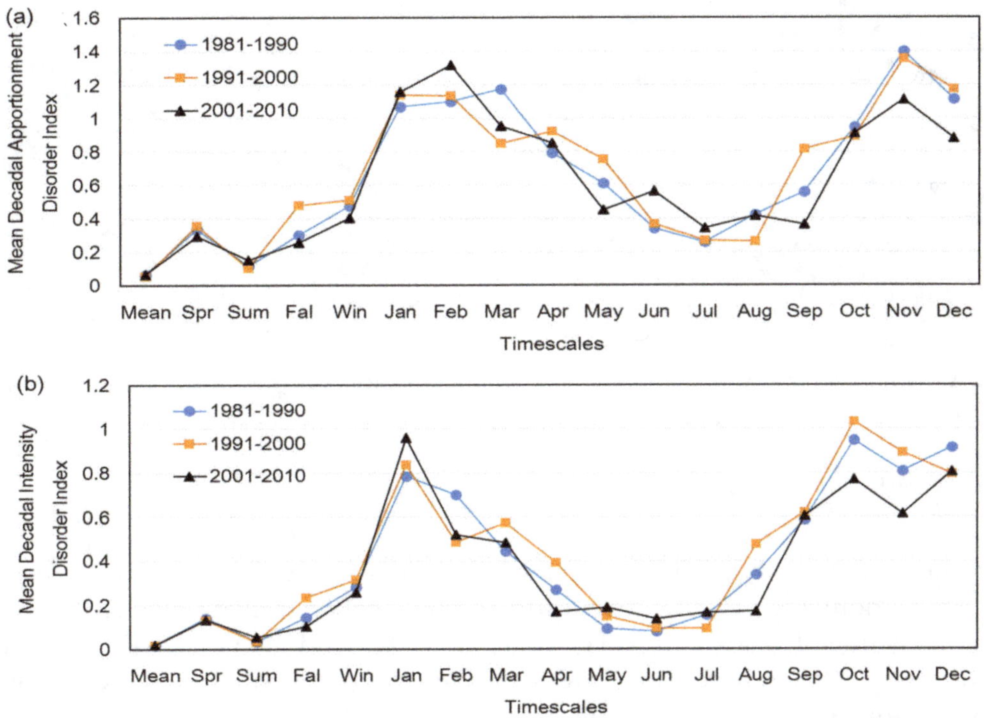

Figure 9. (**a**) Mean decadal apportionment disorder index of precipitation time series at different timescales; (**b**) same as (**a**), but for mean decadal intensity disorder index.

4.4. Implication for Crop Production

Based on the results on the spatial distribution in Figure 8, we found that the entropy-based variability generally decreases from west to east, and the precipitation contributions in terms of both precipitation amount and precipitation days decrease from east to west. Therefore, the middle reaches of the Hexi corridor region has both relatively high variability and precipitation amount/precipitation days. Accordingly, the crop yield in this region may be more sensitive to the evolutions of precipitation.

The crop production rates in four counties in the middle of the Hexi corridor during the period of 2005–2014 were compared with the computed precipitation disorder index and the standard deviations of anuual precipitaion amount/precipitaiton days for the associated stations, as shown in Figure 10. There are mainly four different crop types, namely, corn, wheat, rape, and barley. The crop disorder index (CDI) and standard deviation were computed to represent the variability of crop production. Among them, the standard deviation values were scaled to a comparable magnitude with other entropy index values. It is observed that there is a more consistent trend between the crop disorder index and precipitation intensity disorder index (IDI), compared with the relationship between crop disorder index and apportionment disorder index. This indicates the higher IDI value may introduce more crop variability, which demonstrates the implication of entropy-based variability for regional crop production.

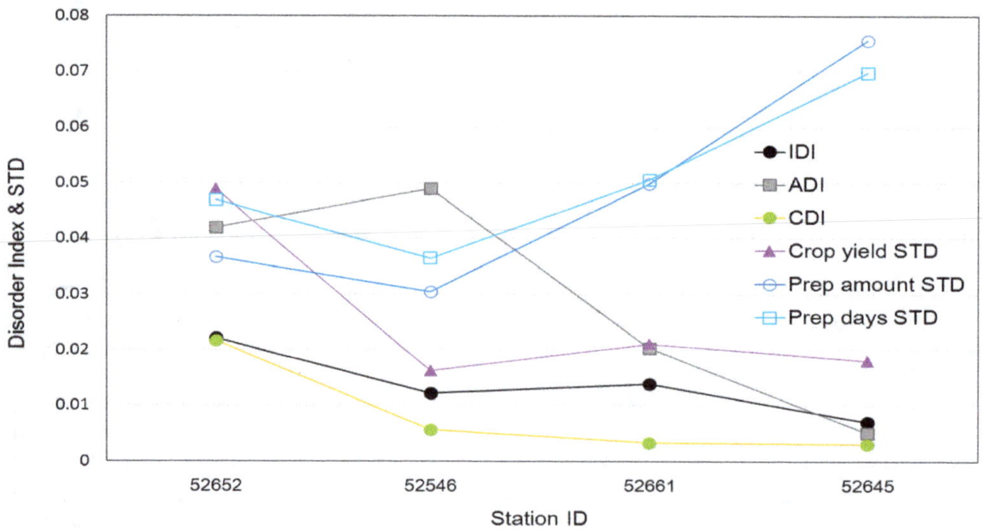

Figure 10. A comparison between crop and precipitation indices for the counties in the middle of the Hexi corridor.

4.5. Co-Variability with Other Meteorological Variables

Screening the co-variability of different meterological or hydrological variations is an important step to revealing the dominant controls in certain geophysical processes, such as for the geochmical hot moments in the study of Arora et al. [21]. To better understand the variability of precipitation and crop production mentioned in Section 4.4, we further checked the entropy-based variabilities of the associated temperature and wind speed. Figure 11 shows the average entropy value for all stations on a monthly scale for three individual decades. More variability appears in May, June, and July for temperature, and in January, November, and December for wind speed. Basically, consistent seasonal distributions are found for past three decades. The relationships between precipitation and temperature on a monthly scale, as shown in Figure 12, displays little precipitation below 0 degrees Celsius, and precipitation reaches its height around 10 degrees Celsius in wet seasons. Furthermore, we compared the crop disorder index CDI in four counties with the computed precipitation disorder index, temperture disorder index (TDI), and wind speed disorder index (WDI) for the associated stations, as shown in Figure 13. The TDI shows relatively steady variability for different stations in the region, and WDI exhibits obvious spatial differences. It is observed that the trend of precipitation intensity disorder index IDI is more consistent with that of CDI, compared with TDI and WDI.

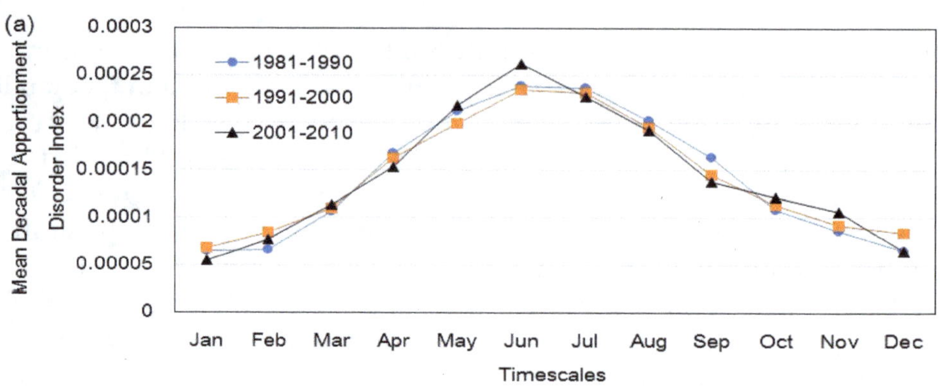

Figure 11. *Cont.*

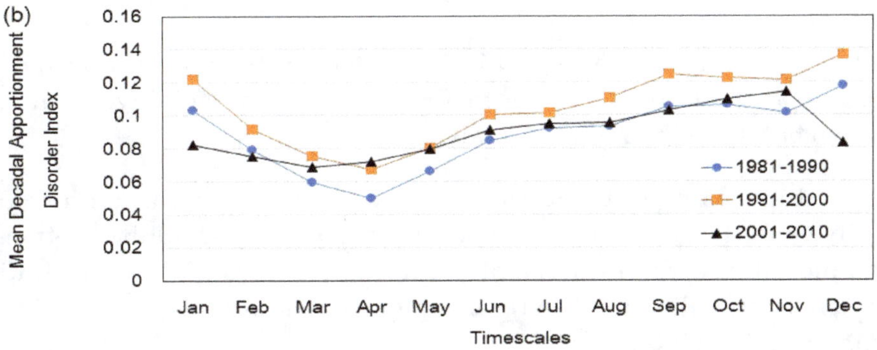

Figure 11. (**a**) Mean decadal apportionment disorder index of monthly temperature time series; (**b**) same as (**a**), but for wind speed.

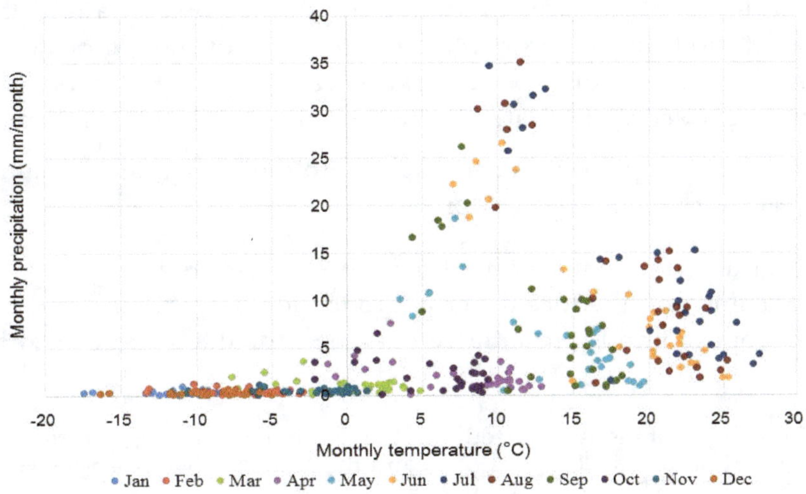

Figure 12. The relationships between temperature and precipitation on a monthly timescale for the studied stations.

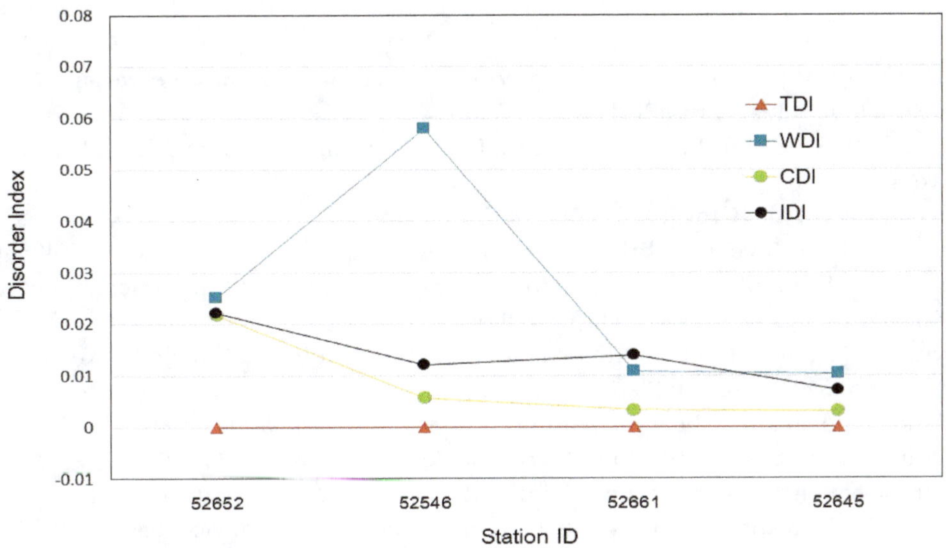

Figure 13. The comparisons between temperature, wind speed, crops, and precipitation disorder index for the counties in the middle of the Hexi corridor.

5. Conclusions

This study undertakes an entropy-based investigation of the spatial and temporal variability of precipitation for the period 1951–2015 over the Hexi corridor in Northwest China. The following conclusions are drawn based on the relevant analyses.

1. The annual variability is mainly contributed to by the winter variability. The precipitation intensity is diversified due to the difference in spatial variability between the monthly precipitation amount and the number of precipitation days. The contributions of different months to seasonal variability are different. A high contribution is detected in March for spring and in December for the winter period.
2. More variability during 2001–2010 is found for the precipitation time series in January, May, June, and July when we consider precipitation days. Regarding the precipitation amount, the variability in February, June, and July has increased in the recent decade.
3. The variability in terms of the precipitation amount and precipitation days decreases from the western part of the Hexi corridor to the eastern part of the region, which indicates more precipitation uncertainty is accommodated in the western part. From the temporal domain, the variability in the recent decade is relatively stronger than the previous decades.
4. The crop yield in historical periods is correlated with the precipitation intensity disorder index for the middle reaches of the Hexi corridor.

Extreme hydrological events may occur for highly disordered regions, which could induce negative effects on agriculture and reduce the crop production. Careful and early water management strategies should be proposed and particularly designed for these areas to mitigate the possible crop losses.

Acknowledgments: This research is supported by the National Training Program of Innovation and Entrepreneurship for Undergraduates (201710019225) and the National Natural Science Foundation of China (51679233).

Author Contributions: The authors designed and performed the research together. Liang Cheng and Dehai Liao collected the data and conducted analysis. Jun Niu and Liang Cheng wrote the draft of the paper. All authors have read and approved the final manuscript.

References

1. Mishra, A.K.; Özger, M.; Singh, V.P. An entropy-based investigation into the variability of precipitation. *J. Hydrol.* **2009**, *370*, 139–154. [CrossRef]
2. Hulme, M.; Doherty, R.; Ngara, T.; New, M.; Lister, D. African climate change: 1900–2100. *Clim. Res.* **2001**, *17*, 145–168. [CrossRef]
3. Easterling, D.R.; Evans, J.L.; Groisman, P.Y.; Karl, T.R.; Kunkel, K.E.; Ambenje, P. Observed variability and trends in extreme climate events: A brief review. *Bull. Am. Meteorol. Soc.* **2000**, *81*, 417–425. [CrossRef]
4. Karl, T.R.; Knight, R.W. Secular trends of precipitation amount, frequency, and intensity in the United States. *Bull. Am. Meteorol. Soc.* **1998**, *79*, 231–241. [CrossRef]
5. Wang, F.; Niu, J. The implication of climate signal for precipitation in the Heihe River basin, Northwest China. *Adv. Meteorol.* **2016**, *2016*, 1078617. [CrossRef]
6. Shannon, C.E. A mathematical theory of communication. *Bell. Syst. Tech. J.* **1948**, *27*, 379–423. [CrossRef]
7. Jaynes, E.T. Information theory and statistical mechanics, I. *Phys. Rev.* **1957**, *106*, 620–630. [CrossRef]
8. Jaynes, E.T. Information and statistical mechanics, II. *Phys. Rev.* **1957**, *108*, 171–190. [CrossRef]
9. Cui, H.; Singh, V.P. On the cumulative distribution function for entropy-based hydrologic modeling. *Trans. ASABE* **2012**, *55*, 429–438. [CrossRef]
10. Cui, H.; Singh, V.P. Computation of suspended sediment discharge in open channels using Tsallis entropy—Based Methods and Empirical Formulas. *J. Hydrol. Eng.* **2013**, *19*, 18–25. [CrossRef]

11. Cui, H.; Singh, V.P. Configurational entropy theory for streamflow forecasting. *J. Hydrol.* **2015**, *521*, 1–7. [CrossRef]
12. Cui, H.; Singh, V.P. Minimum relative entropy theory for streamflow forecasting with frequency as a random variable. *Stoch. Environ. Res. Risk Assess.* **2016**, *30*, 1545–1563. [CrossRef]
13. Singh, V.P. The use of entropy in hydrology and water resources. *Hydrol. Process.* **1997**, *11*, 587–626. [CrossRef]
14. Singh, V.P. *Entropy Theory in Hydrologic Science and Engineering*; McGraw-Hill Education: New York, NY, USA, 2015.
15. Sivakumar, B. Chaos theory in hydrology: Important issues and interpretations. *J. Hydrol.* **2000**, *227*, 1–20. [CrossRef]
16. Sivakumar, B. Chaos Identification and Prediction Methods. In *Chaos in Hydrology*; Springer: Dordrecht, The Netherlands, 2017; pp. 173–197.
17. Dwivedi, D.; Mohanty, B.P. Hot spots and persistence of nitrate in aquifers across scales. *Entropy* **2016**, *18*, 25. [CrossRef]
18. Maruyama, T.; Kawachi, T.; Maeda, S. Entropy-based Assessments of Monthly Rainfall Variability. *J. Rainwater Catchment Syst.* **2002**, *8*, 21–25. [CrossRef]
19. Khan, M.I.; Liu, D.; Fu, Q.; Azmat, M.; Luo, M.; Hu, Y.; Zhang, Y.; Abrar, F.M. Precipitation variability assessment of northeast China: Songhua River basin. *J. Earth Syst. Sci.* **2016**, *125*, 957–968. [CrossRef]
20. Atieh, M.; Rudra, R.; Gharabaghi, B.; Lubitz, D. Investigating the Spatial and Temporal Variability of Precipitation using Entropy Theory. *J. Water Manag. Model.* **2017**. [CrossRef]
21. Arora, B.; Dwivedi, D.; Hubbard, S.S.; Steefel, C.I.; Williams, K.H. Identifying geochemical hot moments and their controls on a contaminated river-floodplain system using wavelet and entropy approaches. *Environ. Model. Softw.* **2016**, *85*, 27–41. [CrossRef]

7

Spatial Optimization of Agricultural Land use based on Cross-Entropy Method

Lina Hao [1], Xiaoling Su [1,*], Vijay P. Singh [2] and Olusola O. Ayantobo [1,3]

1 College of Water Resources and Architectural Engineering, Northwest A & F University, Yangling 712100, China; haolina@nwafu.edu.cn (L.H.); Skollar24k@yahoo.com (O.O.A.)
2 Department of Biological & Agricultural Engineering and Zachry Department of Civil Engineering, Texas A & M University, College Station, TX 77843-2117, USA; vsingh@tamu.edu
3 Department of Water Resources Management and Agricultural-Meteorology, Federal University of Agriculture, PMB 2240 Abeokuta, Nigeria
* Correspondence: xiaolingsu@nwsuaf.edu.cn

Abstract: An integrated optimization model was developed for the spatial distribution of agricultural crops in order to efficiently utilize agricultural water and land resources simultaneously. The model is based on the spatial distribution of crop suitability, spatial distribution of population density, and agricultural land use data. Multi-source remote sensing data are combined with constraints of optimal crop area, which are obtained from agricultural cropping pattern optimization model. Using the middle reaches of the Heihe River basin as an example, the spatial distribution of maize and wheat were optimized by minimizing cross-entropy between crop distribution probabilities and desired but unknown distribution probabilities. Results showed that the area of maize should increase and the area of wheat should decrease in the study area compared with the situation in 2013. The comprehensive suitable area distribution of maize is approximately in accordance with the distribution in the present situation; however, the comprehensive suitable area distribution of wheat is not consistent with the distribution in the present situation. Through optimization, the high proportion of maize and wheat area was more concentrated than before. The maize area with more than 80% allocation concentrates on the south of the study area, and the wheat area with more than 30% allocation concentrates on the central part of the study area. The outcome of this study provides a scientific basis for farmers to select crops that are suitable in a particular area.

Keywords: cross-entropy minimization; land suitability evaluation; spatial optimization

1. Introduction

Scarcity of agricultural water and land resources is becoming severe due to the growing population and continued economic development, and has become a critical issue in formulating sustainable developmental policies [1–3]. Therefore, appropriate and efficient allocation of agricultural water and land resources has become necessary in regional agricultural sustainable development [4]. Agricultural land use allocation is the core issue of agricultural land and water resource allocation optimization [5,6]. Not only can it guide decision makers in assessing land demand for different crop types, but it can also identify the optimum land spatial unit with characteristics that are related to their geographical locations for each crop type [7–9], as well as simultaneously seeking the best land use layout [10,11].

The process of agricultural land-use allocation would be undertaken in three main stages: (i) demand assessment; (ii) agricultural land suitability evaluation; and (iii) spatial distribution of crop types [7].

To assess the agricultural land suitability of crops, the required environmental and socio-economic conditions are considered [12]. The Food and Agricultural Organization (FAO) developed crop-specific maps of crop suitability classes using the spatial data on soil, topography, features, and crop characteristics [13]. Many studies have developed methods for land suitability allocation based on the FAO framework. In these methods, input attributes and suitability indices were classified into different classes, and weights were assigned to the attributes, depending on their relative importance [5,14–17]. However, the main shortcoming of FAO-based methods is that their crop suitability data are usually available at a 5 min (approximately 9 km × 9 km at the equator) grid and such resolution is too coarse to satisfy the research demand at the irrigation district scale. Furthermore, when evaluating land suitability, very few studies have taken into account the actual distribution of various crops. For example, You and Wood [18] improved the pre-allocation by providing the existing crop distribution maps. To a significant extent, land suitability are determined by biophysical and soil conditions, such as organic matter, total nitrogen, total phosphorus. When the observed data of these attributes cannot be completely covered on the entire region, the actual distribution of crops can provide the supplement of crop distribution information. Using the research findings of Peng [19], an integrated model combined with multi-source remote sensing data was developed to generate maps of spatial distribution with 1 km × 1 km resolution suitable for crop planting.

The main objective of agricultural land use allocation is the spatial allocation of crop types to different spatial units having characteristics related to their geographical locations, for the purpose of seeking the best land use layout [11]. There is extensive literature on various methods of optimization for agricultural cropping patterns, such as linear programming [20], non-linear programming [21], multi-objective programming [22], fuzzy programming [23,24], and stochastic optimization [25]. However, these methods have ignored the effective unification of quantity and space, and have merely focused on the quantity optimization. In other words, these models can provide an optimal cropping pattern but cannot analyze the optimal spatial distribution of crops, which has an important guiding significance in actual production work.

Since numerous variables are involved in spatial optimization, conventional mathematical models are deemed unfit to determine the optimal solution within an acceptable timeframe [12]. Various heuristic algorithms for land-use spatial optimization have also been developed, including particle swarm algorithms [26,27], colony algorithms [28], and genetic algorithms [11,29]. Although these algorithms have a significant global optimization capability, they involve complex patch coding, resulting in programming difficulties. Other methods adopted cellular automation models, based on land-use conversion rules for local areas, to generate land use patterns under different conditions using a bottom-up approach [12,29,30]. However, cellular automaton is restricted by neighborhood rules, and cannot search across space. In this study, spatial crop optimization is defined in the framework of minimum cross entropy. The principle of minimum cross-entropy (POMCE) was formulated by Kullback and Leibler [31] and is detailed by Kullback [32]. The cross entropy can measure the variation between different information contents, which seems an ideal approach to resolve the spatial allocation problem [33]. However, the cross entropy method has usually been applied in determining spatial-temporal changes of land use and applied a meso-scale model for the spatial disaggregation of crop production [34–36], but has rarely been coupled with spatial optimization modeling [18]. For example, You and Wood [37] described an entropy-based approach to conduct a spatially disaggregated assessment of the distribution of crop production. Considering the difficulty in coupling spatial variables with non-spatial variables, a loosely coupled model, based on minimum cross entropy and nonlinear optimization, is constructed in this study. The crop spatial allocation can be performed by determining the minimum cross entropy between the prior distribution and the desired distribution. The prior distribution produces crop-related and environment-related information obtained from integrated multi-source analysis in agricultural land suitability.

This study develops an integrated optimization model for the spatial distribution of agricultural cropping. The model is based on the spatial distribution of crop suitability, spatial distribution of population density, and agricultural land use data. It combines multi-source remote sensing data with constraints of optimal crop areas, which are obtained from an agricultural cropping pattern optimization model. Minimization of cross-entropy is applied to build the model. The model determines the suitable planting region for a specified crop well.

2. Materials and Methods

2.1. Study Area

The study area is the middle reaches of the Heihe River basin (98°30′–101° E, 38°30′–40° N), lying in an arid region of Gansu Provence, northwest of China, and covers an area of 11,427 km^2. The Heihe River basin is the second largest inland river basin in China. In this region, crop production mainly depends on agricultural irrigation, because the mean annual precipitation and evaporation are about 117 mm and 1065 mm, respectively. Water consumption from agricultural irrigation accounts for approximately 90% of the total water consumption in this region [38]. Therefore, the optimal distribution of limited irrigation water and land resources is a key factor for agricultural development and sustainability [4,39]. The study area presents a higher terrain in the southeast and low in the northwest, and the elevation is between 1235 m and 3634 m. there is heterogeneity of biophysical and soil condition in this region. The content of organic matter and nitrogen is higher in the southern part than in the northern part [19]. The middle reaches of Heihe River basin is a commodity grain production base, which is made up of 17 irrigation districts (as shown in Figure 1). The main crops in this region are maize, potato, seed maize, cotton, oil crops, and vegetable. Therefore, planting crops in a suitable region, as well as integration between the quantity structure optimization and spatial allocation optimization, are of primary importance to agricultural production management [40].

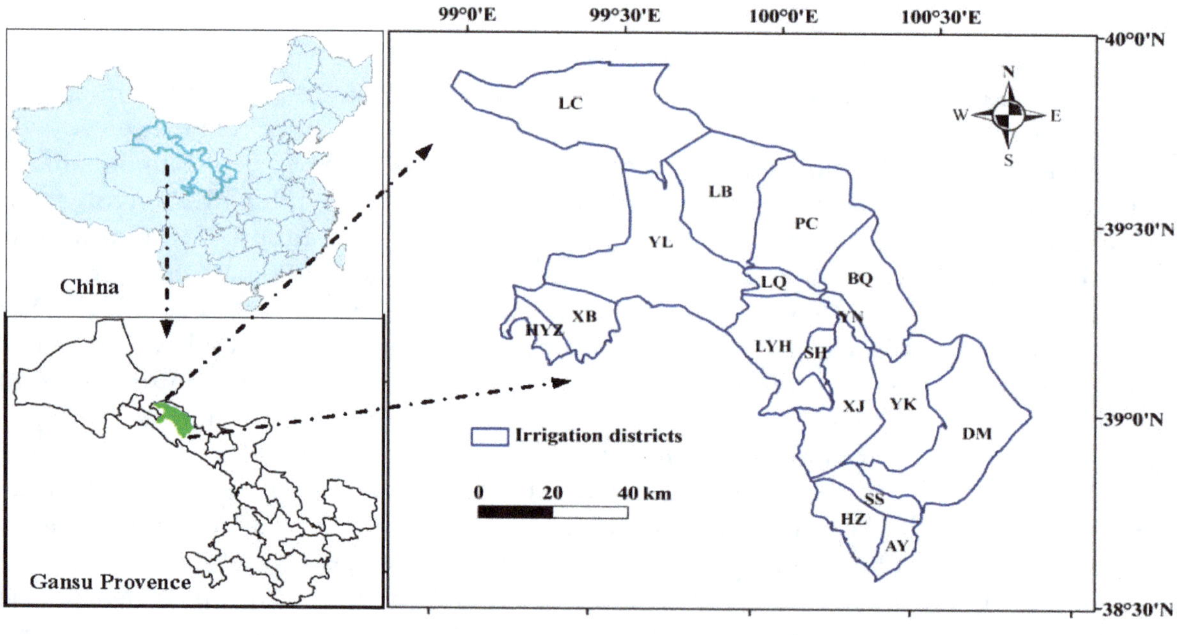

Figure 1. Location of study area in China.

The letters in the figure represent the names of irrigation districts. LC is for Luocheng irrigation district, LB is for Liuba irrigation district, YL is for Youlian irrigation district, XB is for Xinba irrigation district, HYZ is for Hongyazi irrigation district, PC is for Pingchuan irrigation district, LQ is for Liaoquan irrigation district, LYH is for Liyuanhe irrigation district, BQ is for Banqiao irrigation district, YN is for Yanuan irrigation district, SH is for Shahe irrigation district, XJ is for Xijun irrigation district, YK is for Yingke irrigation district, DM is for Daman irrigation district, SS is for Shangsan irrigation district, HZ is for Huazhai irrigation district, and AY is for Anyang irrigation district.

2.2. Methods and Data

Figure 2 shows an overview of the integrated agricultural cropping spatial distribution optimization model. This model is based on minimizing cross-entropy, re-aggregating administrative statistics data, and multi-source remote sensing information data, such as spatial distribution of crop suitability, spatial distribution of population density, and agricultural land use data in a logical framework.

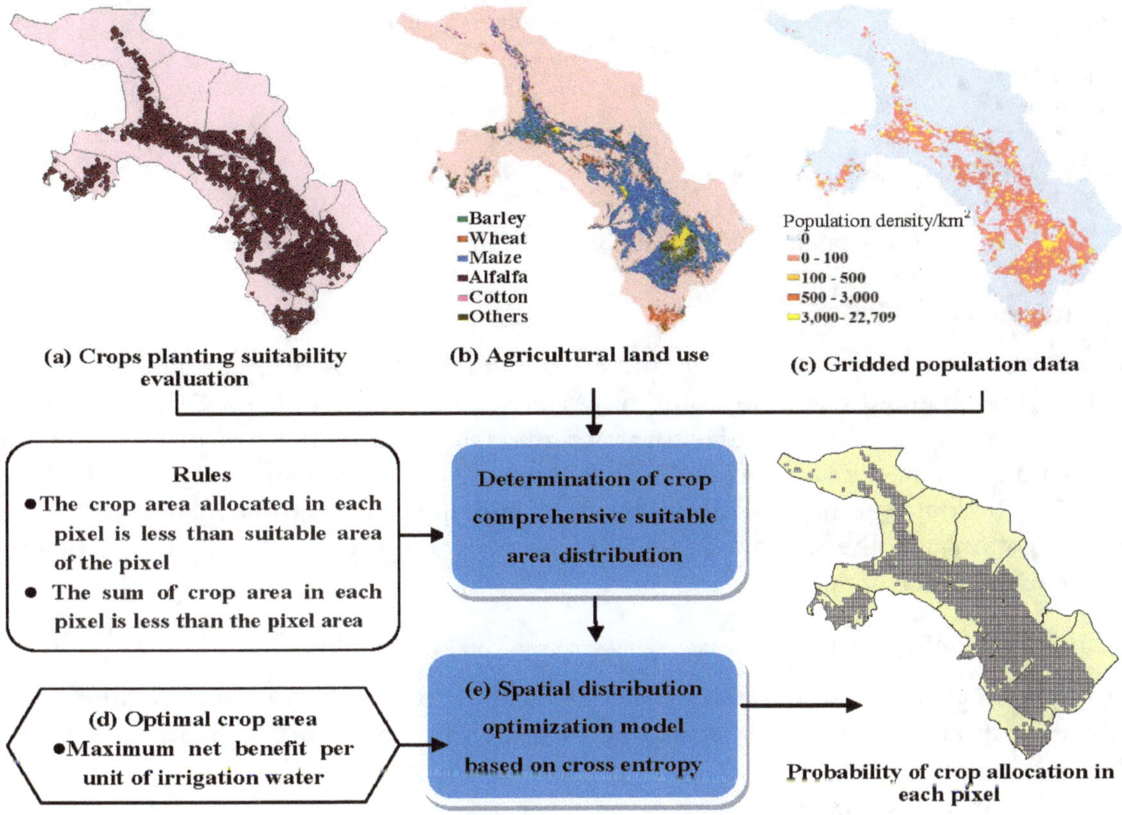

Figure 2. Overview of integrated agricultural cropping spatial distribution optimization model.

2.2.1. Evaluation of Crop Planting Suitability

The evaluation of crop planting suitability is important for agricultural land use allocation. It provides the essential data for the optimization of crop spatial framework in order to realize reasonable utilization of land resources as well as providing references for the scientific management and sustainable utilization of cultivated land resources.

The suitability of crop cultivation in this study is referenced from Peng [19]. Based on the ecological niche fitness theory, Peng [19] selected climatic and environment factors (rainfall, temperature, ET_0), soil characteristics (organic matter, total nitrogen, total phosphorus, total potassium, pH, bulk density), and geographical factors (terrain elevation, slope and aspect). These factors are closely related to crop growth and are used for evaluating the ecological niche and crop planting suitability. Figure 3 shows the spatial distribution of crop planting suitability index. Owing to the limitation of data collection, Peng's [19] study on the maize and wheat planting suitability was employed.

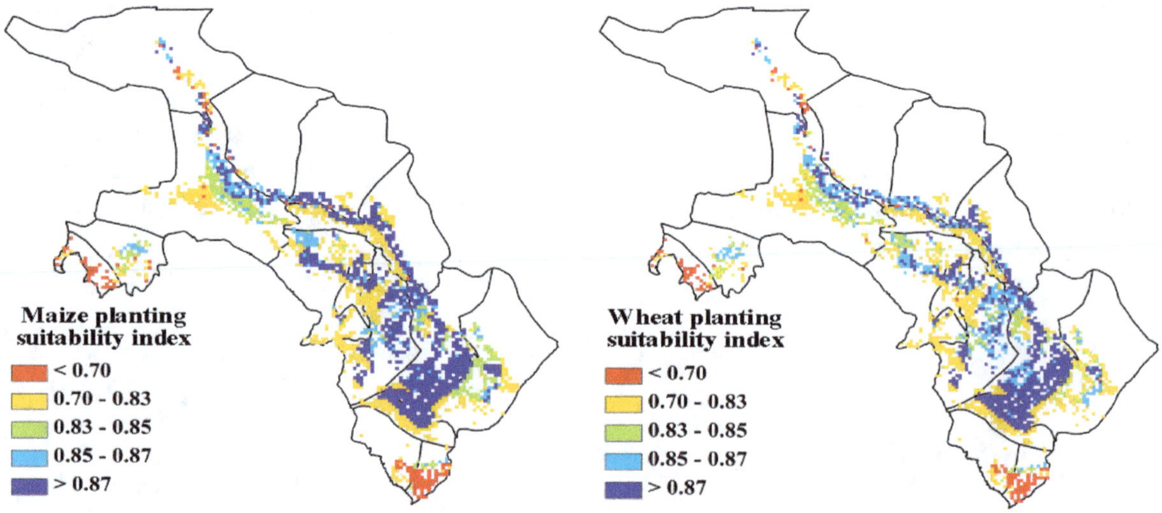

Figure 3. Spatial distribution of crop planting suitability index [19].

2.2.2. Agricultural Land Use

The Heihe River basin land use and land cover data set (HiWATER: Land cover map of Heihe River basin) were obtained from the Cold and Arid Regions Sciences Data Center (http://westdc.westgis.ac.cn/) which provided the 2011–2015 monthly data covering the type of surface. The data set is based on China's domestic satellite HJ/CCD data which has a high temporal and spatial resolution (30 M) [41,42]. This data set increases the classification of cultivated crops, including barley, wheat, maize, alfalfa, cotton and others, and can provide the current crop spatial distribution to determine the crop suitable distribution area.

2.2.3. Gridded Population Data

The spatial distribution of population density data is from the gridded population data of the Heihe River basin and provides the spatial distribution of population density in a 1 km × 1 km grid. The data set is given by the Cold and Arid Regions Sciences Data Center at Lanzhou (http://westdc.westgis.ac.cn) [40].

2.2.4. Optimization Model of Agricultural Cropping Pattern

To realize the appropriate allocation and efficient use of agricultural water and land resources and provide the primary input data for spatial allocation, the optimization model for agricultural cropping pattern is established with the objective of maximum agricultural net benefit per unit of irrigation water under certain agricultural water resources. The planting area of crops in an irrigation district is a decision variable for the model. Crops considered include maize, wheat, potato, maize seed, cotton, oil crop, and vegetable, while the basic data of 2013 used in the model are crop yield, crop prices, cost, agricultural irrigation quota and water availability. This model references the results of the project supported by National Natural Science Fund in China (91,425,302).

(1) Objective function

The agricultural cropping pattern optimization model can be written as follows, the variables x_{ij} is expected area of crop j in irrigation district i (ha); and the objective of optimization model is the maximum agricultural net benefit per unit of irrigation water:

$$\max f = \sum_{i=1}^{17}\sum_{j=1}^{7}((y_{ij}v_{ij} - c_{ij}) \cdot x_{ij}/ET_{ij}) / \sum_{i=1}^{17}\sum_{j=1}^{7} x_{ij} \tag{1}$$

where f is the net benefit per unit of irrigation water (RMB/m^3); i (i = 1, 2, ... , n) is irrigation district identifier, of which there are 17 within the study area; j (j = 1, 2, ... , 7) is the crop type identifier, of which 7 main types are considered in this model (maize, wheat, potato, maize seed, cotton, oil crops, and vegetable); v_{ij} is the price of crop j in the irrigation district i (RMB/kg); y_{ij} is the yield of crop j in the irrigation district i (kg/ha); c_{ij} is the cost of crop j in the irrigation district i (RMB/ha); ET_{ij} is the net irrigation quota of crop j in the irrigation district i (m^3/ha).

(2) Constraints

The irrigation water of irrigation district i should be less than the available water supply:

$$\sum_{i=1}^{n}\sum_{j=1}^{7} m_{ij}x_{ij} \leq Q_i \tag{2}$$

where m_{ij} is the gross irrigation quota of crop j in the irrigation district i (m^3/ha); Q_i is the available water supply in the irrigation district i (m^3).

The irrigation area of irrigation district i would be less than the effective irrigation area X_i (ha):

$$\sum_{i=1}^{n}\sum_{j=1}^{7} x_{ij} \leq X_i \tag{3}$$

The agriculture product would be to meet the local demand:

$$\sum_{i=1}^{n}\sum_{j=1}^{4} x_{ij} \cdot y_{ij} \geq K \cdot P \cdot FN \tag{4}$$

where P is the population in the study area; FN is the per person grain demand, 135 kg/per; and VN is the per person vegetable demand, 140 kg/per; K is demand coefficient, when the agriculture product meet the local demand, K = 1.

$$\sum_{i=1}^{n}\sum_{j=9} x_{ij} \cdot y_{ij} \geq K \cdot P \cdot VN \tag{5}$$

non-negative constraint:

$$x_{ij} \geq 0 \tag{6}$$

2.2.5. Spatial Distribution Optimization Model Based on Cross Entropy

Shannon (1948) introduced information entropy to measure the uncertainty of the expected information. He defined entropy $H(p)$ as a weighted sum of the information [34]. The entropy of a random variable with probability distribution P ($p_1, p_2, \ldots, p_k$) can be expressed using Equation (7) [43]

$$H(P) = -\sum_{i=1}^{k} p_i \ln p_i \tag{7}$$

Jaynes (1957) proposed the maximum entropy principle in statistical inference: the least informative probability distribution P (p_1, p_2, ... , p_k) can be found by maximizing the entropy $H(p)$ [44]. In Equation (7), the solutions are: $p_i = 1/n$, i = 1,2, ... , n, $H(p) = \ln n$ [34,35].

The cross-entropy formulation is based on the Shannon Entropy theory [36]. Cross-entropy was formulated by Kullback and Leibler [31] and is detailed by Kullback [32]. It measures the divergence between the prior distribution and the desired distribution. The principle of minimum cross entropy (POMCE), also referred to as the principle of minimum discrimination information, is obtained by minimizing cross-entropy with respect to the given prior distribution, subject to given constraints [33]. POMCE can be expressed as

$$D(P,Q) = \sum_{i=1}^{k} p_i \ln\left(\frac{p_i}{q_i}\right) \tag{8}$$

where D is the cross-entropy or the discrimination information and the objective is to minimize D. P $(p_1, p_2, \ldots, p_k)$ is the desired distribution, $Q = (q_1, q_2, \ldots, q_k)$ is prior distribution chosen based on all the given information, but does not satisfy the prescribed constraints [45–47].

In this study, we considered a comprehensive crop suitable distribution based on the integrated multi-source data analysis as prior knowledge in the POMCE. Let q_{ij} represent the suitable cultivated land area shares of crop j on pixel i. Therefore,

$$q_{ij} = \frac{Suitable_{ij}}{\sum_i Suitable_{ij}} \tag{9}$$

Based on the spatial distribution of crop suitability, spatial distribution of population density, and agricultural land use data, the spatial distribution optimization model based on cross entropy, subject to the constraints of optimal crop area obtained from agricultural cropping pattern optimization model, is used to determine the optimal spatial distribution of crops.

The objective of crop spatial distribution optimization is to minimize cross-entropy of comprehensive crop distribution probability and desired distribution probability, subject to area constraint on the pixel scale and other related limitations.

The spatial distribution optimization model, based on cross entropy, can be written as follows, variables p_{ij} represent the desired area shares of crop j on pixel i:

$$\min_{p_{ij}} D(p_{ij}, q_{ij}) = \min\left(\sum_i p_{ij} \ln p_{ij} - \sum_i p_{ij} \ln q_{ij}\right) \tag{10}$$

subject to the following constraints:

$$\sum_i p_{ij} = 1 \quad 0 \leq p_{ij} \leq 1 \tag{11}$$

making sure the allocated area of crop j on pixel i would be less than suitable cultivated land area of crop j on pixel i:

$$Area_j \times p_{ij} \leq Suitable_{ij} \tag{12}$$

making sure the allocated area on pixel i would be less than the arable land area on pixel i:

$$\sum_j Area_{ij} \times p_{ij} \leq Available_i \tag{13}$$

where $i = 1, 2, 3, \ldots$, represents the pixel identifier within the study area; $j = 1, 2$, represents the crop identifier within the study area; q_{ij} represent the suitable area shares of crop j on pixel i; $Available_i$ represent the arable land area on pixel i; $Suitable_{ij}$ represent the suitable cultivated land area of crop j on pixel i; and $Area_j$ represent the optimal crop areas, which are obtained from the agricultural cropping pattern optimization model.

3. Results and Discussion

3.1. Crop Demand Assessment

Based on limited data on crop planting suitability, this study focuses only on the spatial allocation of two crops, maize and wheat. Thus, in this section, the maize and wheat demands in the middle reaches of Heihe River basin are analyzed.

As shown in Table 1, seed maize controls the priority of water allocation, due to its higher unit net irrigation benefit and lower water requirement in these irrigation districts compared with other crops. Compared with the actual situation in 2013, the proportion of the total planting areas of corn and wheat through optimization all accounts for about 87%. However, the area of maize should increase, and the area of wheat should decrease in a region-wide range.

Table 1. Comparison of crop structure between present and optimal allocation (ha).

	Maize	Wheat	Potato	Seed Maize	Cotton	Oil Crops	Vegetable	Sum
Actual situation	19,680	14,241	794	75,252	3573	1116	11,258	125,915
Optimization	14,111	12,818	725	82,785	4174	990	10,375	125,977

Table 2 compares maize and wheat areas between present and optimal allocations in different irrigation districts. Comparing the actual crop structure, results revealed that except for the XJ, SS, PC, SH, YL, LC, XB irrigation districts, the maize area in the remaining irrigation districts increased, especially in DM and YK. The change was apparent with 2809 and 2057 ha increases, respectively, while the maize area in YL decreased by 2128 ha. However, DM, YK, PC, BQ, LYH and LC would be more favorable to wheat because of lower water requirement, while in other irrigation districts, the area of wheat decreased. For example, it would reduce to 504 ha and 436 ha in LQ and HYZ district, respectively.

Table 2. Comparison of maize and wheat area between present and optimal allocations in different irrigation district (ha).

Irrigation Districts	Maize		Wheat	
	Actual Situation	Optimization	Actual Situation	Optimization
DM	11,805	14,614	160	167
YK	15,049	17,106	1053	1067
XJ	17,598	17,567	579	400
SS	6474	6421	87	78
AY	13	18	561	545
HZ	0	30	353	199
PC	3013	3000	1180	1268
LQ	2367	2433	1253	749
BQ	3900	3931	167	204
YN	2113	2331	947	679
SH	2873	2210	1007	924
LYH	10,220	10,433	3707	3850
YL	14,540	12,412	867	800
LB	1880	1882	56	52
LC	1400	1303	13	73
XB	1593	1080	640	585
HYZ	93	125	1613	1177
Sum	94,932	96,896	14,241	12,817

3.2. *Crop Comprehensive Suitable Area Distribution*

Using the ArcGIS platform, the whole research area was divided into 9041 1 km × 1 km grids. According to crop planting suitability evaluation, different colors were selected to represent different crop suitability indices, and grids were filled to generate the spatial distribution map of crop planting suitability index.

In light of the spatial distribution of maize and wheat planting suitability obtained above, combined with the spatial distribution of population density and agricultural land use data, the crop comprehensive suitable area distribution can be determined according to the following principles. Figure 4 shows an overview of this process.

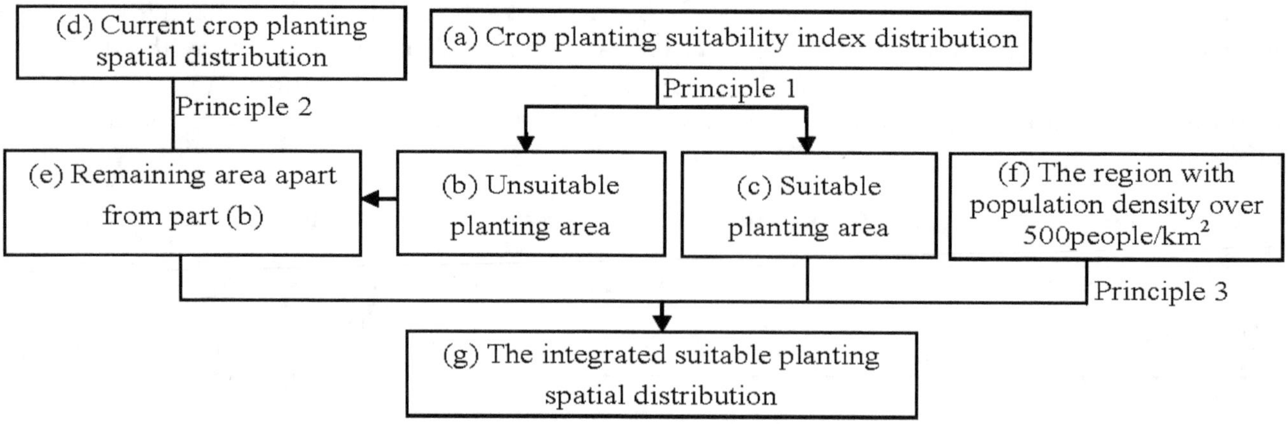

Figure 4. Overview of comprehensive suitable planting spatial distribution.

Principle 1: Preferentially assigning crops to the area with high suitability. First, rank crop planting suitability index (part a in Figure 4) in descending order, then contrast the area corresponding to the crop planting suitability index with the optimized area needed to be allocated, when the former is just larger than the latter, select the corresponding suitability index as a threshold of crops planting suitability index.

For example, Table 3 shows the maize planting suitability index and its corresponding area. When the index is greater than or equal to 0.83, its area is 103,337 ha, which is large enough to allocate the expected area of 96,896 ha obtained from the optimal model. Similarly, Table 4 shows that when the wheat planting suitability index is greater than or equal to 0.83, the area is 20,315 ha which can allocate the expected area of 12,817 ha. As a result, we can distinguish the unsuitable planting area (part b in Figure 4) and the suitable planting area (part c in Figure 4).

Table 3. Area of maize planting suitability index.

Maize Planting Suitability Index	0.83–0.84	0.84–0.85	0.85–0.86	0.86–0.87	0.87–0.88	0.88–0.89	0.89–0.9	0.9–0.92	Sum
Area (ha)	11,295	8242	7006	11,579	30,977	22,851	8899	2486	103,337

Table 4. Area of wheat planting suitability index.

Wheat Planting Suitability Index	0.88–0.89	0.89–0.90	Sum
Area (ha)	16,942	3373	20,315

Principle 2: The evaluation of crop planting suitability index is based on fixed-point sampling, and owing to the limitation of field investigation, there is inevitably some missing information on crop spatial distribution. In order to make up for this deficiency, the current crop planting spatial distribution (part d in Figure 4) is taken into account. Taking part b away from part d, the remaining area (part e in Figure 4) can be considered a supplement of crop distribution information.

Principle 3: According to [18], when the population density exceeds 500 people/km^2, it would be unsuitable for crop growing area. The land would be urban with little agriculture. Combined with the spatial distribution of population density, the region with population density over 500 people/km^2 (part f in Figure 4) is deducted from the suitable planting area.

Figures 5 and 6 show the formation process of maize and wheat comprehensive suitable planting spatial distribution.

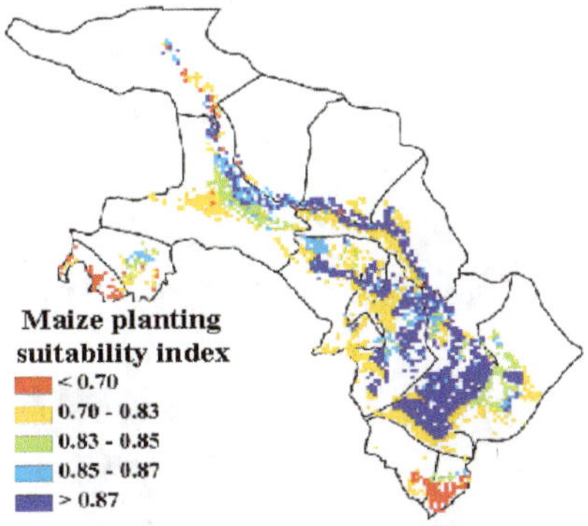

(a) Maize planting suitability index spatial distribution

(b) Maize unsuitable planting area

(c) Maize suitable planting area

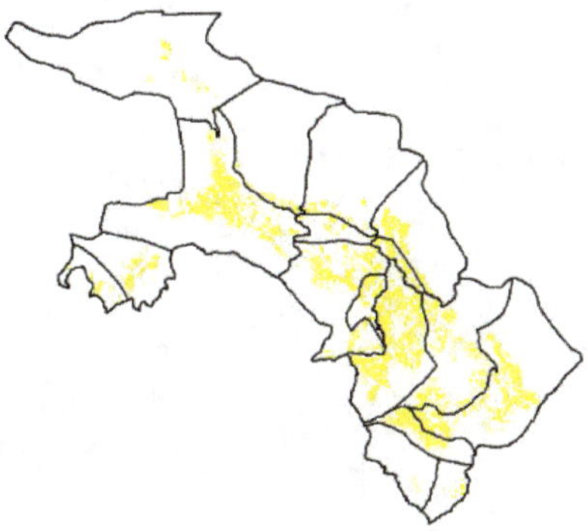

(d) Current maize planting spatial distribution

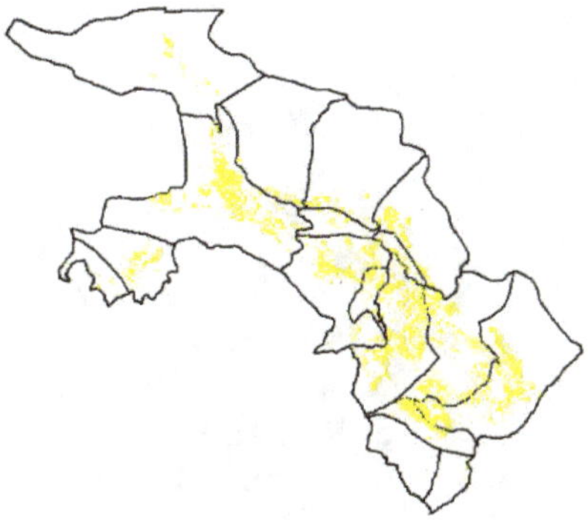

(e) Remaining area apart from part (b)

(f) The region with population density over 500people/km^2

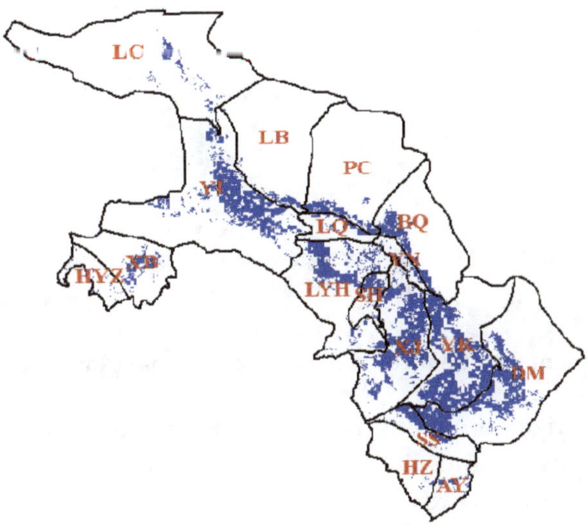

(g) The maize integrated suitable planting spatial distribution

Figure 5. Development of maize comprehensive suitable planting spatial distribution.

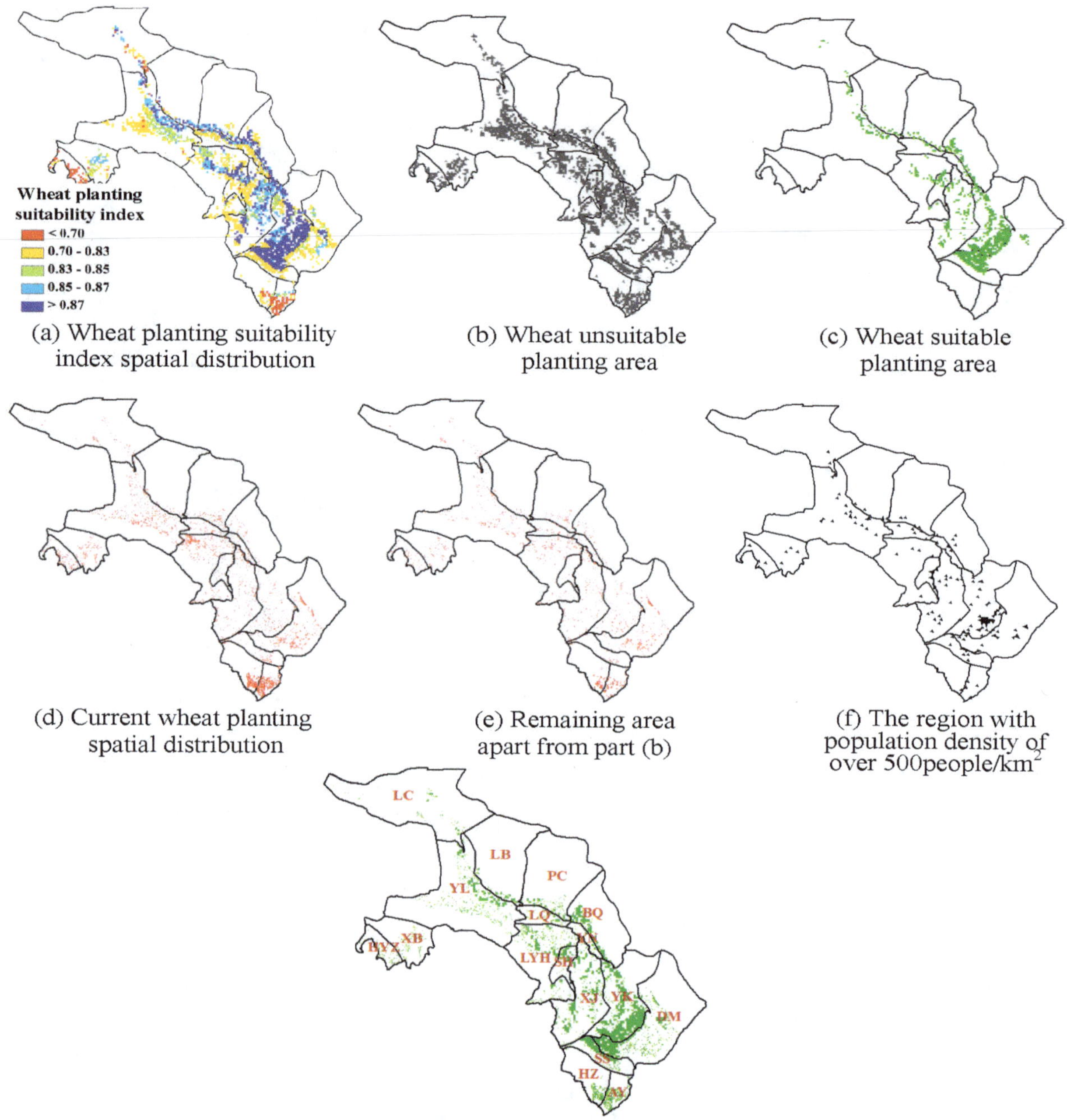

Figure 6. Development of wheat comprehensive suitable planting spatial distribution.

Results show that the maize comprehensive suitable planting spatial distribution is consistent with the current maize planting spatial distribution. The suitable area mainly focuses on the east of the YL irrigation district, the north of the LYH irrigation district, and most areas of the XJ, YK, DM irrigation districts. However, there are fewer areas suitable for growing maize in HYZ, HZ, and LQ.

In YK and DM irrigation district, the suitable area for planting maize is larger than the current area; therefore, an increase in the maize area is suggested in these districts. On the contrary, in YL, XJ and XB irrigation district, the maize area should be reduced because the suitable area is smaller than the current planting area.

Results show that the wheat comprehensive suitable planting spatial distribution has exhibited its obvious characteristics and the regular difference from the current wheat planting spatial distribution.

For example, the planting areas of wheat are relatively centralized in the HZ and AY irrigation districts and evenly distributed in other regions under the existing circumstances. However, the suitable areas for planting wheat are mainly concentrated in the YK, DM and SS irrigation district.

In the LQ, YN and HYZ irrigation districts, the suitable area for planting wheat is smaller than current area; therefore, a reduced area of wheat is suggested in these districts. However, in the BQ, YK, DM and SS irrigation districts, the maize area should be increased, for the suitable area is larger than current planting area.

3.3. Comparison of Crop Area Spatial Distribution between Present and Optimal Allocations

Comparison of maize area spatial distribution between present and optimal allocations is shown in Figure 7. Results show that after optimization, it tends to centralize grids, and the proportion of maize area in each grid is greater than 80% in the south of study area. This mainly focuses on the SS irrigation district, as the suitable area for planting maize in SS irrigation approaches the optimized area needed to be allocated.

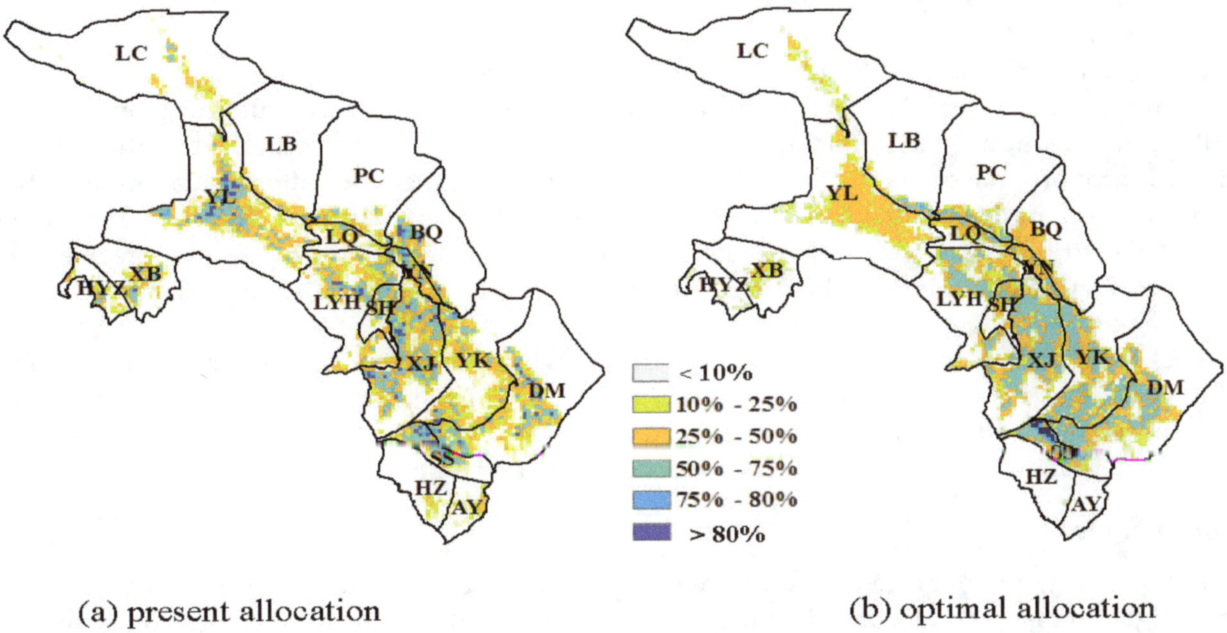

Figure 7. Comparison of maize area spatial distribution between present and optimal allocations.

The number of grids with higher proportion, including the proportion greater than 80% and between 75% and 80% is less than the actual situation, mainly owing to the deduction of unsuitable area for planting maize according to the principles in Section 3.1.

Meanwhile, the optimal spatial distribution shows more centralization than before optimization, because of the number of grids with a proportion of 50–75% increase, while the amount of grids with a proportion of 10–20% and less than a 10% decrease.

Figure 8 shows a comparison of wheat area spatial distribution between present and optimal allocations. Similarly, it indicates that grids with a high proportion of wheat area in each grid are more concentrated than in the actual situation, mainly distributed in LYH, LQ and SH irrigation districts.

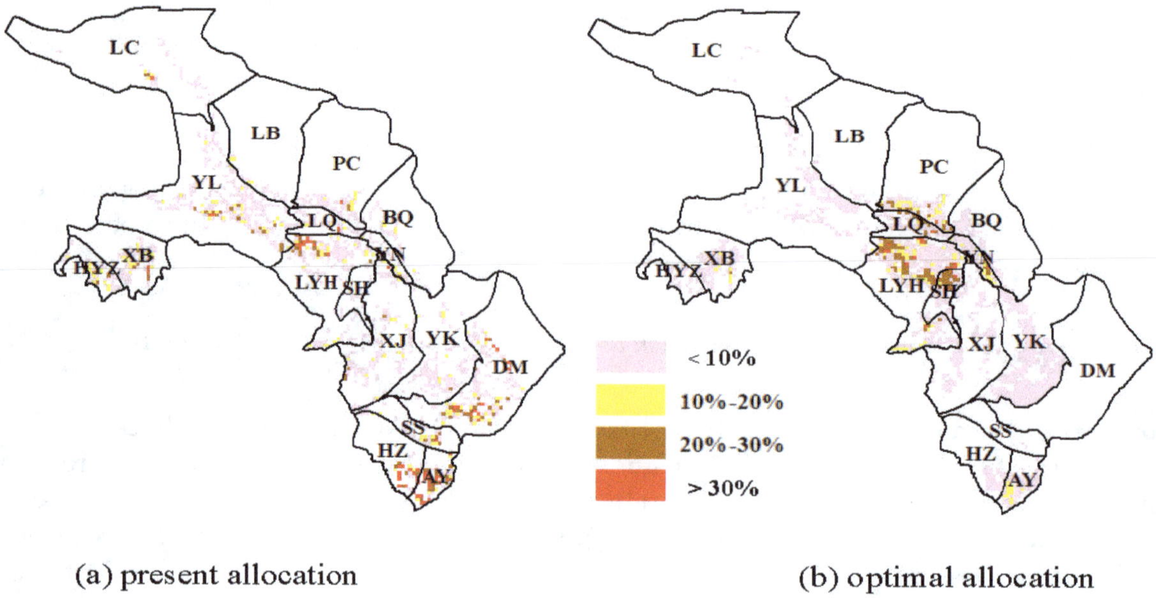

Figure 8. Comparison of wheat area spatial distribution between present and optimal allocations.

Compared with the maize proportion in each grid, the wheat proportion is lower, owing to less optimal wheat area and the decentralization of wheat comprehensive suitable planting spatial distribution. The number of grids with proportion greater than 30% decreases, while the amount of grids with proportion less than 10% and between 10% and 30% increases after spatial optimization.

In the south of the study area (AY and HZ irrigation district), based on the comparison of crop area spatial distribution between present and optimal allocations, the wheat area was more overloaded than its area suitable to plant in the actual situation. Therefore, it is suggested that the wheat area should be reduced in these two districts.

The optimization of crop spatial distribution, which is based on crop planting suitability evaluation and agricultural cropping pattern optimization, can improve the efficient utilization of agricultural water and land resources. The study ensures that crops are planted in suitable areas to provide the agricultural planting results of specialization and visualization.

4. Conclusions

In this study, an integrated agricultural cropping spatial distribution optimization is achieved at the irrigation district scale, based on the combination of multi-source remote sensing information data with optimal crop area. Minimizing cross-entropy was applied to build the model. This study considers the maize and wheat comprehensive suitable area distribution in the middle reaches of Heihe River basin and maize and wheat spatial distribution optimization is obtained. The high proportion of maize and wheat areas is more concentrated than before optimization. The integrated model contributed a new idea to cropping pattern spatial optimization. An optimizing approach based on cross-entropy minimization can lead to the efficient allocation of water resources and appropriate crop spatial distribution simultaneously. This study can ensure that crops are planted in a suitable region and provide a scientific basis for farmers to make crop selection decisions, which has an important guiding significance in the actual production work. However, due to the limitations of data on the planting suitability evaluation of other crops, only the main crops (maize and wheat) were studied. Further studies are therefore recommended using more crops.

Acknowledgments: We are grateful for the grant support from the National Natural Science Fund in China (91425302), National Key Research and Development Program during the 13th Five-year Plan in China (2016YFC0401306) and China Water Resources Industry Research Special Funds for Public Welfare Projects (No. 201301016).

Author Contributions: Lina Hao and Xiaoling Su designed the computations; Lina Hao and Xiaoling Su analyzed the data and wrote the paper; Vijay P. Singh and Olusola O. Ayantobo provides an English editing service and some improvement of style where necessary.

References

1. Li, M.; Guo, P. A coupled random fuzzy two-stage programming model for crop area optimization—A case study of the middle Heihe River basin, China. *Agric. Water Manag.* **2015**, *155*, 53–66. [CrossRef]
2. Dai, Z.Y.; Li, Y.P. A multistage irrigation water allocation model for agricultural land-use planning under uncertainty. *Agric. Water Manag.* **2013**, *129*, 69–79. [CrossRef]
3. Mianabadi, H.; Mostert, E.; Zarghami, M.; van de Giesen, N. A new bankruptcy method for conflict resolution in water resources allocation. *J. Environ. Manag.* **2014**, *144*, 152–159. [CrossRef] [PubMed]
4. Su, X.; Li, J.; Singh, V.P. Optimal Allocation of Agricultural Water Resources Based on Virtual Water Subdivision in Shiyang River Basin. *Water Resour. Manag.* **2014**, *28*, 2243–2257. [CrossRef]
5. Karimi, M.; Sharifi, M.A.; Mesgari, M.S. Modeling land use interaction using linguistic variables. *Int. J. Appl. Earth Obs. Geoinf.* **2012**, *16*, 42–53.
6. Ma, S.; He, J.; Liu, F.; Yu, Y. Model of urban land-use spatial optimization based on particle swarm optimization algorithm. *Trans. Chin. Soc. Agric. Eng.* **2010**, *26*, 321–326.
7. Pilehforooshha, P.; Karimi, M.; Taleai, M. A GIS-based agricultural land-use allocation model coupling increase and decrease in land demand. *Agric. Syst.* **2014**, *130*, 116–125. [CrossRef]
8. Cao, K.; Ye, X. Coarse-grained parallel genetic algorithm applied to a vector based land use allocation optimization problem: The case study of Tongzhou Newtown, Beijing, China. *Stoch. Environ. Res. Risk Assess.* **2012**, *27*, 1133–1142.
9. Cao, K.; Huang, B.; Wang, S.; Lin, H. Sustainable land use optimization using Boundary-based Fast Genetic Algorithm. *Comput. Environ. Urban Syst.* **2012**, *36*, 257–269. [CrossRef]
10. Baja, S.; Arif, S.; Neswati, R. Developing a User Friendly Decision Tool for Agricultural Land Use Allocation at a Regional Scale. *Mod. Appl. Sci.* **2017**, *11*, 11. [CrossRef]
11. Li, X.; Parrott, L. An improved Genetic Algorithm for spatial optimization of multi-objective and multi-site land use allocation. *Comput. Environ. Urban Syst.* **2016**, *59*, 184–194. [CrossRef]
12. Liu, Y.; Tang, W.; He, J.; Liu, Y.; Ai, T.; Liu, D. A land-use spatial optimization model based on genetic optimization and game theory. *Comput. Environ. Urban Syst.* **2015**, *49*, 1–14. [CrossRef]
13. Maleki, F.; Kazemi, H.; Siahmarguee, A.; Kamkar, B. Development of a land use suitability model for saffron (*Crocus sativus* L.) cultivation by multi-criteria evaluation and spatial analysis. *Ecol. Eng.* **2017**, *106*, 140–153. [CrossRef]
14. Reshmidevi, T.V.; Eldho, T.I.; Jana, R. A GIS-integrated fuzzy rule-based inference system for land suitability evaluation in agricultural watersheds. *Agric. Syst.* **2009**, *101*, 101–109. [CrossRef]
15. Sorel, L.; Viaud, V.; Durand, P.; Walter, C. Modeling spatio-temporal crop allocation patterns by a stochastic decision tree method, considering agronomic driving factors. *Agric. Syst.* **2010**, *103*, 647–655. [CrossRef]
16. Akıncı, H.; Özalp, A.Y.; Turgut, B. Agricultural land use suitability analysis using GIS and AHP technique. *Comput. Electron. Agric.* **2013**, *97*, 71–82. [CrossRef]
17. Nouri, H.; Mason, R.J.; Moradi, N. Land suitability evaluation for changing spatial organization in Urmia County towards conservation of Urmia Lake. *Appl. Geogr.* **2017**, *81*, 1–12. [CrossRef]
18. You, L.; Wood, S. Assessing the spatial distribution of crop areas using a cross-entropy method. *Int. J. Appl. Earth Obs. Geoinf.* **2005**, *7*, 310–323. [CrossRef]
19. Peng, C. *Research on the Optimization of Temporal and Spatial Distribution of Regional Crop Evapotranspiration and Its Application in Middle Reaches of Heihe River*; China Agricultural University: Beijing, China, 2016.
20. Sahoo, B.; Lohani, A.K.; Sahu, R.K. Fuzzy Multiobjective and Linear Programming Based Management Models for Optimal Land-Water-Crop System Planning. *Water Resour. Manag.* **2006**, *20*, 931–948. [CrossRef]
21. Henseler, M.; Wirsig, A.; Herrmann, S.; Krimly, T.; Dabbert, S. Modeling the impact of global change on regional agricultural land use through an activity-based non-linear programming approach. *Agric. Syst.* **2009**, *100*, 31–42. [CrossRef]

22. Sarker, R.; Ray, T. An improved evolutionary algorithm for solving multi-objective crop planning models. *Comput. Electron. Agric.* **2009**, *68*, 191–199. [CrossRef]
23. Niu, G.; Li, Y.P.; Huang, G.H.; Liu, J.; Fan, Y.R. Crop planning and water resource allocation for sustainable development of an irrigation region in China under multiple uncertainties. *Agric. Water Manag.* **2016**, *166*, 53–69. [CrossRef]
24. Biswas, A.; Pal, B.B. Application of fuzzy goal programming technique to land use planning in agricultural system. *Omega* **2005**, *33*, 391–398. [CrossRef]
25. Liu, J.; Li, Y.P.; Huang, G.H.; Zhuang, X.W.; Fu, H.Y. Assessment of uncertainty effects on crop planning and irrigation water supply using a Monte Carlo simulation based dual-interval stochastic programming method. *J. Clean. Prod.* **2017**, *149*, 945–967. [CrossRef]
26. Liu, X.; Ou, J.; Li, X.; Ai, B. Combining system dynamics and hybrid particle swarm optimization for land use allocation. *Ecol. Model.* **2013**, *257*, 11–24. [CrossRef]
27. Liu, Y.; Wang, H.; Ji, Y.; Liu, Z.; Zhao, X. Land use zoning at the county level based on a multi-objective particle swarm optimization algorithm: A case study from Yicheng, China. *Int. J. Environ. Res. Public Health* **2012**, *9*, 2801–2826. [CrossRef] [PubMed]
28. Yu, J.; Chen, Y.; Wu, J. Modeling and implementation of classification rule discovery by ant colony optimisation for spatial land-use suitability assessment. *Comput. Environ. Urban Syst.* **2011**, *35*, 308–319. [CrossRef]
29. Liu, Y.; Yuan, M.; He, J.; Liu, Y. Regional land-use allocation with a spatially explicit genetic algorithm. *Landsc. Ecol. Eng.* **2014**, *11*, 209–219. [CrossRef]
30. Li, X.; Yeh, A.G.-O. Neural-network-based cellular automata for simulating multiple land use changes using GIS. *Int. J. Geogr. Inf. Sci.* **2002**, *16*, 323–343. [CrossRef]
31. Kullback, S.; Leibler, R.A. On information and sufficiency. *Ann. Math. Stat.* **1951**, *22*, 79–86. [CrossRef]
32. Kullback, S. *Information Theory and Statistics*; John Wiley: Hoboken, NJ, USA, 1959.
33. Singh, V.P. *Entropy Theory and Its Application in Environmental and Water Engineering*; John Wiley & Sons: Hoboken, NJ, USA, 2013.
34. You, L.; Wood, S.; Wood-Sichra, U. Generating plausible crop distribution maps for Sub-Saharan Africa using a spatially disaggregated data fusion and optimization approach. *Agric. Syst.* **2009**, *99*, 126–140. [CrossRef]
35. You, L.; Wood, S.; Wood-Sichra, U.; Wu, W. Generating global crop distribution maps: From census to grid. *Agric. Syst.* **2014**, *127*, 53–60. [CrossRef]
36. Tan, J.; Yang, P.; Liu, Z.; Wu, W.; Zhang, L.; Li, Z.; You, L.; Tang, H.; Li, Z. Spatio-temporal dynamics of maize cropping system in Northeast China between 1980 and 2010 by using spatial production allocation model. *J. Geogr. Sci.* **2014**, *24*, 397–410. [CrossRef]
37. You, L.; Wood, S. An entropy approach to spatial disaggregation of agricultural production. *Agric. Syst.* **2006**, *90*, 329–347. [CrossRef]
38. Li, M.; Guo, P.; Singh, V.P.; Yang, G. An uncertainty-based framework for agricultural water-land resources allocation and risk evaluation. *Agric. Water Manag.* **2016**, *177*, 10–23. [CrossRef]
39. Hao, L.; Su, X. Determination for suitable scale of oasis and cultivated land in middle reaches of Heihe River basin. *Trans. Chin. Soc. Agric. Eng.* **2015**, *31*, 262–268.
40. Li, X.; Tong, L.; Niu, J.; Kang, S.; Du, T.; Li, S.; Ding, R. Spatio-temporal distribution of irrigation water productivity and its driving factors for cereal crops in Hexi Corridor, Northwest China. *Agric. Water Manag.* **2017**, *179*, 55–63. [CrossRef]
41. Zhong, B.; Ma, P.; Nie, A.; Yang, A.; Yao, Y.; Lü, W.; Zhang, H.; Liu, Q. Land cover mapping using time series HJ-1/CCD data. *Sci. China Earth Sci.* **2014**, *57*, 1790–1799. [CrossRef]
42. Zhong, B.; Yang, A.; Nie, A.; Yao, Y.; Zhang, H.; Wu, S.; Liu, Q. Finer resolution land-cover mapping using multiple classifiers and multisource remotely sensed data in the Heihe river basin. *IEEE J. Sel. Top. Appl. Earth Obs. Remote Sens.* **2015**, 1–20. [CrossRef]
43. Shannon, C.E. A mathematical theory of communication, Part I, Part II. *Bell Syst. Tech. J.* **1948**, *27*, 623–656. [CrossRef]
44. Jaynes, E.T. Information Theory and Statistical Mechanics. *Phys. Rev.* **1957**, *106*, 620–630. [CrossRef]
45. Tan, J.; Li, Z.; Yang, P.; Liu, Z.; Li, Z.; Zhang, L.; Wu, W.; You, L.; Tang, H. Spatiotemporal changes of maize sown area and yield in Northeast China between 1980 and 2010 using spatial production allocation model. *Acta Geogr. Sin.* **2014**, *69*, 353–364.

46. Tang, P.; Yang, P.; Chen, Z. Using cross-entropy method simulates spatial distribution of rice in Northeast China. *Trans. Chin. Soc. Agric. Eng.* **2013**, *29*, 96–104.
47. Liu, Z.; Li, Z.; Tang, P.; Li, Z.; Wu, W.; Yang, P.; You, L.; Tang, H. Change analysis of rice area and production in China during the past three decades. *J. Geogr. Sci.* **2013**, *23*, 1005–1018. [CrossRef]

8

Tsallis Entropy Theory for Modeling in Water Engineering

Vijay P. Singh [1,2,*], **Bellie Sivakumar** [3,4,*] **and Huijuan Cui** [5]

1 Department of Biological and Agricultural Engineering, Texas A&M University, College Station, TX 77843-2117, USA
2 Zachry Department of Civil Engineering, Texas A&M University, College Station, TX 77843-2117, USA
3 School of Civil and Environmental Engineering, The University of New South Wales, Sydney, NSW 2052, Australia
4 Department of Land, Air and Water Resources, University of California, Davis, CA 95616, USA
5 Key Laboratory of Land Surface Pattern and Simulation, Institute of Geographic Sciences and Natural Resources Research, Chinese Academy of Sciences, Beijing 100101, China; cuihj@igsnrr.ac.cn
* Correspondence: vsingh@tamu.edu (V.P.S.); s.bellie@unsw.edu.au (B.S.)

Abstract: Water engineering is an amalgam of engineering (e.g., hydraulics, hydrology, irrigation, ecosystems, environment, water resources) and non-engineering (e.g., social, economic, political) aspects that are needed for planning, designing and managing water systems. These aspects and the associated issues have been dealt with in the literature using different techniques that are based on different concepts and assumptions. A fundamental question that still remains is: Can we develop a unifying theory for addressing these? The second law of thermodynamics permits us to develop a theory that helps address these in a unified manner. This theory can be referred to as the *entropy theory*. The thermodynamic entropy theory is analogous to the Shannon entropy or the information theory. Perhaps, the most popular generalization of the Shannon entropy is the Tsallis entropy. The Tsallis entropy has been applied to a wide spectrum of problems in water engineering. This paper provides an overview of Tsallis entropy theory in water engineering. After some basic description of entropy and Tsallis entropy, a review of its applications in water engineering is presented, based on three types of problems: (1) problems requiring entropy maximization; (2) problems requiring coupling Tsallis entropy theory with another theory; and (3) problems involving physical relations.

Keywords: entropy; water engineering; Tsallis entropy; principle of maximum entropy; Lagrangian function; probability distribution function; flux concentration relation

1. Introduction

Water resources systems serve a multitude of human needs. They are needed for water supply, water transfer, water diversion, irrigation, land reclamation, drainage, flood control, hydropower generation, river training, navigation, coastal protection, pollution abatement, transportation and recreation, among others. Many of the systems (e.g., channels, culverts, impoundments) have been with us since the birth of human civilization. Some (e.g., spillways, small dams, levees) are several centuries old, while some others (e.g., large dams, long-distance water transfer structures) are of more recent origin. In the beginning, systems were designed more or less empirically. Then, engineering and economics constituted the sole foundation of design. About fifty years ago, planning and design of hydraulic structures went through a dramatic metamorphosis. These days, they are based on both engineering and non-engineering aspects. Engineering aspects encompass planning, development, design, operation and management, while non-engineering aspects include environmental impact assessment, socio-economic analysis, policy making and impact on society.

For designing water engineering systems (e.g., channels, levees, bridge piers, drainage structures, dams, reservoirs, spillways), some key questions that need to be addressed relate to the following, among others: peak discharge; velocity distribution; sediment yield, concentration and discharge; pollutant load, concentration and transport; river bed profile, meandering and braiding; downstream and at-a-point hydraulic geometry; flow depth, discharge and velocity routing; and seepage through a dam. For water supply systems, the key questions are concerned with, among others: reliability; loss of energy in the distribution system; and pipe sizes. In addition, pollutant concentration and transport as well as pollution abatement are also now considered as essential components of water resources system design. In a similar manner, because of increased public awareness, primarily triggered by environmental movement, non-engineering aspects of hydraulic design, dubbed under "socio-economic analyses," play a critical role. Besides engineering feasibility and economic viability, issues related to public health, political support, legal and judicial restrictions and social acceptability determine, to a large extent, if the water resources project will go off the ground.

A survey of the water engineering literature shows that there are myriad techniques for addressing questions pertaining to the design of water resources systems. The techniques range from empirical to semi-empirical to physically-based ones. Empirical techniques are data-based; examples are regression, time series analysis and other statistical methods. Semi-empirical methods, also sometimes referred to as conceptual or systems-based techniques, employ mass conservation and some empirical relationship or hypothesis; a good example is the unit hydrograph theory. Physically-based methods employ the laws of conservation of mass, momentum and energy; an example is the use of St. Venant's equations for flow routing or Richards' equation for computing infiltration. Strictly speaking, even physically-based methods also employ empirical parameters and, thus, are not entirely "physically-based." Indeed, all these three types of techniques employ some physics through data or hypotheses or laws. Extensive details of these methods are already well-documented in the literature [1–6]. A more recent and comprehensive account of these methods and applications is also presented in [7].

Because of the large diversity of these techniques, based on different hypotheses and assumptions, it is difficult to present the developments in any subject or field of interest in a unified and coherent manner. This becomes particularly challenging when undertaking water resources system engineering design. There are, of course, some theories that do apply to a wide variety of problems, such as kinematic wave theory and diffusion wave theory [8,9]. These theories can be applied to solve a wide variety of problems where the movement of water, sediment and/or pollutant is involved. However, many problems in water engineering design require a statistical treatment. For addressing such problems, entropy theory can serve as a unifying theory. During the past three decades, entropy theory has been applied to address a wide spectrum of problems in water engineering, including rainfall-runoff [1,3,8], infiltration [2], soil moisture [2], network design [10], velocity distributions [11–14], sediment concentration and discharge [15–17], hydraulic geometry [18–22] and reliability [23], among others. For recent comprehensive accounts of entropy theory applications in water and environmental engineering, see [24–26].

The origin of entropy theory is in the second law of thermodynamics. Koutsoyiannis [27] has presented a nice account of the historical background of entropy. The most commonly used measure of entropy is the Boltzmann-Gibbs-Shannon (BGS) entropy [28], which is often referred simply as the Shannon entropy [29]. Tsallis [30] introduced a more general entropy function for complex systems, which is now referred to as the Tsallis entropy. Tsallis entropy specializes in the Shannon entropy. During the last two decades or so, Tsallis entropy theory has found many applications in water and environmental engineering and there is certainly a great potential to extend the applications to a much wider spectrum of water systems and associated problems.

This paper aims to provide a review of the applications of Tsallis entropy theory in water engineering. It revisits the Tsallis entropy theory, presents a general methodology for application of the theory, shows how entropy theory couples statistical information with physical laws and how it can be employed to derive useful physical constructs in time and/or space and provides a review of physical applications of the Tsallis entropy theory in water engineering.

The rest of the paper is organized as follows. Section 2 reviews the Tsallis entropy theory. Sections 3–6 review the applications of the Tsallis entropy theory in water engineering. Section 3 presents an overview of three types of problems in water engineering. Section 4 reviews problems requiring entropy maximization. Section 5 reviews coupling entropy theory with another theory. Section 6 reviews physical relations. Section 7 draws some conclusions.

2. Tsallis Entropy Theory

2.1. Definition of Entropy

The concept of entropy is closely linked with the concept of uncertainty, information, chaos, disorder, surprise or complexity. Indeed, there are often different interpretations of entropy in different fields [24–26,31–33]. For instance, in statistics, entropy is regarded as a measure of randomness, objectivity or unbiasedness, dependence, or departure from the uniform distribution. In ecology, it is a measure of diversity of species or lack of concentration. In water engineering, it is a measure of information of uncertainty. In industrial engineering, it is a measure of complexity. In manufacturing, it is a measure of interdependence. In management, it is a measure of similarity. In social sciences, it is measure of equality. This may be illustrated as follows.

Consider a random variable X that takes on values x_i, $i = 1, 2, \ldots, N$, that occur with probabilities p_i, $i = 1, 2, \ldots, N$. If $p_i = 1$ (the event is certain to occur), $p_j = 0$ (the event is certain not to occur), $i \neq j$, then it can be said that there is no surprise about the occurrence or non-occurrence of event $X = x_i$ and the occurrence or non-occurrence of this event provides no information. The system that produces such an event has no complexity and is not chaotic or disorderly. On the other hand, if an event x_i occurs with very small probability p_i, say 0.01, then our anticipation of event x_i is highly uncertain and if x_i does indeed occur, then there will be a great deal of surprise about its occurrence. The occurrence of this event provides a great deal of information and the system producing such an event is complex, chaotic and disorderly. Intuitively, the information content of event x_i or the anticipatory uncertainty of x_i prior to the observation decreases as the value of probability $p(x_i)$ increases [24,26]. It may be noted that information is gained only if the variable takes on different values. The value that occurs with a higher probability conveys less information and vice versa.

It is logical to deduce that if a system has more uncertainty, then more information will be needed to characterize it and vice versa. That is, information reduces uncertainty, meaning that, for a system, more information means less uncertainty. Uncertainty increases the need for more information; that is, more uncertainty means more information is needed. Shannon [29] formulated entropy as the expected value of the probabilities of values that a variable or event may take on. The information gained is indirectly measured as the amount of reduction of uncertainty. Thus, entropy is defined as a measure of disorder, chaos, uncertainty, surprise, or information.

From an informational perspective, the information gain from the occurrence of any event x_i, $\Delta H(x_i)$, can be expressed as:

$$\Delta H(x_i) = -\log p_i \tag{1}$$

Equation (1) says that the information gained is minus the logarithm of the inverse of the probability of occurrence. For N events, the average or expected information gain, H_s, is the weighted average of Equation (1):

$$H_s = \sum_{i=1}^{N} H(x_i) = -\sum_{i=1}^{N} p_i \log p_i \tag{2}$$

Equation (2) is the Shannon entropy [29], also called informational entropy. Equation (1) shows that the information gain is directly a function of probability and is, hence, called gain function. Equation (1) can be generalized by expressing it as a power function, given by:

$$\Delta I(x_i) = \frac{1}{m-1}(1 - {p_i}^{m-1}), \sum_{i=1}^{N} p_i = 1 \tag{3}$$

where $\Delta I(x_i)$ is the gain in information from an event i which occurs with probability p_i and m is any real number. The gain function computed from Equation (3) for $m = -1, 0, 1$ and 2, as shown in Figure 1, decreases with an increase in the probability value regardless of the value of m. For increasing value of m, the gain function diminishes for the same probability value. The gain function has a much longer tail, showing very low values of gain as the probability increases.

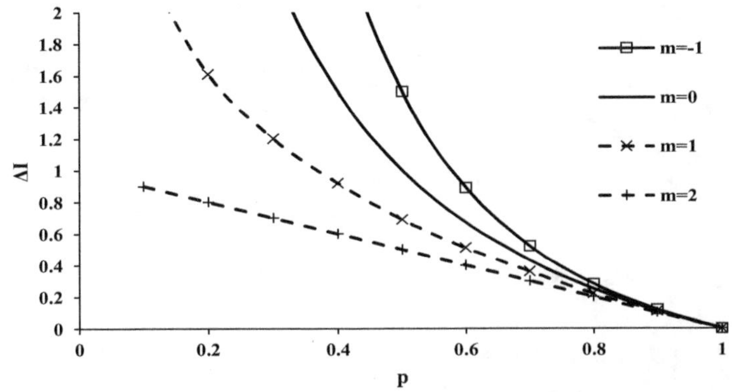

Figure 1. Gain function for $m = -1, 0, 1$ and 2.

Analogous to Shannon entropy, the average or expected gain function for N events, H_m, is the weighted average of Equation (3), given by:

$$H_m = E[\Delta I_i] = \sum_{i=1}^{N} p_i[\frac{1}{m-1}(1 - {p_i}^{m-1})] = \frac{1}{m-1}\sum_{i=1}^{N} p_i[1 - {p_i}^{m-1}] \tag{4}$$

where H_m is designated as the Tsallis entropy or m-entropy [30], which is often referred to as the non-extensive statistic, m-statistic, or Tsallis statistic. It can be shown that as $m \to 1$, Tsallis entropy converges to Shannon entropy. The quantity m is often referred to as the non-extensivity index, Tsallis entropy index, or simply entropy index and its value can be positive or negative. Entropy index m reflects the microscopic dynamics and the degree of nonlinearity of the system. Since almost all real systems (e.g., water systems) are inherently nonlinear in nature, the Tsallis entropy has a clear advantage when compared to the Shannon entropy. Tsallis [34] noted that super-extensivity, extensivity and sub-extensivity occur when $m < 1$, $m = 1$ and $m > 1$, respectively. Interestingly, if $m \geq 0$, $m < 1$ corresponds to the rare events ($0 \leq m < 1$) and $m > 1$ corresponds to frequent events [35,36], pointing to the stretching or compressing of the entropy curve to lower or higher maximum entropy positions.

If random variable X is non-negative continuous with a probability density function (PDF), $f(x)$, then the Shannon entropy can be written as:

$$H_s(X) = H_s(f) = -\int_0^{\infty} f(x) \log f(x) dx \tag{5}$$

Likewise, Tsallis entropy can be expressed as [37–39]:

$$H_m(X) = H_m(f) = \frac{1}{m-1}\int_0^\infty \{f(x) - [f(x)]^m\}dx = \frac{1}{m-1}\{1 - \int_0^\infty [f(x)]^m\}dx \tag{6}$$

From now onwards, subscript m will be deleted and H_m will be simply denoted by H.

A plot of H versus p for $m = -1, -0.5, 0, 0.5, 1$ and 2 is given in Figure 2. For $m < 0$, the Tsallis entropy is concave; and for $m > 0$, it becomes convex. For $m = 0$, $H = (N-1)$ for all p_i's. For $m = 1$, it converges to Shannon entropy. For all cases, the Tsallis entropy decreases as m increases.

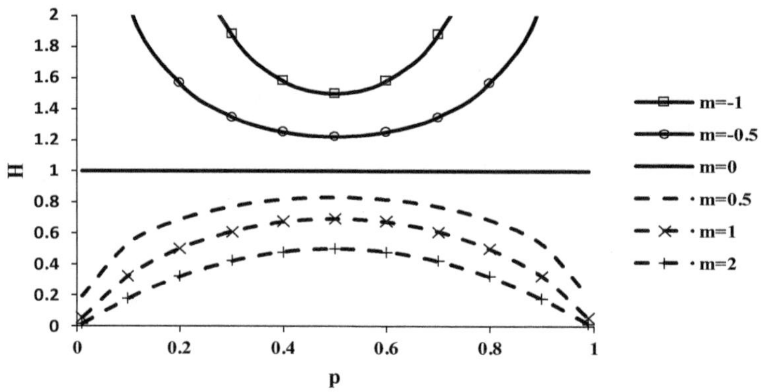

Figure 2. Plot of H for $m = -1, -0.5, 0, 0.5, 1$ and 2.

2.2. Properties of Tsallis Entropy

As mentioned earlier, the concept of entropy is closely linked with the concept of uncertainty, information, chaos, disorder, surprise or complexity. Tsallis entropy has some interesting properties [30,40] that are briefly summarized here.

(1) m-entropy: The m-surprise or m-unexpectedness is defined as $\log_m(1/p_i)$, where the logarithm is the base m. Hence, the m-entropy can be defined as:

$$H = E[\log_m \frac{1}{p_i}] \tag{7}$$

which coincides with the Tsallis entropy:

$$H = E[\frac{1 - {p_i}^{m-1}}{m-1}] \tag{8}$$

in which E is the expectation.

(2) Maximum value: Equation (4) attains an extreme value for all values of m when all p_i's are equal, i.e., $p_i = 1/N$. For $m > 0$, it attains a maximum value and for $m < 0$ it attains a minimum value. The extremum of H becomes

$$H = \frac{N^{m-1} - 1}{1-m} \tag{9}$$

(3) Concavity: For two probability distributions $P = \{p_i, i = 1, 2, \ldots, N\}$ and $Q = \{q_i, i = 1, 2, \ldots, N\}$ corresponding to a unique set of N possibilities, an intermediate probability distribution $G = \{g_i, i = 1, 2, \ldots, N\}$ can be defined for a real a, such that $0 < a < 1$, as:

$$g_i = ap_i + (1-a)q_i \tag{10}$$

for all i's. It can be shown that for $m > 0$,

$$H[G] \geq aH[P] + (1-a)H[Q] \tag{11}$$

and for $m < 0$,

$$H[G] \leq aH[P] + (1-a)H[Q] \tag{12}$$

Functional $H(G) \geq 0$ if $m > 0$ and is, hence, concave; $H(G) = 0$ if $m = 0$; and $H(G) \leq 0$ if $m < 0$ and is, therefore, convex.

(4) Additivity: For two independent systems A and B with ensembles of configurational possibilities $E^A = \{1, 2, \ldots, N\}$ with probability distribution $P^A = \{p_i{}^A, i = 1, 2, \ldots, N\}$ and configurational possibilities $E^B = \{1, 2, \ldots, M\}$ with probability distribution $P^B = \{p_j{}^B, j = 1, 2, \ldots, M\}$, one can express the union of two systems $A \cup B$ and their corresponding ensembles of possibilities $E^{A \cup B} = \{(1, 1), (1, 2), \ldots, (i, j), \ldots, (N, M)\}$. If $p_{ij}{}^{A \cup B}$ represents the corresponding probabilities, then, by virtue of independence, the joint probability will be equal to the product of individual probabilities, i.e., $p_{ij}{}^{A \cup B} = p_i{}^A p_j{}^B$ or $p_{ij}(A+B) = p_i(A)p_j(B)$, for all i and j. One can write the entropy of the union of two systems A and B as:

$$H(A+B) = H(A) + H(B) + (1-m)H(A)H(B) \tag{13}$$

Equation (13) describes the additivity property, which can be extended to any number of systems. In all cases, $H \geq 0$ (non-negativity property). If systems A and B are correlated, then

$$p_{ij}{}^{A \cup B} \neq [\sum_{i=1}^{N} p_{ij}{}^{A \cup B}][\sum_{j=1}^{M} p_{ij}{}^{A \cup B}] \tag{14}$$

for all (i, j). One may define mutual information of the two systems or transinformation S as:

$$T[(p_{ij}{}^{A \cup B})] = H^{A \cup B}[(p_{ij}{}^{A \cup B})] - H^A[(\sum_{i=1}^{N} p_{ij})] - H^B[(\sum_{j=1}^{M} p_{ij})] \tag{15}$$

Considering Equation (15), $T(p_{ij}) = 0$ for all m, if A and B are independent and Equation (15) will reduce to Equation (13). For correlated A and B, $T(p_{ij}) < 0$ for $m = 1$ and $T(p_{ij}) = 0$ for $m = 0$. For arbitrary values of m, it will be sensitive to p_{ij}; it can take on negative or positive values for both $m < 1$ and $m > 1$ with no particular regularity and can exhibit more than one extremum.

(5) Composability: The entropy $H(A + B)$ of a system comprised of two sub-systems A and B can be computed from the entropies of sub-systems, $H(A)$ and $H(B)$ and the entropy index m.

(6) Interacting sub-systems: Consider a set of N possibilities arbitrarily separated into two sub-systems with W_1 and W_2 possibilities, where $W = W_1 + W_2$. Defining $P_{W_1} = \sum_{i=1}^{W_1} p_i$ and $P_{W_2} = \sum_{j=1}^{W_2} p_j$, $P_{W_1} + P_{W_2} = 1 = \sum_{k=1}^{W} p_k$, it can be shown that

$$H(P_W) = H(P_{W_1}, P_{W_2}) + P_{W_1}{}^m H(p_i|P_{W_1}) + P_{N_2}{}^m H(p_j|P_{W_2}) \tag{16}$$

where $p_i|P_{W_1}$ and $p_j|P_{W_2}$ are the conditional probabilities. Note that $p_i^m > p_i$ for $m < 1$ and $p_i^m < p_i$ for $m > 1$. Hence, $m < 1$ corresponds to rare events and $m > 1$ to frequent events [41]. This property can be extended to any number R of interacting sub-systems: $W = \sum_{j=1}^{R} W_j$. Then, defining $w_j = \sum_{i=1}^{N_j} p_i, j = 1, 2, \ldots, N_j$, $\sum_{j=1}^{N} w_j = 1$, Equation (16) can be generalized as:

$$H(\{p_i\}) = H(\{w_j\}) + \sum_{j=1}^{R} w_j{}^m H(\{p_i | w_j\}) \quad (17)$$

Here, $p_j = w_j$.

2.3. *Principle of Maximum Entropy*

Quite often, some information may be known about the random variable X. Then, it seems logical to choose a PDF of X that is consistent with the known information. Since there can be more than one PDF that may satisfy this condition, Jaynes [42,43] formulated the principle of maximum entropy (POME), which states that one should choose the distribution that has the highest entropy, subject to the given information. This distribution will be the least-biased distribution. Furthermore, it is equivalent to the minimum relative (cross) entropy condition when no prior is given, which is also called as the Kullback-Leibler principle [44]. The implication here is that POME takes into account all of the given information and, at the same time, avoids consideration of any information that is not given. This is consistent with Laplace's principle of insufficient reason (or principle of indifference), according to which all outcomes of an experiment should be considered equally likely unless there is information to the contrary. Therefore, POME enables entropy theory to achieve the probability distribution of a given random variable.

The procedures for applications of *POME* entails the following steps [24,26]: (1) definition of Tsallis entropy; (2) specification of constraints; (3) maximization of entropy, in accordance with POME, using of the method of Lagrange multipliers; (4) derivation of the probability distribution in terms of constraints; (5) determination of Lagrange multipliers; and (6) determination of the maximum entropy. Since the definition of entropy has already been reviewed earlier, it will not be repeated here. The remaining steps, (2)–(6), are briefly discussed here.

2.3.1. Specification of Constraints

For deriving the PDF of a random variable X using POME, appropriate constraints need to be defined. Papalexiou and Koutsoyiannis [28] suggested three considerations for defining constraints: (1) Simplicity and physical meaningfulness; (2) little variability in the future; and (3) definition in terms of laws of physics—mass, momentum and energy conservation and constitutive laws—as far as possible. For simplicity, constraints are often defined in terms of statistical moments and, fortunately, the moments are related to the laws of physics.

Let C_i denote the i-th constraint, $i = 0, 1, 2, \ldots, n$, where n is the number of constraints. Let $g_r(x)$ be an arbitrary function of X. Then, constraints C_r, $r = 0, 1, 2, \ldots, n$, can be defined as:

$$C_0 = \int_a^b f(x)dx = 1 \quad (18)$$

$$C_r = \int_a^b g_r(x)\, f(x)\, dx = E[g_r(x)] = \overline{g_r(x)},\ r = 1, 2, \ldots, n \quad (19)$$

where $g_0(x) = 0$; $g_r(x)$, $r = 1, 2, \ldots, n$, represents some function of x and $\overline{g_r(x)}$ is the expectation of $g_r(x)$. Equation (18) states the total probability theorem that the PDF must satisfy. If $r = 1$ (first moment) and $g_1(x) = x$, then Equation (19) represents the mean $\overline{x}$; likewise, for $r = 2$ (second moment) and

$g_2(x) = (x - \overline{x})^2$, it denotes the variance (σ^2) of X. The first moment corresponds to the conservation of mass and the second moment to the conservation of momentum. Similarly, $r = 3$ (third moment), measuring skewness, corresponds to the conservation of energy. It is noticed that the order of moments higher than 3 may be unreliable [45]. In water engineering, two or three constraints usually suffice.

2.3.2. Entropy Maximizing Using Lagrange Multipliers

Tsallis entropy, given by Equation (6), can be maximized, subject to constraints defined by Equations (18) and (19), using the method of Lagrange multipliers [42,43]. Therefore, the Lagrangian function L can be expressed as:

$$L = \int_0^\infty f(x)\frac{1-[f(x)]^{m-1}}{m-1}dx + (\lambda_0 - \frac{1}{m-1})[\int_0^\infty f(x)dx - 1] + \sum_{i=1}^{M}\lambda_i[\int_0^\infty g_i(x)f(x)dx - \overline{g_i(x)}] \quad (20)$$

where λ_i, i = 0, 1, 2, ... , M, are the Lagrange multipliers. Note that $-1/(m-1)$ is added to the zeroth Lagrange multiplier for algebraic simplification. Using the Euler-Lagrange calculus of variation, differentiating Equation (20) with respect to $f(x)$ and equating the derivative to zero, we obtain:

$$\frac{dL}{df(x)} = 0 = \frac{1}{m-1} - [f(x)]^{m-1} + \lambda_0 - \frac{1}{m-1} + \sum_{i=1}^{M}\lambda_i g_i(x) \quad (21)$$

2.3.3. Determination of Probability Distribution

Equation (21) yields the least-biased probability distribution of X:

$$f(x) = [\lambda_0 + \sum_{i=1}^{M}\lambda_i g_i(x)]^{\frac{1}{m-1}} \quad (22)$$

Integrating Equation (22), the cumulative distribution function $F(x)$ is obtained as:

$$F(x) = \int_0^x [\lambda_0 + \sum_{i=1}^{M}\lambda_i g_i(x)]^{\frac{1}{m-1}} dx \quad (23)$$

The properties of the probability distribution will depend on the value of m, Lagrange multipliers, functions $g_i(x)$ and M.

2.3.4. Determination of the Lagrange Multipliers

Equation (22) contains M unknown Lagrange multipliers that can be determined with the use of Equations (18) and (19). Substituting Equation (22) in Equations (18) and (19), the result is, respectively:

$$\int_0^\infty [\lambda_0 + \sum_{i=1}^{M}\lambda_i g_i(x)]^{\frac{1}{m-1}} dx = 1 \quad (24)$$

$$\int_0^\infty g_i(x)[\lambda_0 + \sum_{i=1}^{M}\lambda_i g_i(x)]^{\frac{1}{m-1}} dx = \overline{g_i(x)},\ i = 1, 2, \ldots, M \quad (25)$$

Solution of Equations (24) and (25) yields the unknown Lagrange multipliers λ_i, $i = 1, 2, \ldots, M$. It may be noted from Equation (24) that λ_0 can be expressed as a function of other Lagrange multipliers and, therefore, the unknown multipliers are λ_i, i = 1, 2, ... , M. Except for simple cases, Equations (24) and (25) do not have an analytical solution but can be solved numerically.

2.3.5. Determination of Maximum Entropy

Substitution of Equation (24) in Equation (6) leads to the maximum Tsallis entropy:

$$H = \frac{1}{m-1}\{1 - \int_{0}^{\infty} [\lambda_0 + \sum_{i=1}^{M} \lambda_i g_i(x)]^{\frac{m}{m-1}} dx\} \tag{26}$$

Equation (26) shows that the entropy of the probability distribution of X depends only on the specified constraints, since the Lagrange multipliers themselves can be expressed in terms of the specified constraints.

3. Applications in Water Engineering: Overview

The problems that can be addressed using the Tsallis entropy theory can be classified into three groups. The first group consists of problems that only require the maximization of the Tsallis entropy, which can be accomplished using POME. Examples of such problems include frequency analysis, parameter estimation, network evaluation and design, spatial and inverse spatial analysis, geomorphologic analysis, grain size distribution analysis, flow forecasting, complexity analysis and clustering [46–52].

The types of problems included in the second group require coupling with another theory, such as the theory of stream power or theory of minimum energy dissipation rate. Examples of problems in this group include hydraulic geometry [19,20] and evaporation [53]. Also included in this group are problems wherein first relations between entropy and design variables are derived and then relations between design variables and system characteristics are established. Examples include geomorphologic relations for elevation, slope and fall; and evaluation of water distribution systems [54,55].

The third group includes problems that require deriving a physical relation either in time or in space. This means, the domain of analysis requires a shift from the probability domain to a real domain (space or time), which is accomplished by invoking a relation between the cumulative probability distribution function and the design variable. Examples of such problems are infiltration capacity, soil moisture movement in vadose zone, groundwater head distribution, velocity distribution, rainfall-runoff relation, channel geometry, rating curve, flow-duration curve, erosion and sediment yield, sediment concentration and discharge, debris flow, longitudinal river profile, hydraulic geometry, channel cross-section shape and rating curve [19–22,56–61]. The objective in this class of problems is to derive a relation of the design variable as a function of time or space.

These three kinds of problems are further detailed in the following sections, with examples from water engineering.

4. Problems Requiring Entropy Maximization

Fundamental to solving problems that only require entropy maximization is the derivation of probability distribution. A multitude of problems in water engineering involve essentially data analysis for deriving either a probability distribution or computing entropy.

4.1. Frequency Distributions

The procedure for deriving a maximum entropy-based frequency distribution has been discussed above. The procedure is illustrated here using only two constraints: mean (first moment) and second moment. For any probability density function, $f(x)$, of a random variable X, the total probability must equal one, i.e.,

$$\int_{0}^{\infty} f(x)dx = 1 \tag{27}$$

The first and second moments can be defined, respectively, as:

$$\int_0^\infty xf(x)dx = E(x) = \mu_1 \tag{28}$$

and

$$\int_0^\infty x^2 f(x)dx = E(x^2) = \mu_2 \tag{29}$$

In order to obtain the least-biased $f(x)$, subject to Equations (27)–(29), Equation (6) can be maximized for $m > 0$ using the method of Lagrange multipliers. The Lagrange function L can be written as:

$$L = \int_0^\infty f(x)\frac{1-[f(x)]^{m-1}}{m-1}dx - \lambda_0[\int_0^\infty f(x)dx - 1] - \lambda_1[\int_0^\infty xf(x)dx - \mu_1] - \lambda_2[\int_0^\infty x^2 f(x)dx - \mu_2] \tag{30}$$

Differentiating L with respect to $f(x)$ and equating the derivative to zero, one obtains:

$$f(x) = \{\frac{1}{m} + \frac{(1-m)}{m}[\lambda_0 + \lambda_1 x + \lambda_2 x^2]\}^{1/(m-1)} \tag{31}$$

Defining $k = (1-m)/m$ [37] and $\alpha_i = m\lambda_i$, $i = 0, 1, 2$, Equation (31) can be written as:

$$f(x) = (1+k)^{-1-1/k}[1 + k(\alpha_0 + \alpha_1 x + \alpha_2 x^2)]^{-1-1/k} \tag{32}$$

Equation (32) is the entropy-based probability density function of power type.

The Lagrange multipliers, λ_i, $i = 0, 1, 2$ and, consequently, $\alpha_i = m\lambda_i$, $i = 0, 1, 2$, can be estimated using Equations (27) to (29). For simplification, let λ_2 be assumed as 0. Then, Equation (32) becomes:

$$f(x) = (1+k)^{-1-1/k}[1 + k(\alpha_0 + \alpha_1 x)]^{-1-1/k} \tag{33}$$

whose parameters $\alpha_i = m\lambda_i$, $i = 0, 1$, are estimated using Equations (27) and (28). Substituting Equation (33) in Equation (27), one obtains:

$$\alpha_1 = (1+k)^{-1-1/k}(1 + k\alpha_0)^{-1/k} \tag{34}$$

or

$$\alpha_0 = -1 + \frac{1}{k}[\alpha_1(1+k)^{1+1/k}]^k \tag{35}$$

The other Lagrange multiplier can be determined by inserting Equation (33) in Equation (28) and is given as:

$$-\frac{(1+k)^{-1-1/k}}{k(1-\frac{1}{k})}(1 + k\alpha_0)^{1-1/k} = \alpha_1^2 \overline{x} \tag{36}$$

By solving Equations (34) or (35) with (36), Lagrange multipliers α_1 and α_0 can be determined. Inserting Equation (34) in Equation (33), one obtains

$$f(x) = (1+k)^{-1-1/k}(1 + k\alpha_0)^{-1-1/k}[1 + k(1+k)^{-1-1/k}k(1 + k\alpha_0)^{-1-1/k}x] \tag{37}$$

Let

$$\beta = [(1+k)(1 + k\alpha_0)]^{-1-1/k} \tag{38}$$

Using Equation (38), Equation (37) becomes

$$f(x) = \frac{1}{\beta}[1 + \frac{kx}{\beta}]^{-1-1/k} \tag{39}$$

Equation (39) is a two-parameter generalized Pareto distribution.
Let

$$y = x + \frac{\beta}{k} \quad \Rightarrow \quad x = y - \frac{\beta}{k} \tag{40}$$

Equation (39) then becomes

$$f(y) = \frac{1}{\beta}[\frac{ky}{\beta}]^{-1-1/k} \tag{41}$$

Equation (41) is a two-parameter Pareto distribution. If $k \to 0$, then Equation (41) leads to an exponential distribution:

$$f(y) = \frac{1}{\beta}\exp[\frac{y}{\beta}] \tag{42}$$

Koutsoyiannis [37,38] proposed a generalization, following which Equation (32) can be expressed as:

$$f(x) = (1+k)^{-1-1/k}[1 + k(\alpha_0 + \alpha_1 x + \alpha_2 x^{c_1})]^{-1-1/k} x^{c_2-1} \tag{43}$$

where c_1 and c_2 are shape parameters. Equation (43) has four parameters: scale parameter α_1 and shape parameters k and c_1 and c_2. Note that α_0 is not a parameter, because it is a constant based on the satisfaction of Equation (27). It is, however, not clear as to what led to the generalized form of Equation (43). Koutsoyiannis [37] suggested that random variable X^{c_2} would have Beta Prime (also referred to as Beta of the second kind) distribution [62]. Then, the distribution of X would be referred to as the power-transformed Beta Prime (PBP). Koutsoyiannis [37] showed that Equation (43) can specialize into several exponential and power-type probability distributions, such as PBP-L1 $(k \to 0)$, gamma $(k \to 0,\ c_1 \to 1)$, Weibull $(k \to 0,\ c_2 = c_1)$, Pareto $(c_2 = c_1 = 1)$, Beta Prime $(c_1 = 1)$, PBP-L2 $(k \to \infty,\ k\alpha_0 \to k_0,\ k\alpha_1 \to \alpha_1)$ and others.

4.2. Network Evaluation and Design

Hydrometric data are required for an efficient planning, design, development, operation and management of water engineering systems. Many studies have applied the Shannon entropy theory to assess and optimize data collection networks (e.g., water quality, rainfall, streamflow, hydrometric, elevation, landscape, etc.) [63,64] but not the Tsallis entropy theory. The basic idea for developing a methodology for data collection network design is that it must take into account the information of each gaging station or potential gaging station in the network. A station with a higher information content is given a higher priority over other stations that have lower information content but the information content of a station must be tempered with the degree of use. That is, a station that is used by one user might be given a lower priority than a station that has diverse uses.

A framework for network design or evaluation considers a range of factors, such as: (a) objectives of sampling; (b) variables to be sampled; (c) locations of measurement stations; (d) frequency of sampling; (e) duration of sampling; (f) uses and users of data; and (g) socio-economic considerations. A network design has two modes: (1) number of gages and their locations (space evaluation); and (2) time interval for measurement (time evaluation).

Let there be two stations A and B with ensembles of configurational possibilities $E^A = \{1, 2, \ldots, N\}$ with probability distribution $P^A = \{p_j{}^A,\ i = 1, 2, \ldots, N\}$ and configurational possibilities $E^B = \{1, 2, \ldots, M\}$ with probability distribution $P^B = \{p_j{}^B,\ j = 1, 2, \ldots, M\}$. Then, the union of the two stations $A \cup B$ and their corresponding ensembles of possibilities are $E^{A\cup B} = \{(1,1), (1,2), \ldots, (i,j), \ldots, (N,M)\}$. If $p_{ij}{}^{A\cup B}$ represents the corresponding probabilities, then the mutual information or transinformation S can be expressed as:

$$T[(p_{ij}{}^{A\cup B})] = H^{A\cup B}[(p_{ij}{}^{A\cup B})] - H^{A}[(\sum_{i=1}^{N} p_{ij})] - H^{B}[(\sum_{j=1}^{M} p_{ij})] \tag{44}$$

From Equation (44), $T(p_{ij}) = 0$ for all m, if X and Y are independent. For correlated X and Y, $T(p_{ij}) < 0$ for $m = 1$ and $T(p_{ij}) = 0$ for $m = 0$. For arbitrary values of m, it will be sensitive to p_{ij}; it can take on negative or positive values for both $m < 1$ and $m > 1$ with no particular regularity and can exhibit more than one extremum.

4.3. Directional Information Transfer Index

By dividing by the marginal entropy, the mutual information T can be normalized [65] as:

$$\frac{T}{H} = DIT = \frac{(H - H_{Lost})}{H} = 1 - \frac{H_{Lost}}{H} \tag{45}$$

where H_{Lost} is the amount of information lost. The ratio of T by H is called the Directional Information Transfer (DIT) index. Mogheir and Singh [66] called it as the Information Transfer Index (ITI). The DIT varies from zero to unity and denotes the fraction of information transferred from one station to another. A zero value of DIT corresponds to the case where sites are independent and, therefore, no information is transmitted. A value of unity for DIT corresponds to the case where sites are fully dependent and no information is lost. Since $\text{DIT}_{XY} = T/H(X)$ is not the same as $\text{DIT}_{YX} = T/H(Y)$, DIT is not symmetrical. The term DIT_{XY} describes the fractional information inferred by station X about station Y, whereas DIT_{YX} describes the fractional information inferred by station Y about station X. Between two stations, the station with the higher DIT should be given higher priority because of its greater capability in inferring (predicting) the information at other sites. The DIT can be applied for regionalization of the network or watersheds.

4.4. Reliability of Water Distribution Networks

A water distribution system can be designed by minimizing head losses, costs, risks and departures from specified values of water quantity, pressure and quality and also maximizing reliability [67]. Thus, it becomes a multi-objective optimization problem. However, it is not uncommon to formulate the design problem as a single-objective optimization problem, where the system capital and operational costs are minimized and, at the same time, the laws of hydraulics are satisfied and the targets of water quantity and pressure at demand nodes are met. Fundamental to either type of optimization is reliability [68–70].

To develop a Tsallis entropy-based redundancy measure of the network with N nodes, where the nodes may be considered to constitute sub-systems, the Tsallis entropy of a node j can now be expressed in terms of W_{ij} as:

$$S_j = \frac{1}{m-1}[\sum_{i=1}^{n(j)} W_{ij} - W_{ij}{}^{m})] = \frac{1}{m-1}\{\sum_{i=1}^{n(j)}[\frac{q_{ij}}{Q_j} - (\frac{q_{ij}}{Q_j})^{m}]\} \tag{46}$$

where m is the entropy index and is a real number and S_j is an entropic measure of redundancy at node j and is local redundancy. Maximizing S_j would maximize redundancy of node j and is equivalent to maximizing entropy at node j. The maximum value of S_j is achieved when all W_{ij}'s or q_{ij}/Q_j's are equal. This occurs when all q_{ij}'s are equal. For the entire water distribution network, redundancy is a function of redundancies S_j's of individual nodes in the network.

The overall network redundancy can be assessed in two ways. First, the network redundancy can be assessed by the relative importance of a link to its node and its importance recognized by q_{ij}/Q_j.

In this case, the redundancy is maximized at each node. It may, however, be noted that the network redundancy is not a sum of nodal redundancies. Second, the network redundancy can be assessed by the relative importance of a link to the total flow and its importance recognized by q_{ij}/Q_0. Here, the proposition is that the importance of a link relative to the local flow is not as important as it is to the total flow. In this case too, the network redundancy is not a sum of nodal redundancies. In order to acknowledge the relative importance of a link to the entire network, the nodal redundancy S_{j*} can be expressed as:

$$S_{j*} = \frac{1}{m-1}\sum_{i=1}^{n(j)}\left[\frac{q_{ij}}{Q_0} - \left(\frac{q_{ij}}{Q_0}\right)^m\right] = \frac{1}{m-1}\left[\frac{Q_j}{Q_0} - \sum_{i=1}^{n(j)}\left(\frac{q_{ij}}{Q_0}\right)^m\right] \tag{47}$$

It may be noted that S_{j*}, given by Equation (47), is similar to the S_j given by Equation (46). In this case too, the maximum value of S_{j*} will occur when the q_{ij} values are equal at the j-th node. It can also be shown that the maximum network redundancy will be achieved when all the q_{ij} values are equal. It may, however, be noted that

$$S_{j*} = \frac{1}{m-1}\sum_{i=1}^{n(j)}\left[\frac{q_{ij}}{Q_0} - \left(\frac{q_{ij}}{Q_0}\right)^m\right] \neq \frac{1}{m-1}\left[1 - \sum_{i=1}^{n(j)}\left(\frac{q_{ij}}{Q_0}\right)^{m-1}\right] \tag{48}$$

This is because,

$$\sum_{i=1}^{n(j)} q_{ij} = Q_j \neq Q_0 \text{ and } \sum_{i=1}^{n(j)} \frac{q_{ij}}{Q_0} \neq 1 \tag{49}$$

Therefore, in the second approach, Equation (47) can be used in the spirit of Tsallis entropy or considering it via partial Tsallis entropy [36].

The network redundancy for N nodes is a function of redundancies of individual nodes, S_j's, in the network. However, it will not be a simple summation of these nodal redundancies, because of the non-extensive property of the Tsallis entropy. For the first approach, it can be shown that the network redundancy (with N nodes) can be expressed as:

$$\begin{aligned} S_{1\cup 2\cup \ldots \cup N} = & \sum_{1\le j\le N} S_j + (1-m)\sum_{1\le j1<j2\le N} S_{j1}S_{j2} + (1-m)^2 \sum_{1\le j1<j2<j3\le N} S_{j1}S_{j2}S_{j3} + \ldots \\ & + (1-m)^{N-1}\sum_{1\le j1<j2\ldots<jN\le N} S_{j1}S_{j2}\ldots S_{jN} \end{aligned} \tag{50}$$

In order to develop an appreciation for Equation (47), it will be instructive to expand Equation (47) in terms of flow quantities. Equation (47) is just the sum of nodal redundancies.

For the second approach, the network redundancy (with N nodes: 1, 2, ... , N) can be expressed as:

$$\begin{aligned} S_{1\cup 2\cup 3\cup 4\ldots\cup N*} & = (m-1)^{N-1}\prod_{j=1}^{N} S_{j*} - \frac{(m-1)^{N-2}}{Q_0}[Q_1 S_{2*}\ldots S_{N*} + Q_2 S_{1*} S_{3*}\ldots S_{N*} + \ldots Q_N S_{1*} S_{2*}\ldots S_{N-1*}] \\ & + \ldots + \frac{(m-1)}{Q_0{}^{N-1}}[Q_1 Q_2\ldots Q_{N-1} S_{N*} + Q_1 Q_2 \ldots Q_{N-2} S_{N-1*} + \ldots + Q_2 Q_2 \ldots Q_N Q_1] \end{aligned} \tag{51}$$

It can be seen that Equation (51) for network redundancy with the second approach is significantly different from Equation (50) with the first approach.

5. Problems Requiring Coupling with another Theory

There are some problems where the entropy theory can only be part of the solution methodology and needs to be coupled with another theory. Consider, for example, hydraulic geometry of a river or channel, which is defined by the relations between discharge and each of hydraulic variables (e.g., flow width, depth, velocity under bankfull conditions) and each of geometric variables (e.g., bed roughness, slope). Hydraulic geometry is of two types: (1) downstream; and (2) at-a-point. For either type, the values of hydraulic and geometric variables are average annual values corresponding to the equilibrium or stable condition of the river. At this condition, the river will try to spend the least amount of energy for transporting water and sediment. In response to the influx of water and sediment coming from the watershed, the river will adjust these variables or characteristics in order to attain the stable condition. This means, the river will spread the adjustment as equitably as the environment will allow and will follow the theory of minimum energy dissipation rate. The equal rate of adjustment can be described by the principle of maximum entropy (POME) that the river will follow. In this manner, the theory of minimum energy dissipation rate and entropy theory are combined for determining the hydraulic geometry of a river or designing a stable canal. The advantage of the entropy theory is that it explicitly allows to incorporate the constraints imposed by the watershed and the design. For example, if a river is leveed, then it cannot adjust its width and the adjustment will be shared by depth and velocity. Likewise, if a canal is lined, it cannot adjust its width, slope and roughness and will have to adjust its depth and velocity. In this manner, a whole hierarchy of hydraulic geometry relations can be obtained, depending on the constraints imposed. The other existing theories do not allow this flexibility. In what follows, some specific examples of problems requiring coupling entropy theory with other theories are presented.

5.1. Hydraulic Geometry

Hydraulic geometry is defined by relations between discharge (Q) and each of channel width (B), flow depth (d), flow velocity (V), slope (S) and roughness factor (say Manning's n). Hydraulic geometry is either downstream hydraulic geometry or at-a-station hydraulic geometry [71]. Langbein [72] and Yang et al. [73] reasoned that hydraulic geometry relations correspond to the equilibrium state of the channel. In order to attain this state, the channel adjusts its hydraulic variables and the adjustment is shared equally among these variables. In practice, the channel is seldom in equilibrium, meaning that the adjustment among hydraulic variables will be unequal. This, then, suggests that there will be a family of hydraulic geometry relations, depending on the adjustment of hydraulic variables and the adjustment should explain the variability in the parameters of hydraulic geometry relations. For downstream hydraulic geometry, Singh et al. [19,20] hypothesized that, for a given influx of discharge from the watershed, the channel would minimize its stream power by adjusting three controlling variables: depth, width and friction.

Coupling the theory of minimum energy dissipation rate with the principle of maximum entropy, three possibilities can occur corresponding to the spatial rate of adjustment of friction, the spatial rate of adjustment of width and the spatial rate of adjustment of flow depth. These possibilities then lead to the formulation of, respectively, the proportion of the adjustment of stream power (SP) by friction, the proportion of the adjustment of SP by channel width and the proportion of the adjustment of SP by flow depth and, hence, to four sets of hydraulic geometry relations, as follows: (1) the spatial change in SP is accomplished by an equal spatial adjustment between flow width B and resistance expressed by Manning's n; (2) the spatial variation in SP is accomplished by an equal spatial adjustment between flow depth and flow width; (3) the spatial variation in SP is accomplished by an equal spatial adjustment between flow depth and resistance; and (4) the spatial variation in SP is accomplished by an equal spatial adjustment between flow depth, flow width and resistance. These four possibilities can occur in the same river in different reaches or in the same reach at different times, or in different rivers at the same time or at different times.

The hydraulic geometry relations are expressed as:

$$B = aQ^b,\ d = cQ^f,\ V = kQ^m,\ n = NQ^p,\ S = sQ^y \tag{52}$$

where a, c, k, N and s are parameters; and b, f, m, p and y are exponents. Values of these exponents and parameters depend upon the possibility under consideration and the entropy theory permits explicit expressions for the exponents and parameters. Singh et al. [19,20] showed that most of the downstream hydraulic geometry relations reported in the literature can be derived as special cases of the entropy-based equations.

For at-a-station hydraulic geometry, when discharge changes, a river cross-section can adjust its width, depth, velocity, roughness and slope or a combination thereof. Singh and Zhang [21,22] reasoned that the channel cross-section will adjust or minimize its SP by adjusting these four variables: (1) P_B can be interpreted as the proportion of the temporal change of SP due to the temporal rate of adjustment of width; (2) P_h as the proportion of the temporal change of SP due to the temporal rate of adjustment of depth; (3) P_α as the proportion of the temporal change of SP due to the temporal rate of adjustment of friction; and (4) P_S as the proportion of the temporal change of SP due to the temporal rate of adjustment of slope. These cases involve probabilities of four variables, meaning that any adjustment in hydraulic variables in combinations of two, three or four may occur. These give rise to different configurations of adjustment that do indeed occur in nature [74]. Thus, the equality among four probabilities yields 11 possibilities and, hence, leads to 11 sets of equations: (1) $P_B = P_h$; (2) $P_B = P_\alpha$; (3) $P_B = P_S$; (4) $P_h = P_\alpha$; (5) $P_h = P_S$; (6) $P_\alpha = P_S$; (7) $P_B = P_h = P_\alpha$; (8) $P_B = P_\alpha = P_S$; (9) $P_B = P_h = P_S$; (10) $P_h = P_\alpha = P_S$; and (11) $P_B = P_h = P_\alpha = P_S$. It should be noted that all eleven possibilities can occur in the same river cross-section at different times, or in different river cross-sections at the same time or at different times. Williams [75] explored 11 cases, which are similar to the above 11 possibilities. The resulting hydraulic geometry relations are of the same form as Equation (52) but expression for exponents and parameters therein are different.

5.2. Evaporation

Evaporation is a process by which liquid water is converted into water vapor. The process of evaporation entails four elements [76]: (1) supply of energy; (2) supply of water; (3) tendency of liquid water molecules to escape; and (4) turbulent transport. The source of energy that is needed for evaporation to occur from land surfaces is solar radiation, which can be defined by radiative flux. The supply of water can be from precipitation or irrigation, determining soil wetness, which is characterized by soil moisture. Fugacity refers to the tendency of water molecules to escape and is expressed by the saturated vapor pressure at the liquid-vapor interface. The turbulent transport of water vapor and heat is determined by wind speed and thermal instability of the surface layer and is defined by turbulent sensible heat flux into the atmosphere. Wang et al. [76] argued that, under thermodynamic equilibrium, thermal and hydrologic states of the land surface resulting from the interaction between land and atmospheric processes tend to maximize evaporation.

For a given radiative energy flux, the rate of evaporation depends on the combination of surface soil moisture, surface soil temperature and sensible heat flux, as well as the dynamic feedbacks among them at the land surface. There can be many combinations of ground and sensible heat fluxes, evaporation rate and net radiation that can satisfy the energy balance. Wang et al. [76] hypothesized that the preferred combination is the one that maximizes evaporation. Denoting the rate of evaporation by E, surface soil moisture by w, surface soil temperature by T, sensible heat flux into the atmosphere by H, ground heat flux by G and net radiative by R_n at the surface, maximizing evaporation, subject to the energy balance, the result is:

$$E = \max[E(w, T, H; R)|E + H + G = R_n] \tag{53}$$

for all combinations of independent variables w, T and H. Wang et al. [76] investigated three cases: (1) $R = R_n - G$, representing the turbulent energy budget, as described by the Bowen ratio; (2) $R = R_n$, corresponding to the partitioning of the net radiation into latent, sensible and ground heat fluxes; and (3) $R = R_i$, representing the budget of all surface energy fluxes. These cases express land-atmosphere interactions.

6. Problems Involving Physical Relations

In water engineering, we often need to determine physical relationships, such as infiltration rate as a function of time, runoff or discharge as a function of time, soil moisture as a function of depth from the soil surface, velocity as a function of flow depth measured from the bottom, sediment concentration as a function of flow depth and sediment discharge as a function of time. For deriving physical relationships, the probability domain and the physical domain need to be concatenated and this can be done by hypothesizing the cumulative distribution function (CDF) of a design (dependent) variable (e.g., flux, say discharge) in terms of independent (concentration) variable (e.g., stage of flow).

6.1. Hypotheses on Cumulative Probability Distribution Function

Different types of hypotheses have been formulated when applying entropy theory to derive relationships for design variables. Examples of a linear hypothesis include velocity distribution as a function of flow depth, wind velocity as a function of height, sediment concentration profile along the flow depth, rating curve and groundwater discharge along the horizontal direction of flow. It is noted that the CDF should have a one-to-one relation with the design variable (i.e., random variable) of interest and its value should only be between 0 and 1; 0 for the minimum value of the random variable and 1 for the maximum value. For deriving a two-dimensional velocity (u) distribution, Cui and Singh [77] hypothesized a general form of the CDF as:

$$F(u) = \left[1 - \left(\frac{x}{B}\right)^2\right]^b \left(\frac{y}{D}\right)^a \text{ for all } (x,\, y) \text{ on } I(u) \tag{54}$$

in which y is the vertical dimension, x is the transverse direction, a and b are shape parameters and B and D act as scale parameters or normalizing quantities. The CDF given by Equation (54) has a one-to-one relationship with the velocity value u; in other words, the CDF is unique on each isovel $I(u)$ and has a value of 0 at $I(0)$ and 1 at $I(u_{\max})$. Also, CDF is 0 at $x = B$ or $y = 0$ and is 1 at $x = 0$ and $y = D$ (Figure 3). It is continuous and differentiable in both x and y.

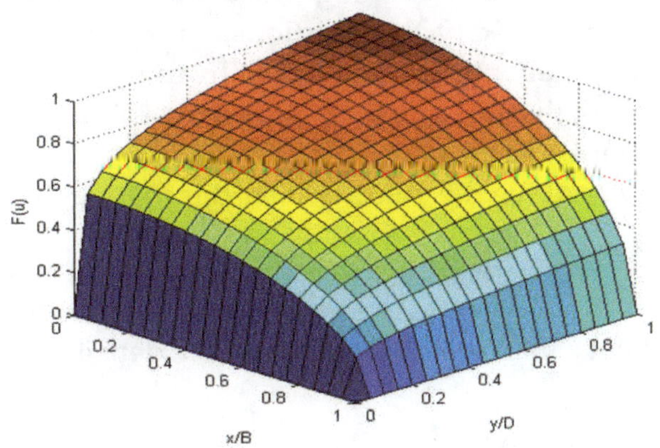

Figure 3. CDF of 2-D velocity distribution with $a = 0.5$, $b = 0.2$ in the idealized cross-section.

The two-dimensional (2-D) velocity distribution can also be expressed by transforming the Cartesian coordinates [vertical (y) and transverse (x)] into curvilinear s-r coordinate system in which r has a unique, one-to-one relation with a value of velocity and s (coordinate) curves are their orthogonal trajectories [56]. Then, the CDF of velocity in a channel cross section can be expressed as:

$$F(u) = \frac{r - r_0}{r_{\max} - r_0} \tag{55}$$

where r is a coordinate between r_0 and r_{max} and corresponds to y.

For the one-dimensional (1-D) form of Equation (54) with $b = 0$, it is seen from Equation (54) that the CDF will remain between 0 and 1 when $0 \leq x \leq B$ and $0 \leq y \leq D$. Thus, the CDF for 1-D velocity distribution can be simplified as:

$$F(u) = \left(\frac{y}{D}\right)^a \text{ for all } (y) \text{ on } I(u) \tag{56}$$

This equation implies that the shape parameter a is critical. If $a = 1$, then

$$F(u) = \left(\frac{y}{D}\right) \text{ for all } (y) \text{ on } I(u) \tag{57}$$

In cases where the design variable is maximum at $y = 0$ and minimum at maximum y (i.e., $y = D$), as is the case for suspended sediment concentration, it is distributed from a maximum value at the channel bed and decreases towards the water surface. Hence, the CDF should be assumed in a way so that it is 1 at the channel bed ($y = 0$) and 0 at the water surface ($y = D$). The CDF of sediment concentration can then be expressed as:

$$F(c) = 1 - \left(\frac{y}{D}\right)^a \tag{58}$$

It is seen from Equation (58) that $F(c)$ will be between 0 and 1 if $0 \leq y \leq D$.

6.2. One-Dimensional Velocity Distribution

Velocity distribution is needed for determining flow discharge, scour around bridge piers, erosion and sediment transport, pollutant transport, energy and momentum distribution coefficients, hydraulic geometry, watershed runoff and river behavior. Velocity distributions have been derived using experimental, hydrodynamic, or entropy methods. Shannon entropy has been applied to derive one- and two-dimensional velocity distributions [26]. Singh and Luo [11,78] and Luo and Singh [12] employed Tsallis entropy to derive the 1-D velocity distribution. The Tsallis entropy-based velocity distribution has been shown to have an advantage over the Shannon entropy-based distribution. However, in these entropy-based velocity distributions, the CDF has been assumed to be linear, meaning the velocity is equally likely along the vertical from the channel bed to the water surface. This assumption is fundamental to the derivation of velocity distributions but has not been adequately scrutinized. Further, this assumption is weak and may partly explain the reason why these velocity distributions do not accurately describe the velocity near the channel bed.

For deriving 1-D velocity distribution, the constraint equation is derived using the continuity equation such that $g_1(u) = u$ and limits of integration are 0 and u_D (maximum velocity at the surface or flow depth D). The entropy theory then yields the probability density function of velocity u:

$$f(u) = \left[\frac{m-1}{m}(\lambda_* + \lambda_1 u)\right]^{\frac{1}{m-1}} \tag{59}$$

where $\lambda_* = \frac{1}{m-1} + \lambda_0$.

Following Cui and Singh [77], the CDF of u can be hypothesized as Equation (56). Then, a general velocity distribution based on the Tsallis entropy theory [14] is obtained as:

$$u = -\frac{\lambda_*}{\lambda_1} + \frac{1}{\lambda_1}\frac{m}{m-1}\left[\lambda_1\left(\frac{y}{D}\right)^a + \left(\frac{m-1}{m}\lambda_*\right)^{\frac{m}{m-1}}\right]^{\frac{m-1}{m}} \tag{60}$$

Equation (60) can be simplified by defining a dimensionless parameter G as:

$$G = \frac{\lambda_1 u_{\max}}{\lambda_1 u_{\max} + \lambda_*} \tag{61}$$

Parameter G can be regarded as an index of the velocity distribution uniformity. Equation (60) can be cast as:

$$\frac{u}{u_{\max}}=1-\frac{1}{G}\left(1-((1-G)^{\frac{m}{m-1}}+\left(1-(1-G)^{\frac{m}{m-1}}\right)\left(\left(\frac{y}{D}\right)^{a}\right)^{\frac{m-1}{m}}\right) \tag{62}$$

Equation (62) shows that for a given m value, the velocity distribution can be obtained with only two parameters: a and G. A bigger G value tends to slow the growth of the velocity from the channel bed to the water surface, while the parameter a has an opposite effect. A lower value of G tends to linearize the velocity distribution and a higher value non-linearize. The opposite is the case for exponent a. The velocity distribution is more sensitive to a than to G.

6.3. Two-Dimensional Velocity Distribution

For two-dimensional (2-D) velocity distribution, Chiu [56] proposed a transformation that converts the Cartesian coordinates [vertical (y) and transverse (z)] into curvilinear s-r coordinate system, in which r has a unique, one-to-one relation with a value of velocity and s (coordinate) curves are orthogonal trajectories to r. Following the coordinate system of Chiu [56], Luo and Singh [12] employed the Tsallis entropy. Then, expressing the CDF of velocity in a channel cross section by Equation (57), the velocity distribution becomes

$$u=\left(\frac{m}{m-1}\right)\frac{1}{\lambda_1}\left\{\left(\lambda_1\frac{r-r_0}{r_{\max}-r_0}\right)+\left[\left(\frac{m-1}{m}\right)\lambda_*\right]^{\frac{m}{m-1}}\right\}^{\frac{m-1}{m}}-\frac{\lambda_*}{\lambda_1} \tag{63}$$

Defining entropy parameter $M=\lambda_1 u_{\max}^2$, the dimensionless velocity distribution can be expressed as:

$$\frac{u}{u_{\max}}=\frac{2}{M}\left[M\left(\frac{r-r_0}{r_{\max}-r_0}\right)+\frac{(4-M)^2}{16}\right]^{\frac{1}{2}}-\frac{4-M}{2M} \tag{64}$$

The Tsallis entropy-based approach of Luo and Singh [12] was either superior or comparable to Chiu's distribution for the data sets used therein for testing. However, due to the complexity of the coordinate system and a large number of parameters used, the application of these methods may be limited. Later, using normal x-y coordinate system, Cui and Singh [13] obtained the 2-D velocity distribution equation as:

$$u=-\frac{\lambda_*}{\lambda_1}+\frac{1}{\lambda_1}\frac{m}{m-1}\left[\lambda_1 F(u)+\left(\frac{m-1}{m}\lambda_*\right)^{\frac{m}{m-1}}\right]^{\frac{m-1}{m}} \tag{65}$$

where the CDF, $F(u)$, is defined as Equation (54). As defined in the case of one-dimensional velocity distribution, the dimensionless entropy parameter G is also used here as:

$$G=\frac{\lambda_1 u_{\max}}{\lambda_1 u_{\max}+\lambda_*} \tag{66}$$

Parameter G is found to be related to the ratio of mean and maximum velocity and a quadratic relation is obtained by computing from observed mean and maximum velocity values as [13]:

$$\frac{\overline{u}}{u_{\max}}=\Psi(G)=0.554G^2-0.077G+0.568 \tag{67}$$

Now, with the use of Equation (67), the general velocity distribution Equation (65) can be cast as:

$$\frac{u}{u_{\max}} = 1 - \frac{1}{G}\left(1 - \left[(1-G)^{\frac{m}{m-1}} + \left(1-(1-G)^{\frac{m}{m-1}}\right)F(u)\right]^{\frac{m-1}{m}}\right) \quad (68)$$

Equation (68) is the general 2-D velocity distribution equation in terms of parameter G and maximum velocity.

6.4. Suspended Sediment Concentration

The sediment concentration distribution can also be derived using Tsallis entropy. For deriving the one-dimensional sediment concentration distribution, the constraint equation is derived using the mean concentration, such that $g_1(c) = c$ and limits of integration are $a = c_0$ (maximum concentration at the bed surface or flow depth $D = 0$) and $b = 0$ at the water surface, as [16]:

$$f(c) = \left[\frac{m-1}{m}(\lambda_* + \lambda_1 c)\right]^{\frac{1}{m-1}} \quad (69)$$

where $\lambda_* = \frac{1}{m-1} + \lambda_0$. Equation (69) is the least-biased entropy-based probability density function of sediment concentration, which is fundamental to determining the sediment concentration distribution. The dimensionless sediment concentration can then be obtained as:

$$\frac{c}{c_0} = 1 - \frac{1}{N}\left(1 - \left\{(1-N)^{\frac{m}{m-1}} + \left[1-(1-N)^{\frac{m}{m-1}}\right]F(c)\right\}^{\frac{m-1}{m}}\right) \quad (70)$$

where N is a dimensionless parameter as a function of maximum concentration and the Lagrange multipliers expressed as:

$$N = \frac{\lambda_1 c_0}{\lambda_1 c_0 + \lambda_*} \quad (71)$$

By plotting the empirical observations of the mean over the maximum concentration values and corresponding N values, the implicit function can be regressed as a quadratic polynomial as:

$$\frac{\bar{c}}{c_0} = 0.176 + 0.5083N - 0.1561N^2 \quad (72)$$

with a coefficient of determination as 0.99. Thus, N can be used for deriving sediment concentration distribution instead of solving nonlinear equations for λ_1 and λ_*.

6.5. Sediment Discharge

The suspended sediment discharge can be computed simultaneously by integrating sediment concentration and velocity over the cross-section where the velocity distribution and sediment concentration can be obtained empirically as well as using the Tsallis entropy theory. That is, suspended sediment discharge can be derived using different combinations of entropy-based and empirical methods for velocity and sediment concentration distributions. Thus, the suspended sediment discharge can be computed with the use of entropy in three ways: (1) velocity distribution by a standard formula and concentration distribution by an entropy-based equation; (2) velocity distribution by an entropy-based equation and concentration distribution by a standard formula; and (3) velocity distribution and concentration distribution both by entropy-based equations. In the discussion that follows, only the combination where both are derived using Tsallis entropy is considered.

The Tsallis entropy-based velocity distribution [13], given by Equation (68) and sediment concentration [16], given by Equation (70), are integrated from bottom of the channel to the water surface to obtain the suspended sediment discharge, as:

$$q_s = u_{\max} c_0 \int_0^D \left[1 - \tfrac{1}{N}\left(1 - ((1 + 0.5\ln N)F(c) - 0.5\ln N)^{2/3}\right)\right] \left[1 - \tfrac{1}{G}(1 - ((1 + 0.5\ln G)F(u) - 0.5\ln G)^{2/3}\right] dy \tag{73}$$

To get an explicit solution of Equation (73) is difficult; however, it can be simplified with the use of mean values. The first term in the integration can be replaced by mean sediment concentration and the second term can be replaced by mean velocity, such that Equation (73) reduces to

$$q_s = D u_{\max} c_0 (0.554G^2 - 0.777G + 0.568)(0.554N^2 - 0.777N + 0.568) \tag{74}$$

which provides a simple method to compute sediment discharge. Entropy parameters G and N are fixed for each channel cross-section [17]. Thus, once the entropy parameters have been obtained for some known cross-section, with observed maximum velocity and sediment concentration, the sediment discharge can be obtained with ease.

6.6. Flow-Duration Curve

The flow-duration curve (FDC) is used for predicting the distribution of future flows, forecasting of future recurrence frequencies, determining low-flow thresholds for defining droughts, generating hydropower, constructing load-duration curves, determining power-duration curves and comparing watersheds. For deriving an FDC, it is assumed that temporally-averaged discharge Q is a random variable, varying from a minimum value Q_{min} to a maximum value Q_{max}, with a probability density function (PDF) denoted as $f(Q)$. The time interval for which the discharge is averaged depends on the purpose of constructing an FDC but it is often taken as one day. Considering $g_1(x)$ = mean $Q = Q_{mean}$, the PDF of Q can be derived as [79]:

$$f(Q) = \left[\frac{m-1}{m}\left(\frac{1}{m-1} - \lambda_0 - \lambda_1 Q\right)\right]^{\frac{1}{m-1}} \tag{75}$$

It is interesting to note that at $Q = 0$, $f(Q)$ becomes $\{[(m-1)/m][(1/(m-1)) - \lambda_0]\}^{1/(m-1)}$. Similar to velocity distribution, a dimensionless parameter M can be defined as:

$$M = \frac{\lambda_1 Q_{\max}}{\lambda_1 Q_{\max} - \lambda_*} \tag{76}$$

It was found [79] that M is linearly related to the ratio between the mean flow and the maximum flow, which, using regression, can be written as:

$$M = 2.246 - 4.891\frac{\overline{Q}}{Q_{\max}} \tag{77}$$

with the squared correlation coefficient of 0.9972. The FDC can be expressed as:

$$\begin{aligned}\frac{Q}{Q_{\max}} &= 1 - \tfrac{1}{M} - (1 - \tfrac{1}{M})\left\{-(\tfrac{1}{\lambda_*})^{\frac{m}{m-1}}\lambda_1(\tfrac{m}{m-1})^{\frac{1}{m-1}}F(Q) + 1\right\}^{\frac{m-1}{m}} \\ &= 1 - \tfrac{1}{M}\left(1 - \left\{(\tfrac{m}{m-1})^{\frac{m}{m-1}}\tfrac{M}{Q_{\max}}[F(Q) - 1] + 1\right\}^{\frac{m-1}{m}}\right)\end{aligned} \tag{78}$$

7. Conclusions

A survey of the literature shows that the Tsallis entropy theory has great potential to address a wide range of problems in water engineering and in many other fields, as reviewed in this paper; see also [80–85] for some more recent studies. As generalization of the Shannon entropy, the Tsallis entropy can be applied in generalized equilibrium and statics in physics. The advantage of using the theory is that it can combine statistical information with physical laws, permits deriving physical relations as functions of time or space and derives probability distributions in terms of the specified constraints. However, analysis using the Tsallis entropy theory becomes complicated when one or two constraints are involved and it is more difficult to obtain the analytical expressions. Besides, the m-statistic must be carefully chosen when applying the Tsallis entropy, since it sometimes involves complex operations. Although entropy is a thermodynamic quantity, development of the thermodynamic basis of entropy-based relations has not been accomplished yet. Until now, the Tsallis entropy has been applied to determine frequency distributions, network evaluation, hydraulic geometry, evaporation, velocity distribution, sediment concentration distribution, flow duration curves and several other problems, as reviewed above. The outcomes are certainly encouraging. Until now, as we have reviewed in this paper, the Tsallis entropy theory has been applied in water engineering particularly with the principle of maximum entropy (POME). There is, however, also potential to combine the Tsallis entropy with the minimum relative (cross) entropy, for the condition that prior assumptions can be made. There is no question that the Tsallis entropy theory has a much greater potential to study a wide spectrum of problems in water engineering. It is hoped that this review will stimulate further interest in this fascinating field.

Acknowledgments: Bellie Sivakumar acknowledges the financial support from the Australian Research Council (ARC) through the Future Fellowship grant (FT110100328). Huijuan Cui acknowledges the financial support from Key Research Program of the Chinese Academy of Sciences (ZDRW-ZS-2016-6-4).

Author Contributions: All three authors substantially contributed to the article. Vijay P. Singh conceived the ideas for this review article; Vijay P. Singh, Bellie Sivakumar and Huijuan Cui had detailed discussions about the contents, structure and organization of the manuscript; Vijay P. Singh prepared the initial draft; Bellie Sivakumar and Huijuan Cui contributed to modifications and improvements; Vijay P. Singh, Bellie Sivakumar and Huijuan Cui contributed to the responses to the review comments and manuscript revision.

Notation

B	channel width
c	sediment concentration
C_r	constraint
D	water depth
E	evaporation
$f(x)$	probability density function
G	ground heat flux
H	entropy
$\Delta I(x_i)$	gain in information
$k = (1 - m)/m$	
m	Tsallis entropy index
n	Manning's n
N	natural number representative
p_i	probability
Q	flow
q_s	sediment discharge
R	net radiative
S_j	entropic measure of redundancy
u	velocity

x	transverse direction
y	vertical dimension
σ^2	variance
λ_i	Langrange multiplier
μ_1	first moment
μ_2	second moment
w	surface soil moisture

References

1. Singh, V.P. *Hydrologic Systems: Vol. 1. Rainfall-Runoff Modeling*; Prentice Hall: Englewood Cliffs, NJ, USA, 1988.
2. Singh, V.P. *Hydrologic Systems: Vol. 2. Watershed Modeling*; Prentice Hall: Englewood Cliffs, NJ, USA, 1989.
3. Singh, V.P. (Ed.) *Computer Models of Watershed Hydrology*; Water Resources Publications: Highland Ranch, CO, USA, 1995.
4. Salas, J.D.; Delleur, J.W.; Yevjevich, V.; Lane, W.L. *Applied Modeling of Hydrologic Time Series*; Water Resources Publications: Littleton, CO, USA, 1995.
5. Beven, K.J. *Rainfall-Runff Modelling: The Primer*; Wiley: Chichester, UK, 2001.
6. Sivakumar, B.; Berndtsson, R. *Advances in Data-Based Approaches for Hydrologic Modeling and Forecasting*; World Scientific Publishing Company: Singapore, 2010.
7. Singh, V.P. *Handbook of Applied Hydrology*, 2nd ed.; McGraw-Hill Education: New York, NY, USA, 2017.
8. Singh, V.P. *Kinematic Wave Modeling in Water Resources: Surface Water Hydrology*; John Wiley & Sons: New York, NY, USA, 1995.
9. Singh, V.P. *Kinematic Wave Modeling in Water Resources: Environmental Hydrology*; John Wiley & Sons: New York, NY, USA, 1996.
10. Harmancioglu, N.B.; Fistikoglu, O.; Ozkul, S.D.; Singh, V.P.; Alpaslan, M.N. *Water Quality Monitoring Network Design*; Kluwer Academic Publishers: Boston, MA, USA, 1999; p. 299.
11. Singh, V.P.; Luo, H. Entropy theory for distribution of one-dimensional velocity in open channels. *J. Hydrol. Eng.* **2011**, *16*, 725–735. [CrossRef]
12. Luo, H.; Singh, V.P. Entropy theory for two-dimensional velocity distribution. *J. Hydrol. Eng.* **2011**, *16*, 303–315. [CrossRef]
13. Cui, H.; Singh, V.P. Two-dimensional velocity distribution in open channels using the Tsallis entropy. *J. Hydrol. Eng.* **2013**, *18*, 331–339. [CrossRef]
14. Cui, H.; Singh, V.P. One-dimensional velocity distribution in open channels using Tsallis entropy. *J. Hydrol. Eng.* **2014**, *19*, 290–298. [CrossRef]
15. Simons, D.B.; Senturk, F. *Sediment Transport Technology*; Water Resources Publications: Highland Ranch, CO, USA, 1976.
16. Cui, H.; Singh, V.P. Suspended sediment concentration in open channels using Tsallis entropy. *J. Hydrol. Eng.* **2014**, *19*, 966–977. [CrossRef]
17. Cui, H.; Singh, V.P. Computation of suspended sediment discharge in open channels by combining Tsallis entropy-based methods and empirical formulas. *J. Hydrol. Eng.* **2014**, *19*, 18–25. [CrossRef]
18. Singh, V.P. On the theories of hydraulic geometry. *Int. J. Sediment Res.* **2003**, *18*, 196–218.
19. Singh, V.P.; Yang, C.T.; Deng, Z.Q. Downstream hydraulic geometry relations: 1. Theoretical development. *Water Resour. Res.* **2003**, *39*, 1337. [CrossRef]
20. Singh, V.P.; Yang, C.T.; Deng, Z.Q. Downstream hydraulic geometry relations: 2. Calibration and testing. *Water Resour. Res.* **2003**, *39*, 1338. [CrossRef]
21. Singh, V.P.; Zhang, L. At-a-station hydraulic geometry: I. Theoretical development. *Hydrol. Process.* **2008**, *22*, 189–215. [CrossRef]
22. Singh, V.P.; Zhang, L. At-a-station hydraulic geometry: II. Calibration and testing. *Hydrol. Process.* **2008**, *22*, 216–228. [CrossRef]
23. Singh, V.P.; Oh, J. A Tsallis entropy-based redundancy measure for water distribution network. *Physica A* **2014**, *421*, 360–376. [CrossRef]
24. Singh, V.P. *Entropy Theory and Its Application in Environmental and Water Engineering*; John Wiley & Sons: Sussex, UK, 2013.

25. Singh, V.P. *Entropy Theory in Hydraulic Engineering: An Introduction*; ASCE Press: Reston, VA, USA, 2014.
26. Singh, V.P. *Entropy Theory in Hydrologic Science and Engineering*; McGraw-Hill Education: New York, NY, USA, 2015.
27. Koutsoyiannis, D. Physics of uncertainty, the Gibbs paradox and indistinguishable particles. *Stud. Hist. Philos. Mod. Phys.* **2013**, *44*, 480–489. [CrossRef]
28. Papalexiou, S.M.; Koutsoyiannis, D. Entropy based derivation of probability distributions: A case study to daily rainfall. *Adv. Water Resour.* **2012**, *45*, 51–57. [CrossRef]
29. Shannon, C.E. The mathematical theory of communications, I and II. *Bell Syst. Tech. J.* **1948**, *27*, 379–423. [CrossRef]
30. Tsallis, C. Possible generalization of Boltzmann-Gibbs statistics. *J. Stat. Phys.* **1988**, *52*, 479–487. [CrossRef]
31. Singh, V.P. The use of entropy in hydrology and water resources. *Hydrol. Process.* **1997**, *11*, 587–626. [CrossRef]
32. Singh, V.P. Hydrologic synthesis using entropy theory: Review. *J. Hydrol. Eng.* **2011**, *16*, 421–433. [CrossRef]
33. Singh, V.P. *Introduction to Tsallis Entropy Theory in Water Engineering*; CRC Press/Taylor and Francis: Boca Raton, FL, USA, 2016.
34. Tsallis, C. Entropic nonextensivity: A possible measure of complexity. *Chaos Solitons Fractals* **2002**, *12*, 371–391. [CrossRef]
35. Tsallis, C. On the fractal dimension of orbits compatible with Tsallis statistics. *Phys. Rev. E* **1998**, *58*, 1442–1445. [CrossRef]
36. Niven, R.K. The constrained entropy and cross-entropy functions. *Physica A* **2004**, *334*, 444–458. [CrossRef]
37. Koutsoyiannis, D. Uncertainty, entropy, scaling and hydrological stochastics. 1. Marginal distributional properties of hydrological processes and state scaling. *Hydrol. Sci. J.* **2005**, *50*, 381–404. [CrossRef]
38. Koutsoyiannis, D. Uncertainty, entropy, scaling and hydrological stochastics. 2. Time dependence of hydrological processes and time scaling. *Hydrol. Sci. J.* **2005**, *50*, 405–426. [CrossRef]
39. Koutsoyiannis, D. A toy model of climatic variability with scaling behavior. *J. Hydrol.* **2005**, *322*, 25–48. [CrossRef]
40. Tsallis, C. Nonextensive statistical mechanics: Construction and physical interpretation. In *Nonextensive Entropy-Interdisciplinary Applications*; Gell-Mann, M., Tsallis, C., Eds.; Oxford University Press: New York, NY, USA, 2004; pp. 1–52.
41. Tsallis, C. Nonextensive statistical mechanics and thermodynamics: Historical background and present status. In *Nonextensive Statistical Mechanics and Its Applications*; Abe, S., Okamoto, Y., Eds.; Series Lecture Notes in Physics; Springer: Berlin, Germany, 2001.
42. Jaynes, E.T. Information theory and statistical mechanics, I. *Phys. Rev.* **1957**, *106*, 620–630. [CrossRef]
43. Jaynes, E.T. Information theory and statistical mechanics, II. *Phys. Rev.* **1957**, *108*, 171–190. [CrossRef]
44. Kullback, S.; Leibler, R.A. On information and sufficiency. *Ann. Math. Stat.* **1951**, *22*, 79–86. [CrossRef]
45. Lombardo, F.; Volpi, E.; Koutsoyiannis, D.; Papalexiou, S.M. Just two moments! A cautionary note against use of high-order moments in multifractal models in hydrology. *Hydrol. Earth Syst. Sci.* **2014**, *18*, 243–255. [CrossRef]
46. Krstanovic, P.F.; Singh, V.P. A univariate model for longterm streamflow forecasting: I. Development. *Stoch. Hydrol. Hydraul.* **1991**, *5*, 173–188. [CrossRef]
47. Krstanovic, P.F.; Singh, V.P. A univariate model for longterm streamflow forecasting: II. Application. *Stoch. Hydrol. Hydraul.* **1991**, *5*, 189–205. [CrossRef]
48. Krstanovic, P.F.; Singh, V.P. A real-time flood forecasting model based on maximum entropy spectral analysis: I. Development. *Water Resour. Manag.* **1993**, *7*, 109–129. [CrossRef]
49. Krstanovic, P.F.; Singh, V.P. A real-time flood forecasting model based on maximum entropy spectral analysis: II. Application. *Water Resour. Manag.* **1993**, *7*, 131–151. [CrossRef]
50. Krasovskaia, I. Entropy-based grouping of river flow regimes. *J. Hydrol.* **1997**, *202*, 173–191. [CrossRef]
51. Krasovskaia, I.; Gottschalk, L. Stability of river flow regimes. *Nordic Hydrol.* **1992**, *23*, 137–154.
52. Singh, V.P. *Entropy-Based Parameter Estimation in Hydrology*; Kluwer Academic Publishers: Boston, MA, USA, 1998; p. 365.
53. Wang, J.; Bras, R.L. A model of evapotranspiration based on the theory of maximum entropy production. *Water Resour. Res.* **2011**, *47*, W03521. [CrossRef]
54. Yang, C.T. Unit stream power and sediment transport. *J. Hydraul. Div. ASCE* **1972**, *98*, 1805–1826.

55. Fiorentino, M.; Claps, P.; Singh, V.P. An entropy-based morphological analysis of river-basin networks. *Water Resour. Res.* **1993**, *29*, 1215–1224. [CrossRef]
56. Chiu, C.L. Entropy and 2-D velocity distribution in open channels. *J. Hydraul. Eng. ASCE* **1988**, *114*, 738–756. [CrossRef]
57. Cao, S.; Knight, D.W. Design of threshold channels. Hydra 2000. In Proceedings of the 26th IAHR Congress, London, UK, 11–15 September 1995.
58. Cao, S.; Knight, D.W. New Concept of hydraulic geometry of threshold channels. In Proceedings of the 2nd Symposium on the Basic Theory of Sedimentation, Beijing, China, 30 October–2 November 1995.
59. Cao, S.; Knight, D.W. Entropy-based approach of threshold alluvial channels. *J. Hydraul. Res.* **1997**, *35*, 505–524. [CrossRef]
60. Singh, V.P. Entropy theory for movement of moisture in soils. *Water Resour. Res.* **2010**, *46*, 1–12. [CrossRef]
61. Singh, V.P. Tsallis entropy theory for derivation of infiltration equations. *Trans. ASABE* **2010**, *53*, 447–463. [CrossRef]
62. Evans, M.; Hastings, N.; Peacock, B. *Statistical Distributions*; John Wiley & Sons: New York, NY, USA, 2000.
63. Krstanovic, P.F.; Singh, V.P. Evaluation of rainfall networks using entropy: 1. Theoretical development. *Water Resour. Manag.* **1992**, *6*, 279–293. [CrossRef]
64. Krstanovic, P.F.; Singh, V.P. Evaluation of rainfall networks using entropy: 2. Application. *Water Resour. Manag.* **1992**, *6*, 295–314. [CrossRef]
65. Yang, Y.; Burn, D.H. An entropy approach to data collection network design. *J. Hydrol.* **1994**, *157*, 307–324. [CrossRef]
66. Mogheir, Y.; Singh, V.P. Application of information theory to groundwater quality monitoring networks. *Water Resour. Manag.* **2002**, *16*, 37–49. [CrossRef]
67. Perelman, L.; Ostfeld, A.; Salomons, E. Cross entropy multiobjective optimization for water distribution systems design. *Water Resour. Res.* **2008**, *44*, W09413. [CrossRef]
68. Goulter, I.C. Current and future use of systems analysis in water distribution network design. *Civ. Eng. Syst.* **1987**, *4*, 175–184. [CrossRef]
69. Goulter, I.C. Assessing the reliability of water distribution networks using entropy based measures of network redundancy. In *Entropy and Energy Dissipation in Water Resources*; Singh, V.P., Fiorentino, M., Eds.; Kluwer Academic Publishers: Dordrecht, The Netherlands, 1992; pp. 217–238.
70. Walters, G. Optimal design of pipe networks: A review. In Proceedings of the 1st International Conference Computational Water Resources, Rabat, Morocco, 14–18 March 1988; Computational Mechanics Publications: Southampton, UK, 1988; pp. 21–31.
71. Leopold, L.B.; Maddock, T.J. *Hydraulic Geometry of Stream Channels and Some Physiographic Implications*; US Government Printing Office: Washington, DC, USA, 1953; p. 55.
72. Langbein, W.B. Geometry of river channels. *J. Hydraul. Div. ASCE* **1964**, *90*, 301–311.
73. Yang, C.T.; Song, C.C.S.; Woldenberg, M.J. Hydraulic geometry and minimum rate of energy dissipation. *Water Resour. Res.* **1981**, *17*, 1014–1018. [CrossRef]
74. Wolman, M.G. *The Natural Channel of Brandywine Creek, Pennsylvania*; US Government Printing Office: Washington, DC, USA, 1955.
75. Williams, G.P. *Hydraulic Geometry of River Cross-Sections-Theory of Minimum Variance*; US Government Printing Office: Washington, DC, USA, 1978.
76. Wang, J.; Salvucci, G.D.; Bras, R.L. A extremum principle of evaporation. *Water Resour. Res.* **2004**, *40*, W09303. [CrossRef]
77. Cui, H.; Singh, V.P. ON the cumulative distribution function for entropy-based hydrologic modeling. *Trans. ASABE* **2012**, *55*, 429–438. [CrossRef]
78. Singh, V.P.; Luo, H. Derivation of velocity distribution using entropy. In Proceedings of the IAHR Congress, Vancouver, BC, Canada, 9–14 August 2009; pp. 31–38.
79. Singh, V.P.; Byrd, A.; Cui, H. Flow duration curve using entropy theory. *J. Hydrol. Eng.* **2014**, *19*, 1340–1348. [CrossRef]
80. Weijs, S.V.; van de Giesen, N.; Parlange, M.B. HydroZIP: How hydrological knowledge can be used to improve compression of hydrological data. *Entropy* **2013**, *15*, 1289–1310. [CrossRef]
81. Chen, J.; Li, G. Tsallis wavelet entropy and its application in power signal analysis. *Entropy* **2014**, *16*, 3009–3025. [CrossRef]

82. Furuichi, S.; Mitroi-Symeonidis, F.-C.; Symeonidis, E. On some properties of Tsallis hypoentropies and hypodivergences. *Entropy* **2014**, *16*, 5377–5399. [CrossRef]
83. Lenzi, E.K.; da Silva, L.R.; Lenzi, M.K.; dos Santos, M.A.F.; Ribeiro, H.V.; Evangelista, L.R. Intermittent motion, nonlinear diffusion equation and Tsallis formalism. *Entropy* **2017**, *19*, 42. [CrossRef]
84. Evren, A.; Ustaoğlu, E. Measures of qualitative variation in the case of maximum entropy. *Entropy* **2017**, *19*, 204. [CrossRef]
85. Kalogeropoulos, N. The Legendre transform in non-additive thermodynamics and complexity. *Entropy* **2017**, *19*, 298. [CrossRef]

9

Spatio-Temporal Variability of Soil Water Content under Different Crop Covers in Irrigation Districts of Northwest China

Sufen Wang [1,*] **and Vijay P. Singh** [2]

1 Center for Agricultural Water Research in China, China Agricultural University, Beijing 100083, China
2 Department of Biological & Agricultural Engineering, and Zachry Department of Civil Engineering, Texas A & M University, College Station, TX 77843-2117, USA; vsingh@tamu.edu
* Correspondence: wangsuf@cau.edu.cn

Abstract: The relationship between soil water content (SWC) and vegetation, topography, and climatic conditions is critical for developing effective agricultural water management practices and improving agricultural water use efficiency in arid areas. The purpose of this study was to determine how crop cover influenced spatial and temporal variation of soil water. During a study, SWC was measured under maize and wheat for two years in northwest China. Statistical methods and entropy analysis were applied to investigate the spatio-temporal variability of SWC and the interaction between SWC and its influencing factors. The SWC variability changed within the field plot, with the standard deviation reaching a maximum value under intermediate mean SWC in different layers under various conditions (climatic conditions, soil conditions, crop type conditions). The spatial-temporal-distribution of the SWC reflects the variability of precipitation and potential evapotranspiration (ET_0) under different crop covers. The mutual entropy values between SWC and precipitation were similar in two years under wheat cover but were different under maize cover. However, the mutual entropy values at different depths were different under different crop covers. The entropy values changed with SWC following an exponential trend. The informational correlation coefficient (R_0) between the SWC and the precipitation was higher than that between SWC and other factors at different soil depths. Precipitation was the dominant factor controlling the SWC variability, and the crop efficient was the second dominant factor. This study highlights the precipitation is a paramount factor for investigating the spatio-temporal variability of soil water content in Northwest China.

Keywords: spatio-temporal variability; soil water content; entropy; arid region

1. Introduction

Soil water content (SWC) varies both spatially and temporally. Characterizing the spatio-temporal variability of SWC is important for water management in agricultural irrigation [1]. The spatio-temporal variability of SWC is related to a number of factors, such as certain soil properties, topography, vegetation, and climate. The interactions among these factors control the spatio-temporal evolution of soil water content at different scales, including gully scale [2], plot scale [3–5], hillslope scale [6–8], catchment scale [9,10] and regional scale [11–13]. The factors that determine SWC depend on the environment of the area of study and the presence or absence of vegetative canopy [14] and the spatio-temporal variability of SWC affect vegetation coverage and density [15]. Investigations on agricultural watersheds have shown that topography, rainfall, and soil texture have mixed effects on SWC at the watershed and regional scales, and that land cover influences runoff, interception and evapotranspiration processes, and in turn, the soil water dynamics [5,16–19]. The spatial-temporal pattern of SWC can be obtained from measurements carried out in different experimental plots [3,4]. Differences in SWC can be small or significant at different soil depths [20,21].

Because of practical limitations, measurements of SWC for deep soil profiles are time consuming and costly [22]. Hence, only the near surface soil moisture has been investigated [3,4,16,23]. Examining the SWC variability in the upper 30 cm of the soil on a Mediterranean abandoned terrace in north-east Spain [24] showed that the spatial variability of SWC followed a bimodal pattern with increasing soil water content. Only a few studies have taken into account the SWC variability at different soil depths [23,25,26] and develop new high throughput approaches to measure soil water content in the field [27]. Soil water contents vary with depth due to the heterogeneity of environmental factors, such as water and energy input, soil texture [28] and genetic control of root architecture [29].

For investigating the spatio-temporal variability of SWC, both statistical and geostatistical methods have been applied using ground-based and remote sensing data [17,19,24,26]. A few studies have applied the entropy theory to investigate the spatio-temporal variability of SWC and dominant factors [30]. Singh developed an entropy theory for describing the one-dimensional movement of soil moisture in unsaturated soils, and tested the theory using experimental observations reported in the hydrologic literature [31]. Gaur and Mohanty used the Shannon entropy to assess the effect of different physical controls on the spatial mean and variability of soil moisture in Oklahoma and Iowa and found that for most soil moisture conditions [30], soil texture as opposed to vegetation and topography is the dominant physical control at both the point and airborne scales. There is little information on the relationship between SWC and climate conditions or vegetation types in crop growing seasons in arid and semi-arid agricultural areas. This relationship can provide a basis for improving agricultural water use efficiency in arid regions.

Maize (being a C4 plant) and wheat (being a C3 plant) are the predominant crops grown in northwest China, where population density and exploitation of water resources are high [32]. How SWC is controlled in the maize and wheat fields is important for agricultural water management in arid regions in northwest China. To address this question, a field scale experiment was conducted to measure soil water content over canopies in northwest China in 2013 and 2014. The objectives of this study therefore were to: (1) determine how environmental variables alter the spatial and temporal variability of SWC; (2) determine the effect of crop growth stage on SWC at different depths; and (3) quantify the environmental control on the variation of SWC using entropy.

2. Materials and Methods

2.1. Experimental Site and Design

A two-year experiment was conducted in two fields at the Shiyanghe Experimental Station for Water-Saving in Agriculture and Ecology of China Agricultural University, located in Wuwei (Gansu Province of Northwest China, 37°52′ N, 102°50′ E, altitude 1581 m) during the crop growing seasons in 2013 and 2014 (Figure 1). The experimental site is located in a typical continental temperate climate zone, where the mean annual temperature is 8 °C, the annual accumulated temperature (>0 °C) is 3550 °C d^{-1}, the annual precipitation is 164.4 mm, the mean annual pan evaporation is about 2000 mm, the aridity index (the ratio of mean annual evaporation to precipitation) is 15–25, the average annual sunshine duration is 3000 h and frost-free days number 150 d. The groundwater table is below 40–50 m [33].

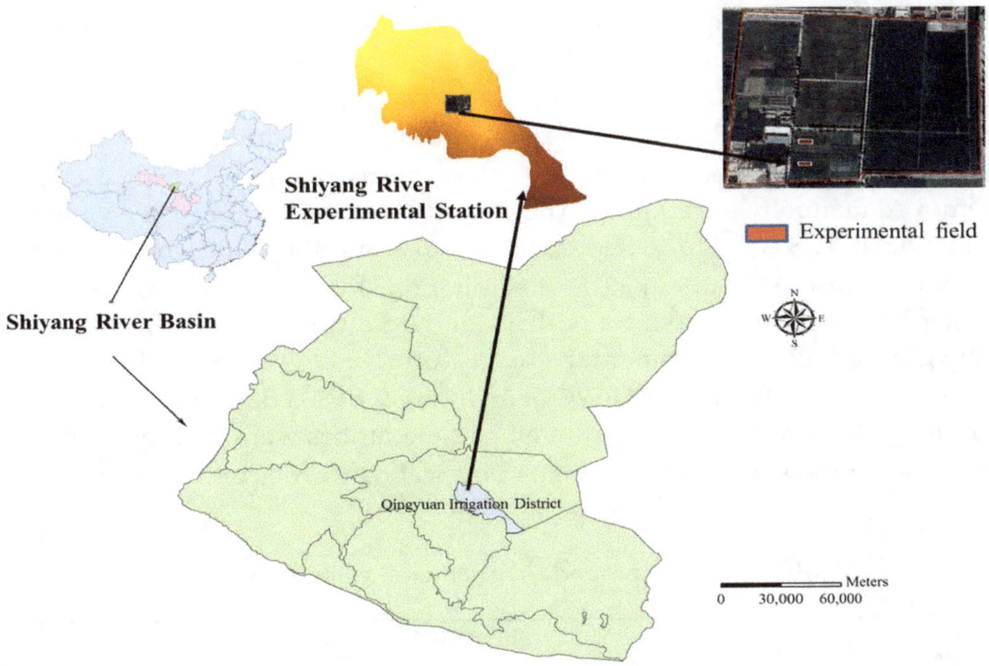

Figure 1. Location of the study area.

The two study sites (field I and field II) have a similar area of about 40 m long and 12 m wide. The SWC (%cm^3cm^{-3}) was monitored under two different crops at an interval of 10 cm providing estimates covering 0 to 120 cm depth. A total of 46 sample points for the study site in 2013 (Figure 2) and 24 sample points in 2014 (Figure 3) were monitored.

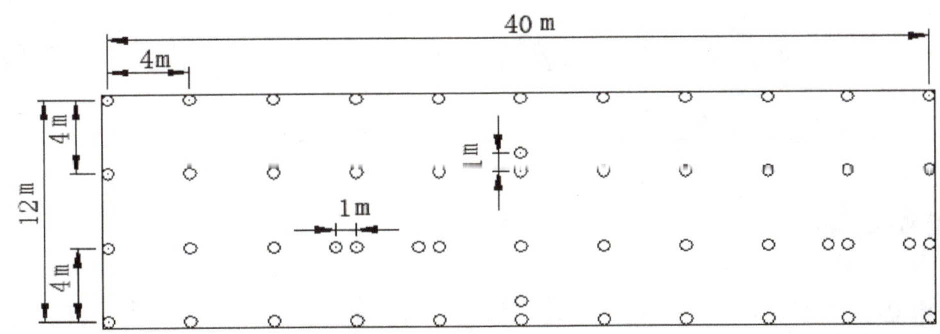

Figure 2. Layout of soil water content measuring probes in the experiment field under maize and wheat covers in 2013.

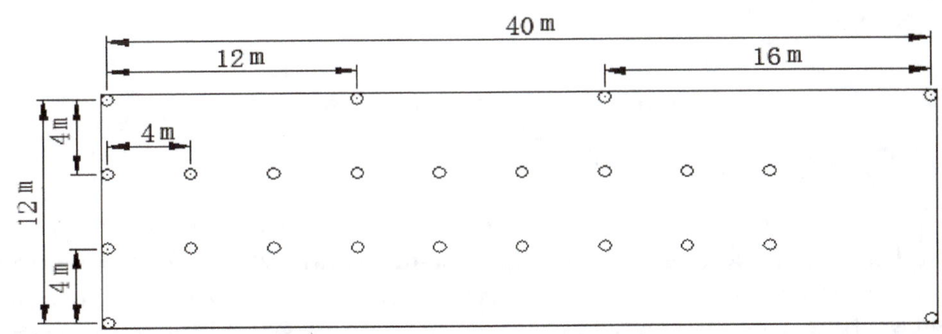

Figure 3. Layout of soil water content measuring probes in the experiment field under maize and wheat covers in 2014.

Spring maize and spring wheat were used in crop rotation, that is, the plot on which wheat was planted in 2013 was planted with maize in 2014. Maize seed was sown through 5.0 cm diameter holes, with a row spacing of 50 cm and a plant spacing of 23.8 cm. Spring wheat was sown in holes of 3.0 cm diameter, with a row spacing of 15 cm and a plant spacing of 10 cm for the two years using a hole sowing machine. Spring wheat was sown with 10–15 seeds in each hole. During each growing season, weeds were removed manually as required. The observation periods for maize were from 14 May to 17 September in 2013 and 28 April to 18 September in 2014; and those for wheat were from 12 May to 27 July in 2013 and 7 April to 23 July in 2014. Precipitation, solar radiation, air temperature, relative humidity and wind speed were measured with a standard automatic weather station at a height of 2.0 m above the ground at the experimental site. A portable Diviner 2000 device (Sentek Pty Ltd., Stepney, Australia) was used to measure SWC at an interval of 3–5 days. Soil samples were collected before and after irrigation events and after rainfall and the probes were calibrated against gravimetric samples for the observation sites. Table 1 lists the physical properties of the soil profile.

Table 1. Physical properties of soil profile at different sites.

Field Site	Soil Depth (cm)	Soil Particle Content (g/g)			Soil Type
		Sand	Silt	Clay	
I	10	0.629	0.268	0.103	sandy loam
	20	0.649	0.251	0.0996	sandy loam
	40	0.656	0.241	0.103	sandy loam
	60	0.443	0.412	0.146	loam
	80	0.653	0.260	0.087	sandy loam
	100	0.931	0.052	0.017	sandy soil
	120	0.616	0.301	0.083	sandy loam
II	10	0.620	0.271	0.109	sandy loam
	20	0.663	0.243	0.095	sandy loam
	40	0.531	0.343	0.127	sandy loam
	60	0.518	0.349	0.134	sandy loam
	80	0.482	0.401	0.117	loam
	100	0.505	0.362	0.133	sandy loam
	120	0.706	0.219	0.074	sandy loam

2.1.1. Statistical Analysis

For each soil depth the coefficient of variation (CV_i) was used for spatial analysis, and temporal standard deviation ($STDV_i$) for temporal analysis. The CV_i and the $STDV_i$ were calculated as:

$$CV_i = \frac{S_i}{\overline{SWC_i}} \tag{1}$$

$$STDV_i = \sqrt{\frac{1}{N}\sum_{i=1}^{n}(SWC_i - \overline{SWC_i})} \tag{2}$$

where S_i is mean squared error. $\overline{SWC_i}$ is the average soil water content for the each soil depth, i represents the soil depth. Heterogeneity was considered weak when $CV \leq 10\%$, moderate when $10\% < CV < 100\%$, and strong when $CV \geq 100\%$ [34]. Descriptive statistics of soil water content under different crop covers based on 10 cm-steps are given in Tables 2 and 3. The temporal average versus the standard deviation of SWC in different soil layers under maize and wheat cover in 2013 and 2014 are shown in Figures 4 and 5, respectively.

Table 2. Statistics of soil water content under different land covers in 2013.

Depth (cm)	Land Cover	Field Site	Minimum ($\%cm^3cm^{-3}$)	Maximum ($\%cm^3cm^{-3}$)	Mean ($\%cm^3cm^{-3}$)	Std. Deviation ($\%cm^3cm^{-3}$)	*CV* (%)
0–10	maize	I	0.62	32.19	11.86	5.30	44.65
	wheat	II	2.13	32.47	15.17	7.39	58.70
10–20	maize	I	1.53	31.94	15.44	4.71	30.53
	wheat	II	3.29	32.73	17.01	6.46	37.98
20–40	maize	I	3.63	36.39	24.12	4.35	18.10
	wheat	II	7.42	30.00	20.66	4.30	20.81
40–100	maize	I	8.65	41.84	28.52	6.41	22.49
	wheat	II	12.86	34.31	22.33	4.41	19.75
100–120	maize	I	8.19	35.61	20.39	5.75	28.20
	wheat	II	10.48	37.89	23.88	5.81	24.33

Table 3. Statistics of soil water content under different land covers in 2014.

Depth (cm)	Land Cover	Field Site	Minimum ($\%cm^3cm^{-3}$)	Maximum ($\%cm^3cm^{-3}$)	Mean ($\%cm^3cm^{-3}$)	Std. Deviation ($\%cm^3cm^{-3}$)	*CV* (%)
0–10	maize	II	4.76	31.59	16.74	5.03	30.05
	wheat	I	6.46	29.09	16.66	5.33	31.99
10–20	maize	II	8.51	31.94	18.69	4.09	21.88
	wheat	I	10.09	30.81	20.74	4.40	21.22
20–40	maize	II	5.14	36.39	32.23	5.15	15.80
	wheat	I	12.76	33.66	22.71	3.46	15.24
40–100	maize	II	4.46	33.16	23.04	5.45	23.66
	wheat	I	13.72	35.89	25.48	4.75	18.64
100–120	maize	II	4.78	38.60	26.79	5.97	22.84
	wheat	I	8.37	30.90	18.33	5.22	28.48

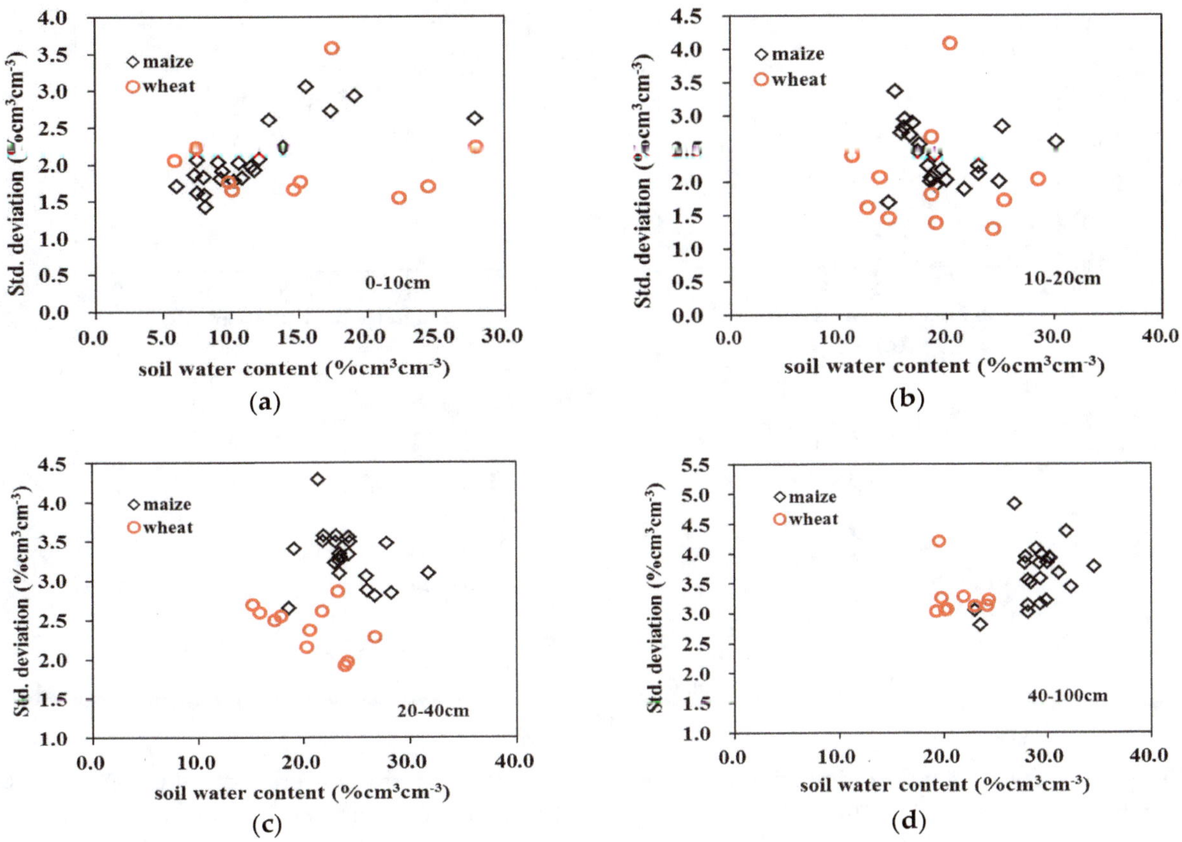

Figure 4. *Cont.*

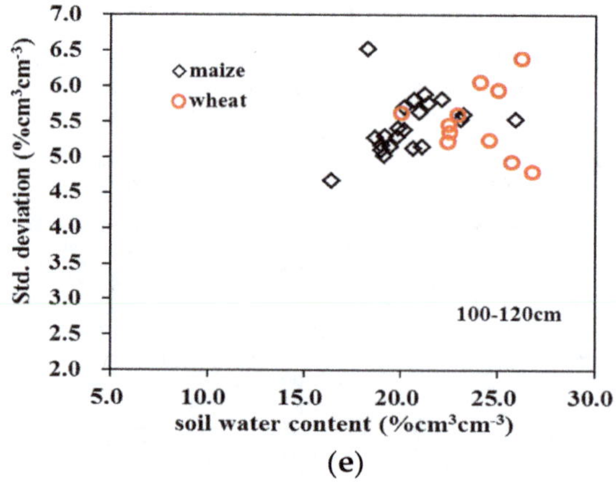

Figure 4. Temporal standard deviation (*STDV*) in time versus temporally averaged soil water content for different soil depths in 2013. *STDV* values for (**a**) 0–10 cm, (**b**) 10–20 cm, (**c**) 20–40 cm, (**d**) 40–100 cm and (**e**) 100–120 cm soil depths.

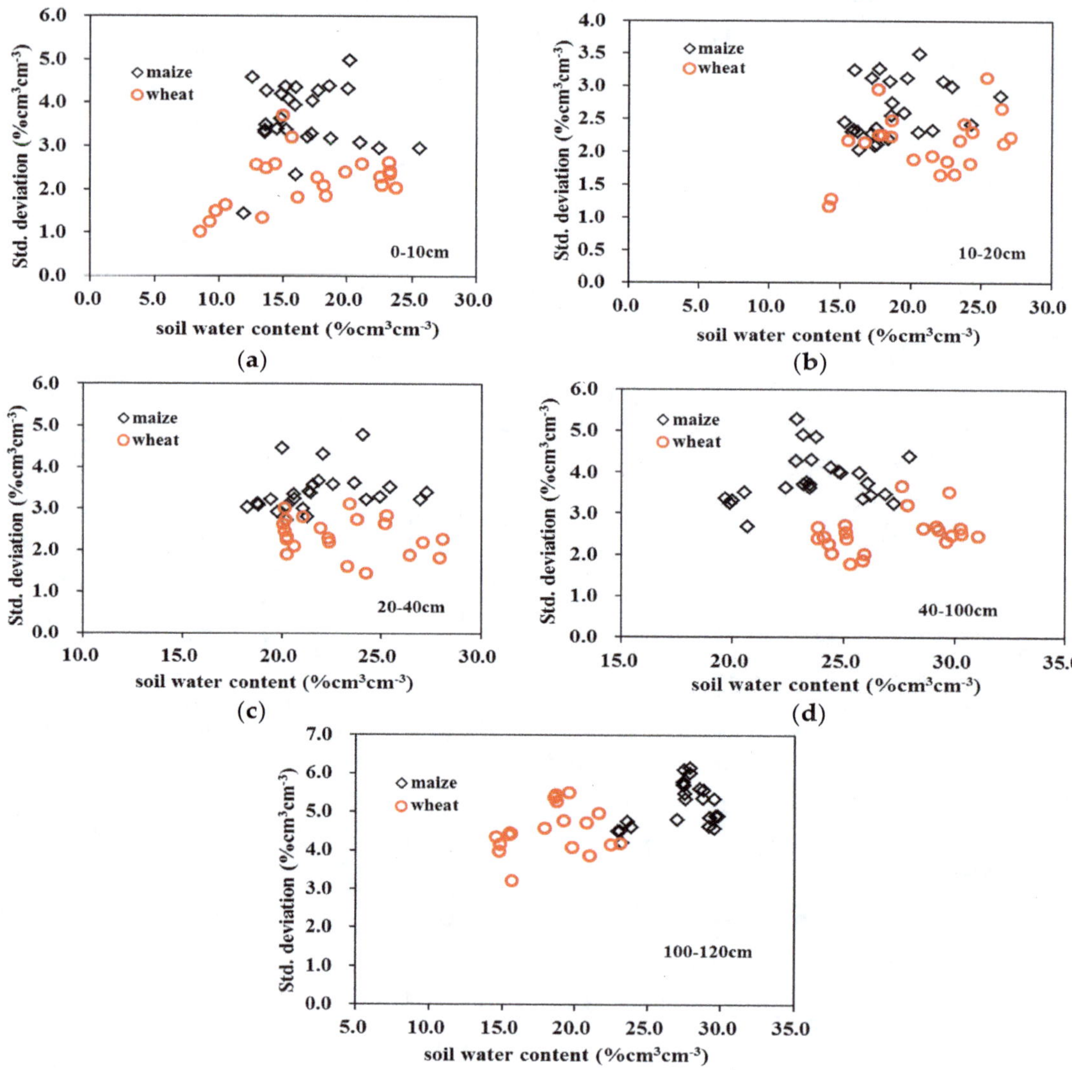

Figure 5. Temporal standard deviation (*STDV*) in time versus temporally averaged soil water content for different soil depths in 2014. *STDV* values for (**a**) 0–10 cm, (**b**) 10–20 cm, (**c**) 20–40 cm, (**d**) 40–100 cm and (**e**) 100–120 cm soil depths.

2.1.2. Entropy

In this study, the Shannon entropy [35,36] was used to investigate the spatial variability and dominant controls on the variability of SWC during the crop growing period. Entropy has an advantage in that it is capable of incorporating the effect of dependent or independent factors on the SWC spatial distribution, and can be used for short or long data sets. Entropy (H) is a statistical quantity representing a measure of information that may be extracted from a system or analogously the uncertainty that the system comprises. For a probability distribution $P = \{p_1, p_2, \cdots, p_n\}$, where $p_1, p_2, \cdots, p_n$ are the probabilities of N outcomes (x_i, $I = 1, 2, \ldots, N$) of a random variable X or a random experiment, Shannon defined a measure H as a function of probabilities as [35]:

$$H(p_1, p_2, \cdots, p_N) = -\sum_{i=1}^{N} p_i \log p_i \qquad \sum_{i=1}^{N} p_i = 1 \tag{3}$$

According to the frequency histograms of SWC for each soil depth constructed with empirical frequencies, a probability p_i is assigned to each bin and calculated. We substituted p_i in Equation (3) to find out the marginal entropies.

For two or more independent random variables X and Y, their respective entropy values may be added. However, if there is dependence between the random variables, then the dependence can be accounted for through mutual information $T(X,Y)$, which is the amount of information common to both the random variables. Consider two simultaneous experiments whose outcomes are represented by X and Y. The mutual entropy of X and Y, denoted as $T(X, Y)$, defines the amount of uncertainty reduced in X when Y is known. It equals the difference between the sum of two marginal entropies and the total entropy:

$$T(X,Y) = H(X) + H(Y) - H(X,Y) \tag{4}$$

Mutual information has an advantage over other measures of information, as it can provide a quantitative measure of a description of the relationship among variables based on their information transmission characteristics. Larger values of T correspond to greater amounts of information transferred. Mutual information is superior to the Pearson correlation coefficient, since it captures both linear and nonlinear dependence, whereas the Pearson correlation coefficient is only suitable for linear relationships, or more generally, for spherical and elliptical dependence structures [37–39]. In reality, the spatial patterns of SWC may be non-stationary and the related processes may be nonlinear [22].

2.1.3. Calculation of Marginal Entropy and Mutual Entropy for Soil Water Content

In this study, the random variables under consideration are the SWC, crop coefficients (K_C), reference crop evapotranspiration (ET_0), and precipitation (P). The highest mutual entropy between soil water content and other factors can be considered to be the most dominant factor that influences the SWC variability.

We arranged SWC values $SWC_{i,d,n}$, where i represents the soil depth, d represents the days (1, 2, ..., d), and n represents the number of soil water content values (1, 2, ..., n), P is daily precipitation, and ET_0 is the reference crop evapotranspiration. In this study, ET_0 was calculated according to the FAO (Food and Agriculture Organization) Penman–Monteith method using daily observed climate data [40,41]. K_C is the daily crop coefficients and was calculated by the single crop coefficient method Equation (5) [40,42–45] as:

$$K_C = \frac{ET_C}{ET_0} \tag{5}$$

where ET_C is the crop evapotranspiration. ET_C was calculated according to the FAO Penman–Monteith method [46].

The frequency histograms of $SWC_{i,d,n}$ were constructed with empirical frequencies obtained from observed the data with no zero counts [47], and hence the Shannon entropy H of the random variable

$SWC_{i,d}$ from the corresponding observed counts $SWC_{i,d,n}$. The marginal entropy of SWC at different soil depths was computed from Equation (2). The mutual entropy between $SWC_{i,d}$ and ET_0, $SWC_{i,d}$ and K_C, $SWC_{i,d}$ and P were calculated using Equation (3).

2.1.4. Informational Correlation Coefficient

The informational correlation coefficient R_0 is a measure of transferable information between random variables X and Y and measures their mutual dependence and does not assume any type of relationship between them. It is expressed in terms of transinformation as:

$$R_0 = \sqrt{1 - \exp(-2T_0)} \tag{6}$$

where T_0 is the transinformation or mutual information representing the upper limit of transferable information between two variables [38]. In this study, three groups were identified: $SWC_{i,d}$ and ET_0, $SWC_{i,d}$ and K_C, $SWC_{i,d}$ and P. From results one can infer the main dominating factors for the SWC variability.

3. Results and Discussion

3.1. Temporal Variation of Soil Water Content under Different Crops

The two-year sampling in this study can be considered as representative of the usual conditions in Northwest China. The accumulated rainfall during the maize growing period was 54.4 mm in 2013 and 196.6 mm in 2014, respectively, and the accumulated rainfall during the wheat growing period was 29.6 mm in 2013 and 128.8 mm in 2014, respectively. Due to the meteorological forcing (precipitation and ET_0), the SWC observed during the study period showed a succession of increases and decreases in response to the succession of precipitation events (Figures 6a and 7a). This led to a steady increase in SWC at different soil depths from June to mid-July for the maize cover. The mean SWC at different depths increased when precipitation took place. The SWC of layer (20–40 cm) can represent the average soil water content (0–120 cm) under different crop covers. During the growing seasons, the SWC changed only slightly during the early growing period (April to mid-May) under different crop covers (Figures 6 and 7) due to little water demand during this stage. The SWC in the 0–120 cm layer varied greatly over the crop growing period. The temporal variability of SWC changed slightly with the increase of soil depth in the two years. Tables 2 and 3 show that the SWC at different soil depths changed typically with crop and soil conditions.

During the observation growing period of maize (from May to late September) in 2013 (dry year) (Table 2). The minimum SWC varied from 0.62%cm^3cm^{-3} to 8.65%cm^3cm^{-3} and the maximum SWC varied from 30%cm^3cm^{-3} to 41.84%cm^3cm^{-3} for different soil depths. Corresponding to the same soil depths, the ranges were 31.57%cm^3cm^{-3}, 30.41%cm^3cm^{-3}, 32.76%cm^3cm^{-3}, 33.19%cm^3cm^{-3} and 27.42%cm^3cm^{-3}, respectively. During the observation period, the range for the 40–100 cm was greater due to the soil texture, where the fraction of sand particles soil in the 40–100 cm layer range from 0.443 to 0.931 (Table 1).

For wheat growing period (from April to late July), The minimum SWC varied from 2.13%cm^3cm^{-3} to 12.86%cm^3cm^{-3} and the maximum SWC varied from 30%cm^3cm^{-3} to 37.79%cm^3cm^{-3} for different soil depths. Corresponding to the same soil depths, the ranges were 30.34%cm^3cm^{-3}, 29.44%cm^3cm^{-3}, 22.58%cm^3cm^{-3}, 21.45%cm^3cm^{-3} and 27.41%cm^3cm^{-3}, respectively. The range of SWC under wheat cover was greater for the 0–10 cm soil layer (Table 2). During the growing period of maize in 2014 (wet year), the minimum SWC varied from 4.46%cm^3cm^{-3} to 8.51%cm^3cm^{-3} and the maximum SWC varied from 30%cm^3cm^{-3} to 38.6%cm^3cm^{-3} for different soil depths. Corresponding to the same soil depths, the ranges were 26.83%cm^3cm^{-3}, 23.43%cm^3cm^{-3}, 31.25%cm^3cm^{-3}, 28.7%cm^3cm^{-3} and 33.82%cm^3cm^{-3}, respectively. The range of SWC under maize cover was greater for the 100–120 cm soil layer. In this year, maize was arranged in field II. Table 1 shows that the fraction of sand particles in 100–120 cm layer ranged from 0.507 to 0.706. Two years of results of SWC under maize cover showed that soil texture had a direct influence on the SWC variability.

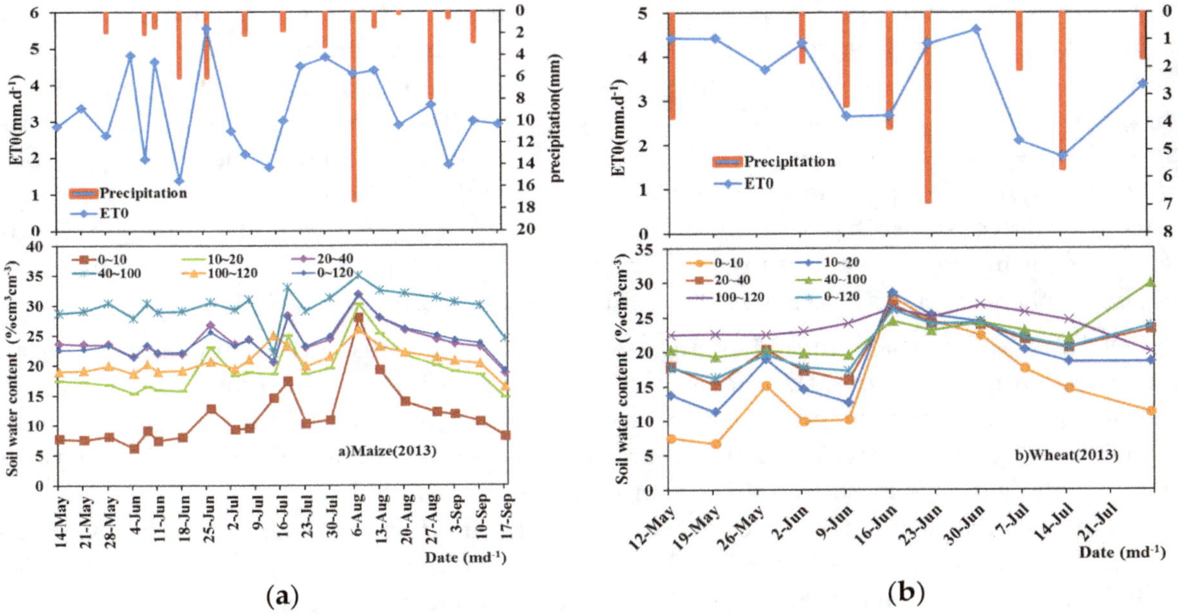

Figure 6. Temporal variation of soil water content under maize cover and wheat cover in 2013. ET_0, P and SWC values for (**a**) maize and (**b**) wheat in 2013.

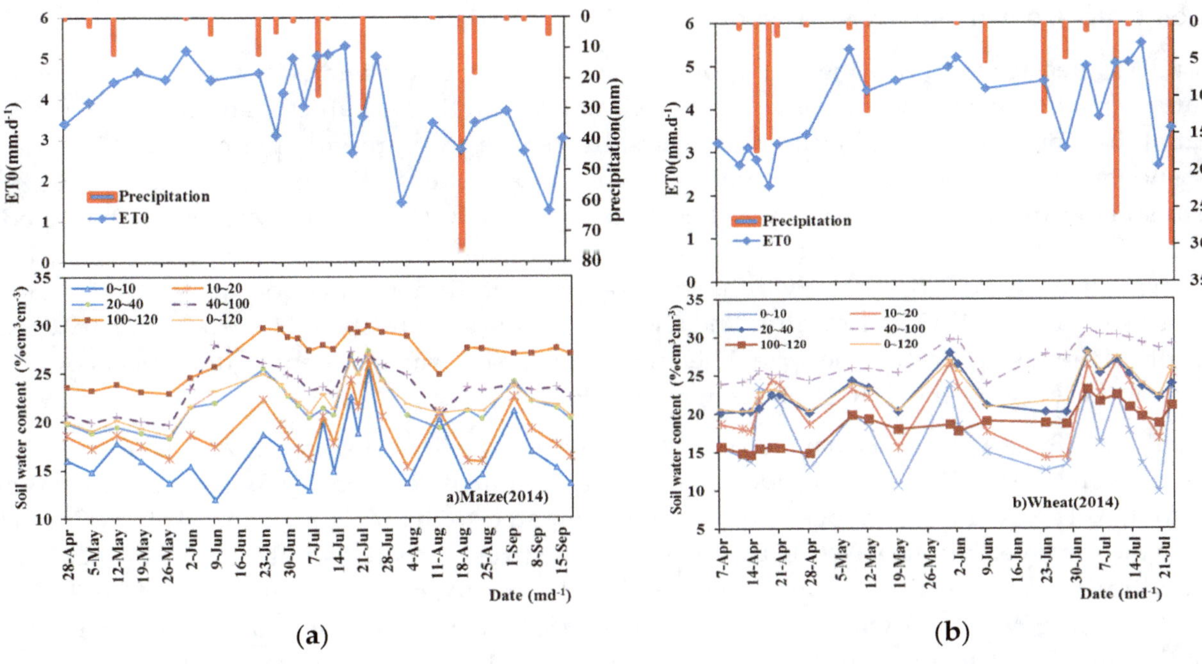

Figure 7. Temporal variation of soil water content under maize cover and wheat cover in 2014. ET_0, P and SWC values for (**a**) maize and (**b**) wheat in 2014.

During the observation period under wheat cover, the minimum SWC ranged from 6.46%cm^{3}cm^{-3} to 13.72%cm^{3}cm^{-3} and the maximum SWC ranged from 29.09%cm^{3}cm^{-3} to 35.89%cm^{3}cm^{-3} for different soil depths. Corresponding to the same soil depths, the ranges were 22.63%cm^{3}cm^{-3}, 20.72%cm^{3}cm^{-3}, 20.9%cm^{3}cm^{-3}, 22.17%cm^{3}cm^{-3} and 22.53%cm^{3}cm^{-3}, respectively. The range of SWC under wheat cover was greater for the 0–10 cm soil layer. The result was the same as that of 2013 in spite of the field plot in two years. Table 1 shows that the soil texture of 0–10 cm in field I was similar to that in field II. The fraction of sand particles in the 0–10 cm layer was 0.6291 in field I, and 0.62 in field II.

The temporal variability of SWC in different layers under different crop covers displayed the same trends. It can be seen that the range under wheat cover in different field plots in the growing seasons in the surface layer was higher than in other layers, because the top layer was more sensitive to the environment. The fluctuation of the upper soil layer (0–10 cm) under wheat was greater in the observation period in 2013 and 2014. This result is similar to what found by Odhiambo and Bomke (2007) with no differences in soil water content in depths 20–40 and 40–60 cm among cropping treatments in south coastal British Columbia. However, the range of deep soil layer (≥40 cm) under maize was greater in the two year study period due to different soil textures. The fraction of sand particles above 40 cm layer was from 0.443 to 0.931 in field I, and 0.482 to 0.706 in field II.

Figures 4 and 5 show the temporal average versus the temporal standard deviation of soil water content of maize and wheat in 2013 and in 2014. In contrast with the top layers, a higher *STDV* was generally found for the deeper layers under two crops. The *STDV* for deep layers changed from 2.5%cm^3cm^{-3} to 6.5%cm^3cm^{-3} in two years. The temporal standard deviation (*STDV*) of SWC increased with increasing depth. In the 20–40 cm layer, the value of wheat was smaller than that of maize. This implies that roots in this depth had a significant effect. Temporally averaged values in deeper layers (40–100 cm and 100–120 cm) had higher standard deviations. The result is similar to that of [48] that SWC in crop covered deep layer (80–100 cm) exhibited a large degree of change, which was affected by crop root water uptake in the same region.

3.2. Spatial Variation of Soil Water Content under Different Crops

3.2.1. Soil Water Content Analysis Based on Coefficient of Variation

For SWC measurements during crop growth, the coefficient of variation (*CV*) was used to evaluate the spatial variation in soil water content in the experimental fields. Tables 2 and 3 show the *CV* of the spatial distribution of SWC in different layers during the crop growing period in 2013 and 2014. Table 2 shows the *CV* of soil water content under the maize cover varied from 18.10% to 44.65% for different soil depths in 2013. The *CV* was between 10% and 100%, so that SWC was considered to exhibit moderate heterogeneity. The *CV* of SWC was the highest in the top layer due to the SWC in the top layer was strongly spatially influenced by precipitation and evapotranspiration, whereas for deeper soil layers, SWC was steady. The *CV* of SWC under wheat cover varied from 19.75% to 58.70% for different soil depths. The spatial heterogeneity of soil water content under wheat cover was moderate and the *CV* for the 0–10 cm layer was the highest.

From Table 3, it can be seen that *CV* of SWC under maize cover changed from 15.80% to 30.05%, for different soil layers. SWC was considered to have moderate heterogeneity due to *CV* was between 10% and 100%. The extent of heterogeneity was less compared to that in 2013. The *CV* of SWC was the highest in the top layer, because SWC in the top layer was strongly spatially influenced by precipitation and evapotranspiration during the growing period, whereas for deeper layers, SWC changed more slowly. The *CV* of SWC under wheat cover varied from 15.24% to 31.99% for different soil layers. It can be seen that the *CV* range was similar to that under maize. The spatial heterogeneity of SWC under wheat cover was moderate and *CV* for the 0–10 cm layer was the highest.

Table 2 showed that the *CV* (from 18.10% to 44.65%) of the spatial distribution of SWC in different layers under maize cover was less than that (from 19.75% to 58.70%) under wheat cover during 2013. In 2014, it can be seen from Table 3 that *CV* (from 15.80% to 30.05%) of the spatial distribution of SWC in different layers under maize cover was similar to that (from 15.24% to 31.99%) under wheat cover. From Figure 8, it can be seen that *CV* of spatial SWC decreased with increasing mean SWC under different crop covers, because the evapotranspiration rate decreased with decreasing soil water content

(Figure 5). Results for *CV* of the spatial SWC can be found in other studies [3,23], and the SWC varied with depth due to the heterogeneity of environmental factors [28,48].

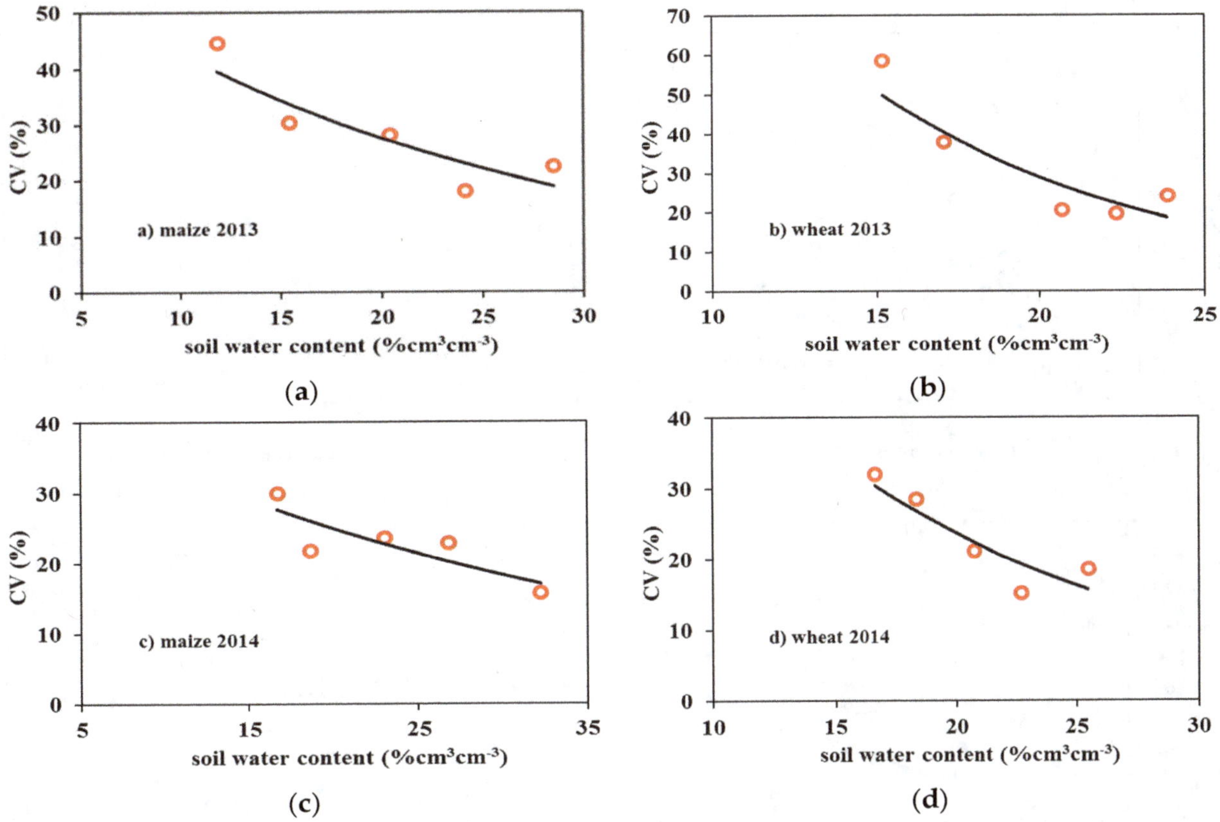

Figure 8. Mean soil water content versus *CV* for maize and wheat in 2013–2014. *CV* values for (**a**) maize, (**b**) wheat cover in 2013, *CV* values for (**c**) maize and (**d**) wheat cover in 2014.

3.2.2. Soil Water Content Analysis Using Entropy

The marginal entropy values of daily soil water content were plotted against the daily mean soil water content for different soil depths under different crop covers in 2013 (Figure 9) and 2014 (Figure 10). Figure 9 shows that entropy was maximum when SWC was the maximum value, and entropy was distributed in the intermediate range under the maize cover in field I in 2013. In the upper soil layer (0–10 cm and 10–20 cm), entropy changed widely during the maize growing period. Entropy was the maximum in the intermediate range under wheat cover for field II in 2013. In the upper soil layer (0–10 cm and 10–20 cm), entropy changed widely during the maize growing period, but the entropy value for the deeper layer changed not so much. The soil texture and crop growth patterns led to the difference under the same weather conditions [49]. In different layers under different crop covers, the marginal entropy of spatial variability and range increased with increasing SWC. In this study, data of marginal entropy and mean SWC were fitted with an exponential model separately for the 0 to 10, 10–20, 20–40, 40–100 and 100–120 cm soil layers, and the results of fitting are shown in Figures 9 and 10.

With all data points included in the analysis, the relationship between marginal entropy and mean SWC apparently was different for different soil layers under the maize cover and the wheat cover in 2013–2014. Entropy changed with soil water content as an exponential model, in which the model parameters indicated the magnitude of the proportional effect on the variability and the dependence of variability on the mean SWC, respectively. Figures 9 and 10 show that the absolute values of both parameters for different soil layers under wheat cover were larger than those of the soil layers under maize cover in 2013–2014. It seems that the soil layer had a relatively greater overall spatial variability under wheat cover and entropy increased relatively faster for the soil layers with increasing SWC.

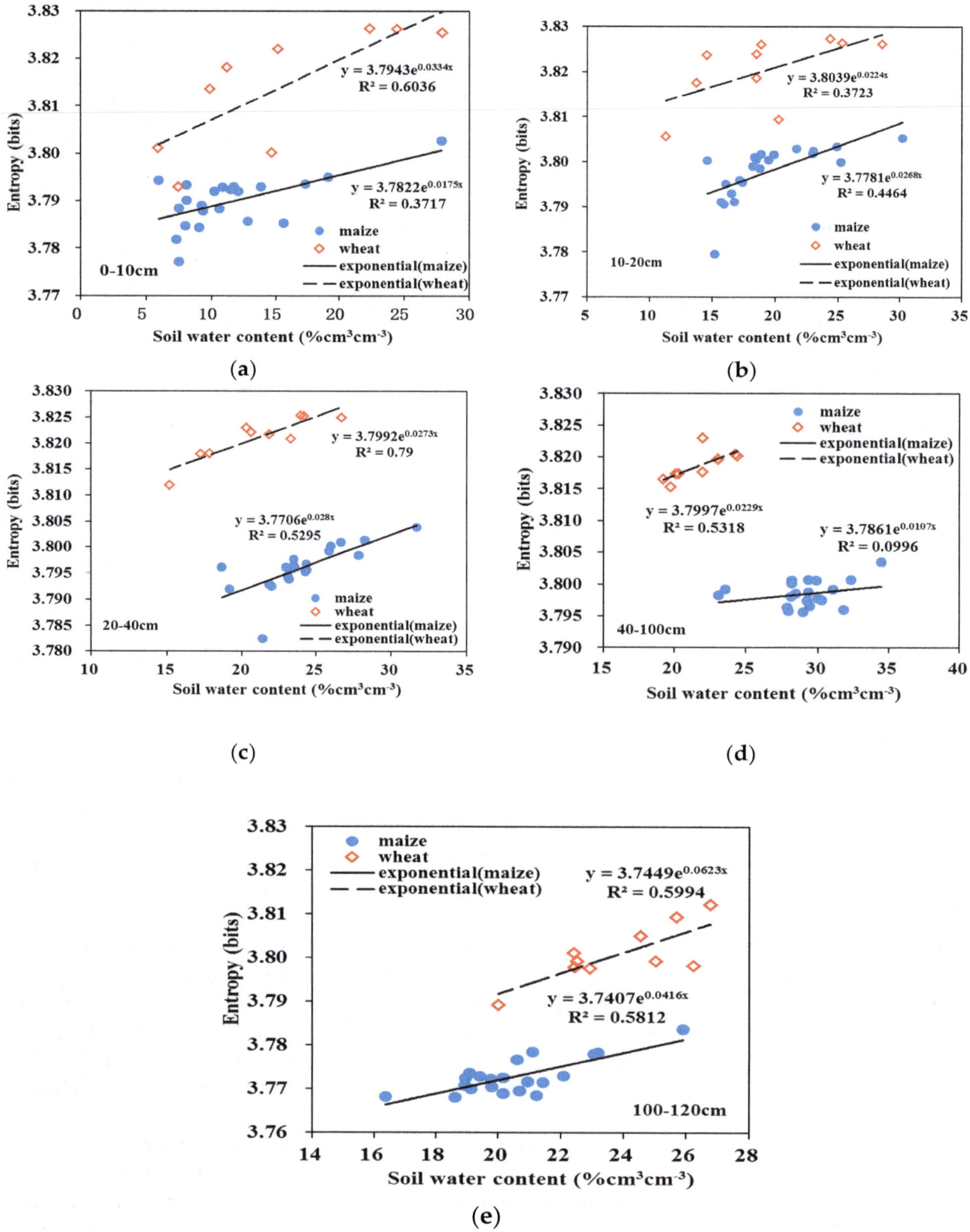

Figure 9. Mean soil water content versus entropy for maize and wheat in 2013. Entropy values for (**a**) 0–10 cm, (**b**) 10–20 cm, (**c**) 20–40 cm, (**d**) 40–100 cm and (**e**) 100–120 cm soil depths.

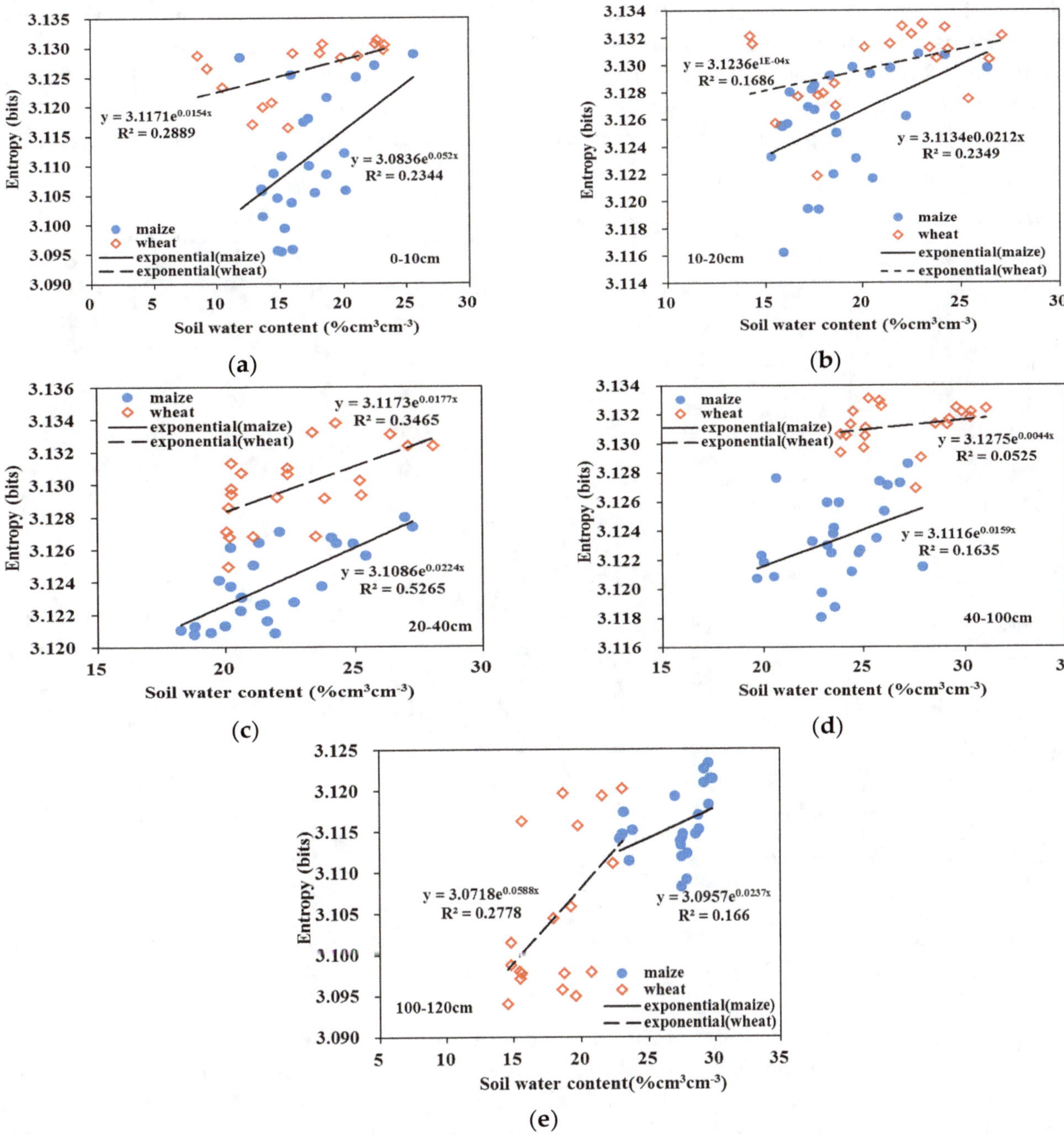

Figure 10. Mean soil water content versus entropy under maize and wheat cover in 2014. Entropy values for (**a**) 0–10 cm, (**b**) 10–20 cm, (**c**) 20–40 cm, (**d**) 40–100 cm and (**e**) 100–120 cm soil depths.

3.3. Calculation of Mutual Entropy for Influencing Factors of Soil Water Content

From Figure 11, it can be seen that a larger value of mutual entropy corresponds to a greater amount of information transferred because the value of mutual entropy represents the amount of information transmitted. In 2013 and 2014, the mutual entropy value of SWC and precipitation was much higher than that of ET_0 and SWC, SWC and K_C under different field plots, which is consistent with previous studies [50]. Crop type was the second contributor that affects the soil water content. Under wheat cover, it was observed that even though the marginal entropy values in 2013 were higher than those in year 2014 (Figures 9 and 10), the mutual entropy values between SWC and precipitation were similar to the value in 2014 (Figure 11). The results reflect that the influence of precipitation to SWC is consistent in spite of the amount. The results reflects the entropy theory dose well in communicating with nature, it helps better understand soil water content and its influence factors in natural systems. The mutual entropy values for different layers were different under wheat cover in different years. For the top soil layers (0–10 cm and 10–20 cm) and observed deepest layer (100–120 cm), the value of mutual entropy between SWC and precipitation (Figure 11b) in dry year (2013) was much higher than that (Figure 11c) in wet year (2014); results were similar to the value between SWC and ET_0.

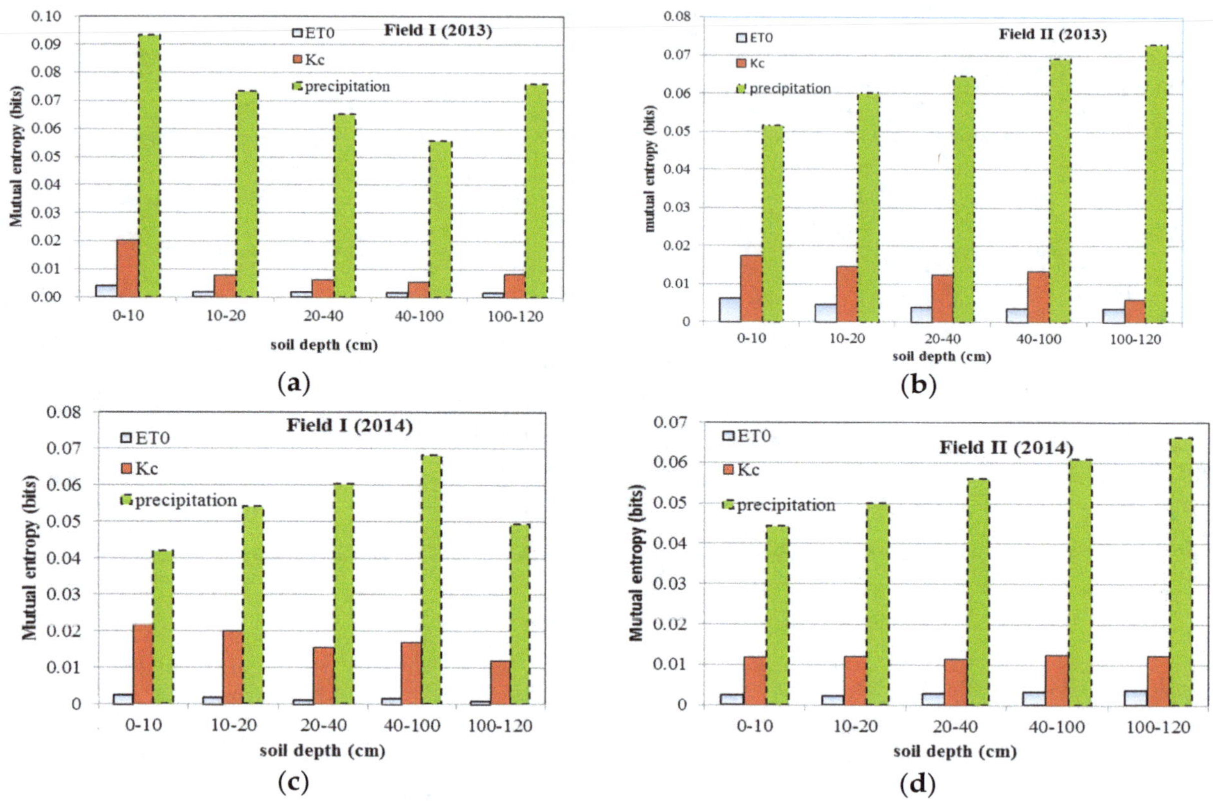

Figure 11. Mutual entropy under different crop covers in 2013–2014. Mutual entropy values for (**a**) Field I and (**b**) Field II in 2013, mutual entropy values for (**c**) Field I and (**d**) Field II in 2014.

Figures 9 and 10 show that the marginal entropy values in 2013 were higher than those in 2014 under maize cover. The mutual entropy values between SWC and precipitation were from 0.055 to 0.091 in 2013 (Figure 11a), which were also bit higher than those (0.045 to 0.067) in 2014 (Figure 11d). The mutual entropy values for different layers were different under wheat cover in the two year period. For the top soil layers (0–10 cm and 10–20 cm), the value of mutual entropy between SWC and precipitation (Figure 11a) in the dry year (2013) was much higher than that (Figure 11d) in the wet year (2014). The value between SWC and ET_0 in 2013 was similar to that in 2014. This result is different from that under wheat cover. The difference resulted from the crop types, wheat (being a C3 plant) extracts more water than maize (being a C4 plant).

Figure 12 shows that informational correlation coefficient (R_0) between SWC and other factors for different soil layers. The value of R_0 between SWC and precipitation was from 0.35 to 0.41 under maize cover (Figure 12a) and from 0.31 to 0.37 under wheat cover (Figure 12b) in 2013. The R_0 value between SWC and precipitation was from 0.29 to 0.36 under maize cover (Figure 12c) and from 0.28 to 0.36 under wheat cover (Figure 12d) in 2014. For the 0–10 cm soil layer, the R_0 value between SWC and precipitation under maize cover was 0.41 and 0.29 in 2013 and 2014, respectively, which was bit higher than the value (0.31 in 2013, 0.28 in 2014) under wheat cover. For the 10–20 cm soil layer, the R_0 value between SWC and precipitation under maize cover was 0.37 and 0.31 in 2013 and 2014, respectively, and the value under wheat cover was 0.34 and 0.32 in 2013 and 2014, respectively. The result was similar to that for the 0–10 cm soil layer.

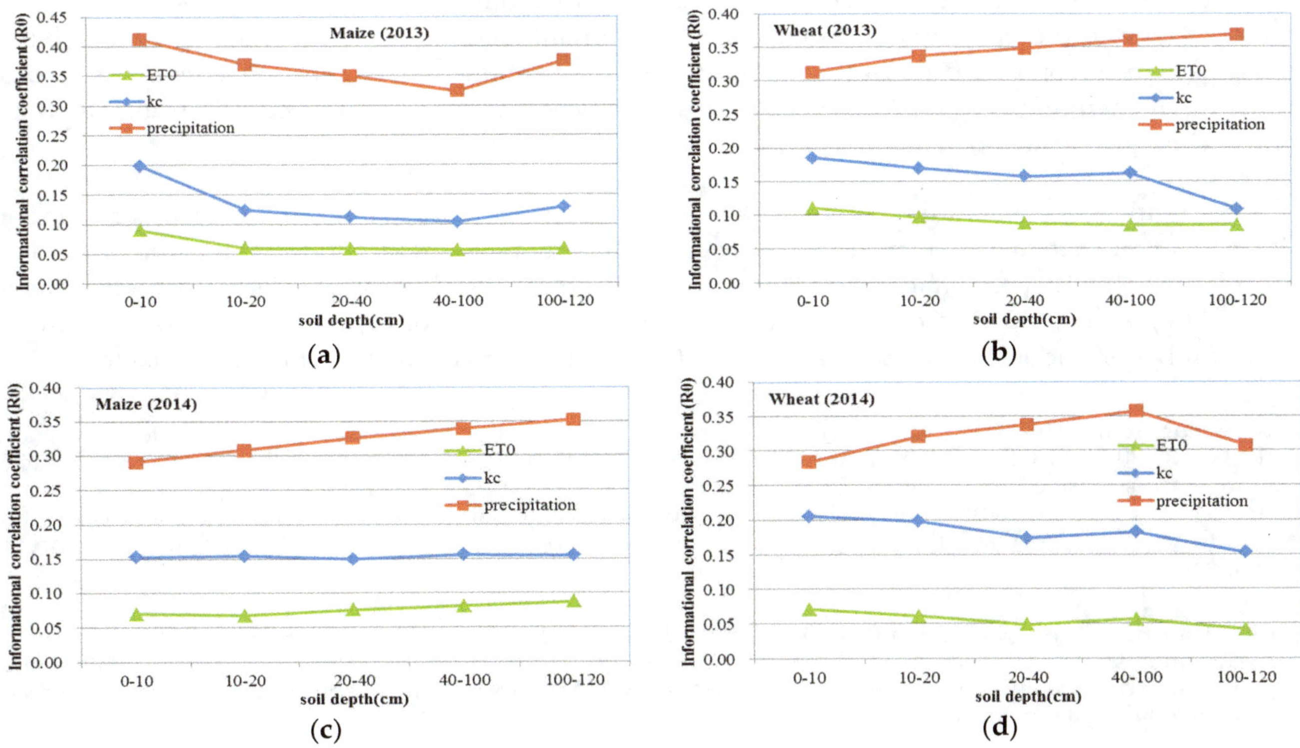

Figure 12. Informational correlation coefficient (R_0) under different crop covers in 2013–2014. R_0 values for (**a**) maize and (**b**) wheat in 2013, R_0 values for (**c**) maize and (**d**) wheat in 2014.

From Figure 12, it can be seen that the transferable information of precipitation for SWC was more than ET_0 and K_C. The value between SWC and K_C was the smallest for different soil layers. The R_0 value was similar under different crop covers in the two years. The influence of K_C on SWC in different layers was steady under different cover crops in the two years. The R_0 value between SWC and K_C under maize cover was from 0.10 to 0.19 in 2013 and from 0.152 to 0.156 in 2014. For wheat cover, the R_0 value between SWC and K_C was from 0.09 to 0.11 and from 0.04 to 0.06 in 2013 and 2014, respectively. In spite of crop type, climate was the dominant factor that affected the SWC. In northwest China, the precipitation is prior to ET_0. So, the irrigation forecast can be according to the surface soil water content and weather conditions for the certain crop cover.

4. Conclusions

In this study, entropy was employed, which was proved as an efficient tool for analyzing the spatio-temporal patterns of SWC and the related environmental conditions. In order to examine the spatial-temporal variability of SWC over the growing season in relation to the environmental variables and to analyze how the influence of these factors vary with the crop growth in different soil depths, both statistical and entropy methods were used to quantify the environmental control on the seasonal variation of SWC based on two-year observations in a field site of Shiyang River basin in Northwest China.

From the distribution of SWC in time, it was found that the temporal variability of SWC in different layers under different cover crops displays the same change trends. The temporal variability of SWC changed slightly with increase of soil depth in the two years. The *CV* of spatial SWC decreased with increasing mean SWC under different covers. The marginal entropy of spatial variability and range increased with increasing SWC for different layers under different crop covers. Entropy changed with SWC following an exponential model.

Mutual entropy results showed that precipitation is the dominant factor on SWC regardless of the crop type. The crop coefficient (K_C) is the second dominant factors for SWC in the two years. So the crop type is the second contributor that affects the SWC variability in different soil layers. The results implied that changing crop type is a necessary adaptation to regional water scarcity exacerbated by climate change in arid and semi-arid areas.

The mutual entropy values between SWC and precipitation are similar in different years under wheat cover, but the marginal entropy values for different layers are different. The value of mutual entropy of SWC with precipitation and that with ET_0 for the top soil layer and observed deepest layer are much higher in 2013 (dry year). This work opens a new avenue for further research to use entropy theory for better understanding the variability of soil water contents and its influencing factors.

Acknowledgments: This work was financially supported by the national key Research and Development Plan (2016YFC0400207), the Chinese National Natural Science Fund (91425302, 51439006, 51109211) and the Special Fund for Agro-scientific Research in the Public Interest (201503125). We are grateful for Zhongyi Liu, Xin Liu and Congcong Peng's help with field data collection. We would also like to thank Nandita Gaur and Zhenlei Yang for valuable discussions. We thank the anonymous reviewers for their constructive comments which helped improve this paper.

Author Contributions: Sufen Wang conceived and designed the experiments; Sufen Wang performed the experiments; Sufen Wang analyzed the data; Sufen Wang and Vijay P. Singh contributed reagents/materials/analysis tools; Sufen Wang and Vijay P. Singh wrote the paper. Both authors have read and approved the final manuscript.

References

1. Vereecken, H.; Huisman, J.A.; Bogena, H.; Vanderborght, J.; Vrugt, J.A.; Hopmans, J.W. On the value of soil moisture measurements in vadose zone hydrology: A review. *Water Resour. Res.* **2008**, *44*, W00D06. [CrossRef]
2. Gao, X.; Wu, P.; Zhao, X.; Zhang, B.; Wang, J.; Shi, Y. Estimating the spatial means and variability of root-zone soil moisture in gullies using measurements from nearby uplands. *J. Hydrol.* **2013**, *476*, 28–41. [CrossRef]
3. Brocca, L.; Morbidelli, R.; Melone, F.; Moramarco, T. Soil moisture spatial variability in experimental areas of central Italy. *J. Hydrol.* **2007**, *333*, 356–373. [CrossRef]
4. Brocca, L.; Melone, F.; Moramarco, T.; Morbidelli, R. Soil moisture temporal stability over experimental areas in Central Italy. *Geoderma* **2009**, *148*, 364–374. [CrossRef]
5. Wijewardana, Y.G.N.S.; Galagedara, L.W. Estimation of spatio-temporal variability of soil water content in agricultural fields with ground penetrating radar. *J. Hydrol.* **2010**, *391*, 24–33. [CrossRef]
6. Famiglietti, J.S.; Rudnick, J.W.; Rodelli, M. Variability in surface moisture content along a hillslope transect: Rattlesnake Hill, Texas. *J. Hydrol.* **1998**, *210*, 259–281. [CrossRef]
7. Famiglietti, J.; Devereaux, J.; Laymon, C.A.; Tsegaye, T.; Houser, P.R.; Jackson, T.J.; Graham, S.T.; Rodell, M.; van Oevelen, P.J. Ground-based investigation of soil moisture variability within remote sensing footprints during the Southern Great Plains 1997 (SGP 97) Hydrology Experiment. *Water Resour. Res.* **1999**, *35*, 1839–1851. [CrossRef]
8. Penna, D.; Borga, M.; Norbiato, D.; Dalla, F.G. Hillslope scale soil moisture variability in a steep alpine terrain. *J. Hydrol.* **2009**, *364*, 311–327. [CrossRef]
9. Liang, W.L.; Hung, F.X.; Chan, M.C.; Lu, T.H. Spatial structure of surface soil water content in a natural forested headwater catchment with a subtropical monsoon climate. *J. Hydrol.* **2014**, *516*, 210–221. [CrossRef]
10. Brocca, L.; Tullo, T.; Melone, F.; Moramarco, T.; Morbidelli, R. Catchment scale soil moisture spatial-temporal variability. *J. Hydrol.* **2012**, *422*, 63–75. [CrossRef]
11. Joshi, C.; Mohanty, B.P.; Jacobs, J.M.; Ines, A.V. Spatiotemporal analyses of soil moisture from point to footprint scale in two different hydroclimatic regions. *Water Resour. Res.* **2011**, *47*, W01508. [CrossRef]

12. Rosenbaum, U.; Bogena, H.R.; Herbst, M.; Huisman, J.A.; Peterson, T.J.; Weuthen, A. Seasonal and event dynamics of spatial soil moisture patterns at the small catchment scale. *Water Resour. Res.* **2012**, *48*, W10544. [CrossRef]
13. Joshi, C. Understanding Spatio-Temporal Variability and Associated Physical Controls of Near-Surface Soil Moisture in Different Hydro-Climates. Ph.D. Thesis, Texas A&M University, College Station, TX, USA, 2013.
14. Gómez-Plaza, A.; Mart Nez-Mena, M.; lbaladejo, J.; Castillo, V.M. Factors regulating spatial distribution of soil water content in small semiarid catchments. *J. Hydrol.* **2001**, *253*, 211–226. [CrossRef]
15. Li, H.; Shen, W.; Zou, C.; Jiang, J.; Fu, L.; She, G. Spatio-temporal variability of soil moisture and its effect on vegetation in a desertified aeolian riparian ecotone on the Tibetan Plateau, China. *J. Hydrol.* **2013**, *479*, 215–225. [CrossRef]
16. Baroni, G.; Ortuani, B.; Facchi, A.; Gandolfi, C. The role of vegetation and soil properties on the spatio-temporal variability of the surface soil moisture in a maize-cropped field. *J. Hydrol.* **2013**, *489*, 148–159. [CrossRef]
17. Joshi, C.; Mohanty, B.P. Physical controls of near surface soil moisture across varying spatial scales in an agricultural landscape during SMEX02. *Water Resour. Res.* **2010**, *46*, W12503. [CrossRef]
18. Mohanty, B.P.; Skaggs, T.H. Spatio-temporal evolution and time-stable characteristics of soil moisture within remote sensing footprints with varying soil, slope, and vegetation. *Adv. Water Resour.* **2001**, *24*, 1051–1067. [CrossRef]
19. Zhou, J.; Fu, B.; Gao, G.; Lü, N.; Lü, Y.; Wang, S. Temporal stability of surface soil moisture of different vegetation types in the Loess Plateau of China. *CATENA* **2015**, *128*, 1–15. [CrossRef]
20. Song, X.Y.; Kang, S.Z.; Shen, B. Dynamic Regularity of Soil Moisture in Farmland with Different Planting on Loess Hilly and Gully Region in North-West of China. *J. Soil Water Conserv.* **2003**, *2*, 130–133.
21. Odhiambo, J.J.O.; Bomke, A.A. Cover crop effects on spring soil water content and the implications for cover crop management in south coastal British Columbia. *Agric. Water Manag.* **2007**, *88*, 92–98. [CrossRef]
22. Hu, W.; Si, B.C. Revealing the relative influence of soil and topographic properties on soil water content distribution at the watershed scale in two sites. *J. Hydrol.* **2014**, *516*, 107–118. [CrossRef]
23. Li, T.; Hao, X.M.; Kang, S.Z. Spatiotemporal variability of soil moisture as affected by soil properties during irrigation cycles. *Soil Sci. Soc. Am. J.* **2014**, 598–608. [CrossRef]
24. Molina, A.J.; Latron, J.; Rubio, C.M.; Gallart, F.; Llorens, P. Spatio-temporal variability of soil water content on the local scale in a Mediterranean mountain area (Vallcebre, North Eastern Spain). How different spatio-temporal scales reflect mean soil water content. *J. Hydrol.* **2014**, *516*, 182–192. [CrossRef]
25. Bogena, H.R.; Herbst, M.; Huisman, J.A.; Rosenbaum, U.; Weuthen, A.; Vereecken, H. Potential of wireless sensor networks for measuring soil water content variability. *Vadose Zone J.* **2010**, *9*, 1002–1013. [CrossRef]
26. Hu, W.; Si, B.C. Soil water prediction based on its scale-specific control using multivariate empirical mode decomposition. *Geoderma* **2013**, *193*, 180–188. [CrossRef]
27. Whalley, W.R.; Binley, A.; Watts, C.W.; Shanahan, P.; Dodd, L.C.; Ober, E.S.; Ashton, R.W.; Webster, C.P.; White, R.P.; Hawkesford, M.J. Methods to estimate changes in soil water for phenotyping root activity in the field. *Plant Soil* **2017**, *415*, 407–422. [CrossRef]
28. Wang, Y.Q.; Shao, M.A.; Liu, Z.P. Vertical distribution and influencing factors of soil water content within 21-m profile on the Chinese Loess Plateau. *Geoderma* **2013**, *193*, 300–310. [CrossRef]
29. Dodd, I.C.; Whalley, W.R.; Obe, R.E.S.; Parry, M.A.J. Genetic and management approaches to boost UK wheat yields by ameliorating water deficits. *J. Exp. Bot.* **2011**, *62*, 5241–5248. [CrossRef] [PubMed]
30. Gaur, N.; Mohanty, B.P. Evolution of physical controls for soil moisture in humid and subhumid watersheds. *Water Resour. Res.* **2013**, *9*, 1244–1258. [CrossRef]
31. Singh, V.P. Entropy theory for movement of moisture in soils. *Water Resour. Res.* **2010**, *46*, W03516. [CrossRef]
32. Kang, S.Z.; Su, X.L.; Tong, L.; Zhang, J.H.; Zhang, L.; Davies, W.J. A warning from an ancient oasis: Intensive human activities are leading to potential ecological and social catastrophe. *Int. J. Sustain. Dev. World* **2008**, *15*, 440–447. [CrossRef]
33. Li, S.; Kang, S.Z.; Zhang, L.; Ortega-Farias, S.; Li, F.; Du, T.; Tong, L.; Wang, S.; Ingman, M.; Guo, W. Measuring and modeling maize evapotranspiration under plastic film-mulching condition. *J. Hydrol.* **2013**, *503*, 153–168. [CrossRef]
34. Warrick, A.W.; Nielsen, D.R. *Spatial Variability of Soil Physical Properties in the Field*; Academic Press: New York, NY, USA, 1980; pp. 319–344.

35. Shannon, C.E. A mathematical theory of communication. *Bell Syst. Tech. J.* **1948**, *27*, 623–656. [CrossRef]
36. Shannon, C.E. A mathematical theory of communication. *ACM SIGMOBILE Mob. Comput. Commun. Rev.* **2001**, *5*, 3–55. [CrossRef]
37. Li, C.; Singh, V.P.; Mishra, A.K. Entropy theory-based criterion for hydrometric network evaluation and design: Maximum information minimum redundancy. *Water Resour. Res.* **2012**, *48*, W05521. [CrossRef]
38. Singh, V.P. *Entropy Theory and Its Application in Environmental and Water Engineering*; Wiley: Hoboken, NJ, USA, 2013.
39. Singh, V.P. *Entropy Theory in Hydrologic Science and Engineering*; McGraw-Hill: New York, NY, USA, 2015.
40. Kang, S.Z.; Gu, B.J.; Du, T.S.; Zhang, J.H. Crop coefficient and ratio of transpiration to evapotranspiration of winter wheat and maize in a semi-humid region. *Agric. Water Manag.* **2003**, *59*, 239–254. [CrossRef]
41. Utset, A.; Farre, I.; Martinez-Cob, A.; Cavero, J. Comparing Penman–Monteith and Priestley–Taylor approaches as reference-evapotranspiration inputs for modeling maize water-use under Mediterranean conditions. *Agric. Water Manag.* **2004**, *66*, 205–219. [CrossRef]
42. Ding, R.S.; Kang, S.Z.; Li, F.S.; Zhang, Y.Q.; Tong, L.; Sun, Q.Y. Evaluating eddy covariance method by large-scale weighing lysimeter in a maize field of north-west China. *Agric. Water Manag.* **2010**, *98*, 87–95. [CrossRef]
43. Ding, R.S. Mechanism of Water and Heat Transfer and Evaportranspitation Modeling over Maize Field in an Arid Inland Region. Ph.D. Thesis, China Agricultrual University, Beijing, China, 2012.
44. Ding, R.S.; Kang, S.Z.; Li, F.S.; Zhang, Y.Q.; Tong, L. Evapotranspiration measurement and estimation using modified Priestley–Taylor model in an irrigated maize field with mulching. *Agric. For. Meteorol.* **2013**, *168*, 140–148. [CrossRef]
45. Lian, C.Y.; Ma, Z.M.; Lu, X.D.; Cao, S.Y. Research on main crops water requirement and crop Coefficient in Oasis irrigation Region. *J. Irrig. Drain.* **2012**, *31*, 136–139. (In Chinese)
46. Allen, R.G.; Pereira, L.S.; Raes, D.; Smith, M. *Crop Evapotranspiration: Guidelines for Computing Crop Water Requirements*; Irrigation and Drainage Paper No.56; FAO: Rome, Italy, 1998.
47. Hausser, J.; Strimmer, K. Entropy inference and the James-Stein estimator, with application to nonlinear gene association networks. *J. Mach. Learn. Res.* **2009**, *10*, 1469–1484.
48. Wang, L. Evaluation of Effectiveness of Soil Moisture in Shiyang River Basin Farmland and Its Dynamic Simulation. Ph.D. Thesis, Northwest A&F University, Xianyang, China, 2014.
49. Henning, H.; Olga, T.; Heinz-Josef, K.; Bernward, M. Subsoil properties and cereal growth as affected by a single pass of heavy machinery and two tillage systemson a Luvisol. *J. Plant Nutr. Soil Sci.* **2008**, *171*, 580–590.
50. Ryu, D.; Famiglietti, J.S. Multi-scale spatial correlation and scaling behavior of surface soil moisture. *Geophys. Res. Lett.* **2006**, *33*, L08404. [CrossRef]

10

A Connection Entropy Approach to Water Resources Vulnerability Analysis in a Changing Environment

Zhengwei Pan [1,2], Juliang Jin [3,*], Chunhui Li [4], Shaowei Ning [3] and Rongxing Zhou [1]

1 School of Civil Engineering and Environmental Engineering, Anhui Xinhua University, Hefei 230088, China; pzhwei1023@163.com (Z.P.); zhourx11@163.com (R.Z.)
2 Institute of Safety and Environmental Assessment, Anhui Xinhua University, Hefei 230088, China
3 School of Civil Engineering, Hefei University of Technology, Hefei 230009, China; ning@hfut.edu.cn
4 Ministry of Education Key Lab of Water and Sand Science, School of Environment, Beijing Normal University, Beijing 100875, China; chunhuili@bnu.edu.cn
* Correspondence: jinjl66@126.com

Abstract: This paper establishes a water resources vulnerability framework based on sensitivity, natural resilience and artificial adaptation, through the analyses of the four states of the water system and its accompanying transformation processes. Furthermore, it proposes an analysis method for water resources vulnerability based on connection entropy, which extends the concept of contact entropy. An example is given of the water resources vulnerability in Anhui Province of China, which analysis illustrates that, overall, vulnerability levels fluctuated and showed apparent improvement trends from 2001 to 2015. Some suggestions are also provided for the improvement of the level of water resources vulnerability in Anhui Province, considering the viewpoint of the vulnerability index.

Keywords: water resources vulnerability; connection entropy; changing environment; set pair analysis; Anhui Province

---

1. Introduction

The concept of vulnerability has its roots in geography and natural hazards research. Moreover, the original concept is the degree to which a system is likely to experience harm because of exposure to a hazard [1,2]. Vulnerability is now considered a central concept in a variety of other research contexts including anthropology, sociology, economics, aerography, ecology, management and sustainability science [3]. There are two key research perspectives in water resources vulnerability: a major role key research perspective is vulnerability of single purpose water systems, such as surface water vulnerability, groundwater vulnerability, and vulnerability of the drinking water supplies, etc. Padowski and Gorelick [4] presented a global analysis of urban water supply vulnerability in 70 surface water-supplied cities in a baseline (2010) condition and a future scenario (2040), which considered increased demand from urban population growth and projected agricultural demand under normal climate conditions, but did not account for climate change. Allouchea et al. [5] assessed groundwater vulnerability to contamination from anthropogenic activities and sea water intrusion based on a more robust "global risk index", and the DRASTIC and GALDIT parametric methods were then linked to a novel land use index. Khakhar et al. [6] derived the intrinsic vulnerability of groundwater against contamination using the GIS platform, and applied DRASTIC model for Ahmedabad district in Gujarat (India), that also contributes to validating the existence of higher concentrations of contaminants/indicators with respect to groundwater vulnerability status in the study area. Rushforth and Ruddell [7] spatially mapped and analyzed the Water Footprint of Flagstaff (AZ, USA), using a county-level database of the U.S. hydro-economy, NWED, which can empower city managers to operationalize a city's water footprint information to reduce vulnerability and increase resilience. Martinez et al. [8] assessed the potential effects of climate change and the vulnerability of water sources to support informed decision-making in Mexico City. Padowski and Jawitz [9] presented a quantitative national assessment of urban water availability and vulnerability for 225 U.S. cities with population greater than 100,000. Additionally, urban vulnerability measures developed here were validated using a media text analysis. Sullivan [10] brought forth the water vulnerability index whereby supply-driven vulnerability and demand-driven vulnerability, through the combination of these various dimensions in a mathematical manner.

Another major role key research perspective concerns vulnerability and the factors influencing water systems. Socio-economy and ecology have also been considered, that is the coupled human-environmental system. A major role of the vulnerability concept is the degree to which a system is either susceptible to, or unable to tackle the adverse effects of climate change. This includes climate variability as well as those extremes listed by the Intergovernmental Panel for Climate Change (IPCC) [11]. Furthermore, vulnerability to climate change has been stated as a function of the character, magnitude and rate of climate variation to which a system is exposed, its responsiveness and its adaptive capacity. Safi et al. [12] developed a climate change vulnerability index as a function of physical vulnerability, sensitivity, and adaptive capacity, and discussed its relationships with climate change mitigation policy support. Liersch et al. [13] established a vulnerability framework through the assessment of vulnerability (V) as a function of exposure (E), sensitivity (S) and adaptive capacity (AC), where the impacts on V by E and S can be summarized as external impacts (EI). The assessment result indicates the difference between the current situation and a chosen scenario of change. Yang et al. [14] presented a multifunctional hierarchy indicator system based on sensitivity and adaptability, and established an evaluation model, the Analytic Hierarchy Process Combining Set Pair Analysis (AHPSPA) model. This AHPSPA model was used to assess the water resource vulnerability of Beijing (China). Gain et al. [15] established a generalized methodological framework for vulnerability assessment to support a participatory decision making mechanism in the field of water resources management, with a particular focus on climate change adaptation. This was in addition to facilitating the work of those active in the field of water management in developing countries by moving towards operational solutions. Ionescu et al. [16] presented the contours of a formal framework of vulnerability to climate change developed on the basis of a grammatical probe that stemmed from the everyday meaning of vulnerability to technical consumption in the context of climate change. Al-Saidia et al. [17] developed the Country Vulnerability Index of Water Resources, which is a composite of socioeconomic and natural components. This approach integrates a country-level standpoint while taking into consideration the associated energy and food issues that are capable of reducing water resources vulnerability. Shabbir and Ahmad [18] analyzed the vulnerability status of the water resources system in Rawalpindi and Islamabad (Pakistan) with the help of an analytic hierarchy process, keeping in mind the intricate, integrated, comprehensive, and hierarchical nature of the vulnerability evaluation of water resources. Füssel [3] presented a conceptual framework and a terminology of vulnerability that makes it possible to develop a concise characterization of any vulnerability concept and the key differences between different concepts. Thus, he thereby bridged the gap between various traditions of vulnerability research.

This paper brings forth a synthesis of the two research views discussed above. First, we will identify the factors of the water system by looking at the human socioeconomic system and the coupled human-environmental system. Second, we will analyze the four transformation processes of the water system in a changing environment: sensitive state, damaged state, recovery state, and equilibrium state. The key impact of natural factors has been defined as natural resilience that deals with the change from a damaged state to a recovery state. The major effect of artificial factors is defined as artificial adaptation that aims for the change from a recovery state to an equilibrium state. Research on water resources vulnerability by synthesis sensitivity (S), natural resilience (R) and artificial adaptation (A) will also be performed.

2. Mechanism of Water Resources Vulnerability in a Changing Environment

With the backdrop of natural hazards science, vulnerability research places significant emphasis on the resilience, response, and resilience of the human socioeconomic system in response to disasters. Accordingly, the concept of vulnerability is required to include exposure, sensitivity, adaptation, and resilience when the system is disturbed, destroyed, or affected. This paper presents the concept of water resources vulnerability, which is the nature and state of a water resources system, wherein, the normal structure and function are compromised by a changing environment. Vulnerability also includes, the sensitivity and adaptation to the disturbance and destruction under a changing environment, the ability to bear and cope with such change, and the level of resilience to damage [12–14,19].

In this transformation mechanism, the water system can experience four states: the sensitive state, the damaged state, the recovery state, and the equilibrium state. The water system undergoes change because of the interference and destruction of the changing environment, in addition to the impacts stemming from environmental change factors. In this way, the water system can be said to be in a sensitive state. The degree of change in the sensitive state is the resulting water resources sensitivity [20], expressed by sensitive factors. A water system can be said to be in a damaged state as the result of environmental interference and destruction. In this scenario, the water system is able to self-regulate and repair through the system factors, for instance water resources endowment, and natural ecological environment. This ability to self-regulate and repair is termed natural resilience, which is adaptation in response to experienced environment and its effects, without planning explicitly or consciously focused on addressing changing environment [20].

Natural resilience is the capacity of the water system, but note that it still possesses relative weakness. The system is capable of self-repair when the degree of damage is weak, but while still in the damaged state it cannot meet the social and economic development and ecological environment requirements if the damage exceeds the natural recovery capacity. It is for this reason we are required to reduce water demand with socioeconomic development and the ecological environment through the adjustment of human behavior and the socioeconomic development mode. This is done, with the improvement of, for example, the ecological environment to adapt to the damaged state, which is defined as artificial or social adaptation. Artificial adaptation is an ability of water systems, humans, social economy and ecology to adjust to potential damage, to take advantage of opportunities, or to respond to consequences [21]. The degree of damage to the water system is steadily alleviated by the combined action of natural recovery and artificial adaptation, in addition to being a system of relative equilibrium. This equilibrium state is dynamic, with the system in a changing atmosphere. The best state for the recovery of the damaged water system is prior to the occurrence of environment changes. In general, it is quite difficult to attain the optimal state, so it may be easier to form a new water system in a changing environment. In accordance with the analysis of the mechanism of water resources, vulnerability in changing environment is presented in Figure 1.

2.1. *Sensitivity*

Water systems are influenced by environmental interference and destruction. These impact factors primarily include water resources utilization, water environment pollution, drought, and floods. Water resources sensitivity is caused and expressed by a sensitive index that looks at water utilization rate (i.e., the ratio of water consumption to water resources availabilities), emission intensity of industrial wastewater (i.e., industrial wastewater discharge to GDP), per capita sewage discharge (i.e., domestic sewage discharge to total population), and the proportion of economic losses of water disasters to GDP (i.e., economic losses of drought and flood disaster to GDP, see Table 1).

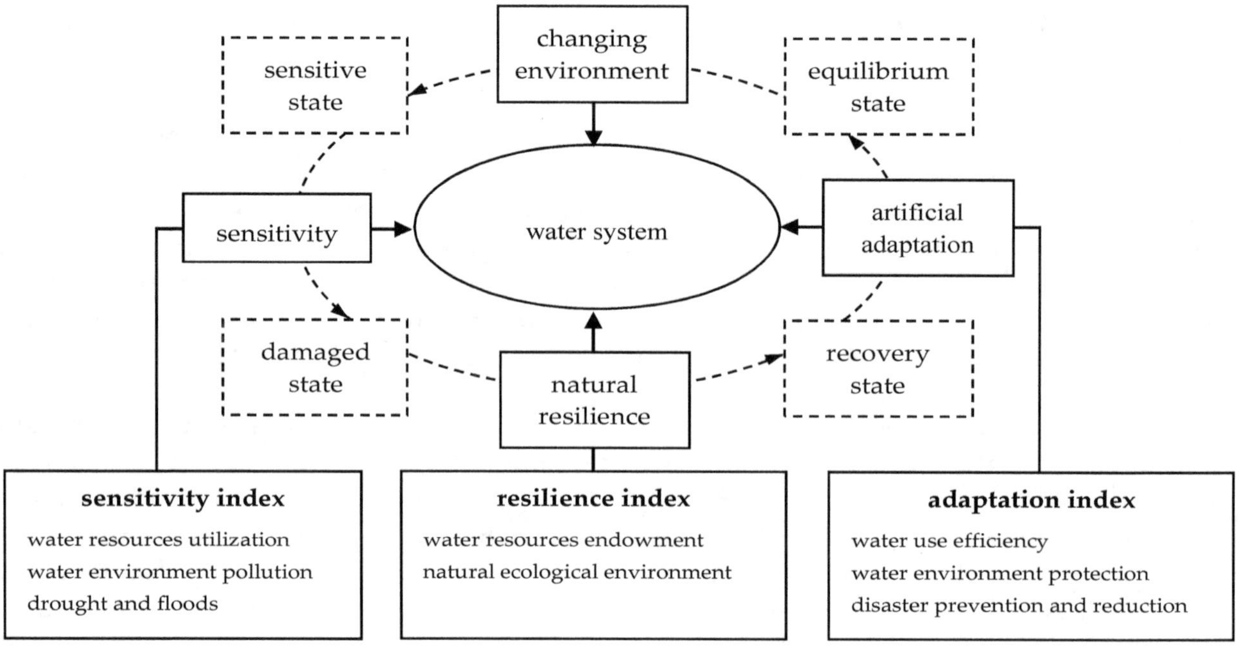

Figure 1. Mechanism diagram of water resources vulnerability in changing environment.

Table 1. Sensitive index of water resources.

Sensitive Factor		Sensitive Index
Water resources utilization	x_1	Water utilization rate (%)
Water environment pollution	x_2	Emission intensity of industrial wastewater (t/10^4¥)
	x_3	Per capita sewage discharge (t/person)
Drought and floods	x_4	Proportion of economic losses of water disasters to GDP (%)

The sensitivity function of the water resources $S(x)$ is proposed to quantitate sensitivity and the sensitive index, which is expressed as follows:

$$S(x) = S(x_1, x_2, x_3, x_4) \tag{1}$$

where $S(x)$ denotes water resources sensitivity, x_1 is the water utilization rate, and x_2 is the emission intensity of industrial wastewater. Furthermore, x_3 depicts the per capita sewage discharge, and x_4 represents the proportion of economic losses of water disasters to GDP.

2.2. Natural Resilience

A water system enters a damaged state because of environmental interference and destruction. The damaged water system is able to self-regulate and repair with the help of water resources endowment, water environmental quality, eco-environmental quality and other factors that aim to encourage the adaption of or repair to the damaged water system. This ability to self-regulate and repair the damaged water system is caused and expressed by resilient index, which looks at the per unit area of available water resources (i.e., the ratio of water resources availabilities to national territorial area), surface water environment quality (i.e., water attainment ratio of monitoring section), and forest cover (i.e., forest coverage, see Table 2).

Table 2. Resilient index of water resources.

Resilient Factor		Resilient Index
Water resources endowment	x_5	Per unit area available amount of water resources ($10^4 \cdot m^3/km^2$)
Water environmental quality	x_6	Surface water environment quality (%)
Eco-environmental quality	x_7	Forest cover (%)

The natural resilience of the water resources function $R(x)$ is proposed to quantify natural resilience and the resilient index, and is presented as follows:

$$R(x) = R(x_5, x_6, x_7) \tag{2}$$

where $R(x)$ indicates water resources natural resilience, x_5 is the per unit area of available water resources, and x_6 represents surface water environment quality. The term x_7 is the forest cover.

2.3. *Artificial Adaptation*

The natural resilience of water resources is constrained. The water system remains in a damaged state where the damage exceeds the natural recovery capacity. In this situation, we must take the initiative to adapt to the environmental changes by, for example, increasing water use efficiency, strengthening water environment protection, disaster prevention, and reduction for the purpose of adapting to the damaged state. This artificial adaptation is caused and expressed by the adaptive index, which includes per unit GDP of water consumption (i.e., the ratio of water consumption to GDP), compliance rate of industrial wastewater and treatment rate of urban sewage (i.e., discharge standard of industrial wastewater and urban sewage), and economic benefits of flood control and fight drought (i.e., economic benefits of disaster prevention and mitigation, see Table 3).

Table 3. Adaptive index of water resources.

Adaptive Factor		Adaptive Index
Water use efficiency	x_8	Per unit GDP of water consumption ($m^3/10^4$¥)
Water environment protection	x_9	Compliance rate of industrial wastewater (%)
	x_{10}	Treatment rate of urban sewage (%)
Disaster prevention and reduction	x_{11}	Economic benefits of flood control and fighting drought (10^8¥)

The proposed artificial adaptation function of water resources $A(x)$ to quantify artificial adaptation and the adaptive index, and is presented as follows:

$$A(x) = A(x_8, x_9, x_{10}, x_{11}) \tag{3}$$

where $A(x)$ denotes water resources artificial adaptation, x_8 is the per unit GDP of water consumption, x_9 indicates the compliance rate of industrial wastewater, x_{10} represents treatment rate of urban sewage, and x_{11} depicts the economic benefits of flood control and fight drought.

2.4. *Water Resources Vulnerability in a Changing Environment*

Water resources vulnerability in a changing environment (V) is the synthetic effect of three elements, namely, water resources sensitivity (S), natural resilience (R) and artificial adaptation (A), and is given as follows:

$$V(x) = f[S(x), R(x), S(x)] \tag{4}$$

where V indicates water resources vulnerability in a changing environment, $S(x)$ denotes sensitivity, $R(x)$ indicates natural resilience, and $A(x)$ presents artificial adaptation.

3. Methodology

Entropy is an concept extensively used in natural and social sciences. It has been developed and perfected in these fields, despite being a simple description of a microcosmic thermodynamic concept with roots in physics research. The evolution of entropy is essentially an extension of "state", which is "state function" in thermodynamics. Entropy is required to possess the attributes of the state function of all, for having used the "mathematic analogy" methods of inference in this type of expansion of the "state". Thereafter, connection entropy is also proposed by the method of "mathematical analogies".

3.1. Development Course

The concept of entropy was primarily introduced in thermodynamics by Clausius in 1854. He initially used the term "equivalence value" to define the concept. In respect of statistical mechanics, the entropy of a system is equal to the logarithm of the number Ω of accessible microstates corresponding to a macroscopical state of this system:

$$S = k \ln \Omega \tag{5}$$

where k denotes Boltzmann's constant and Ω represents the number of microstates having consistency with the given equilibrium macro state.

The use of an intuitive logarithmic measure for information, primarily introduced by Hartley [22], puts forth the suggestion that the self-information of an event increases with the growth of its uncertainty, further implying that the probability of occurrence reduces. In this respect, "S" is termed as a measure of uncertainty.

Schrödinger [23] put forth a proposition regarding a local decrease of entropy for living systems when $(1/D)$ represents the number of states that are prevented from random distribution:

$$-S = k \ln(1/D) \tag{6}$$

where D indicates the number of possible energy states in the system that can be randomly filled with energy. Furthermore, D suggests a measure of disorder, and its reciprocal $1/D$ can be regarded as a direct measure of order.

In 1948, Shannon [24] published his famous paper titled "A Mathematical Theory of Communication", wherein he introduced the entropy of a discrete probability distribution ($p_1, p_2, \ldots,$ p_n), providing an H function of the following form:

$$H = -K \sum_{i=1}^{n} p_i \log p_i \tag{7}$$

where K denotes a positive constant. Moreover, K, merely implying choice of a unit of measurement, plays a focal function in the information theory as measures of information, choice, and uncertainty [24].

In 1972, De Luca and Termini [25] promulgated the concept of entropy again for finite fuzzy sets, similar to Shannon entropy although quite different conceptually, where the range is a set of nonnegative real numbers.

In 1992, Zhao [26] proposed the concepts of identical entropy, discrepancy entropy, contrary entropy, and connection entropy based on the definition of traditional entropy, to measure identity, diversity, opposites, and associative of systems which contains n set pairs.

3.2. Connection Entropy

Set pair analysis (SPA) is a new system analysis approach advanced by the Chinese scholar Zhao [1]. SPA considers both certainty and uncertainty as a system for the purpose of conducting identical-discrepancy-contrary analysis as well as further quantitative mathematical process using a connection number. Because of its realistic approach to deal with uncertainty, SPA has been put to extensive applications in sociology, economics, engineering technology, and management, and has produced numerous research results [14,27–30].

In the case of particular questions, analyses have been conducted on the identity, discrepancy, and opposition of a set pair, the formula of the connection number of set pair "H" for given conditions that describe the relationship between certainty and uncertainty. The expression with respect to set pair "H" is presented as follows [1]:

$$u = a + bI + cJ \tag{8}$$

where $a,\ b,\ c \in [0,\ 1]$, and $a + b + c = 1$. Furthermore, a denotes the identical degree of set pair H, b indicates the discrepancy degree, c is the contrary degree, and I is the coefficient of discrepancy and $I \in [-1,\ 1]$; sometimes I may only refer to discrepancy. Finally, J is the coefficient of contrary, and has been ruled to be -1; sometimes it may only be considered as a mark of the opposites.

Here, n connection degrees are attained: $u_1 = a_1 + b_1I_1 + c_1J_1$, $u_2 = a_2 + b_2I_2 + c_2J_2, \ldots$, and $u_n = a_n + b_nI_n + c_nJ_n$ through the analysis of n set pairs.

Identical entropy is defined as follows: $S_s = \sum\limits_{i=1}^{n} a_i \ln a_i$, discrepancy entropy as follows: $S_f = \sum\limits_{i=1}^{n} b_i \ln b_i \times I$, contrary entropy as follows: $S_p = \sum\limits_{i=1}^{n} c_i \ln c_i \times J$, and connection entropy as follows [1]:

$$S = S_s + S_f + S_p = \sum_{i=1}^{n} a_i \ln a_i + \sum_{i=1}^{n} b_i \ln b_i \times I + \sum_{i=1}^{n} c_i \ln c_i \times J \tag{9}$$

Entropy is the "state function" of a system that is considered a measure of randomness or disorder. The variation of thermodynamic entropy was considered as a measure of variation of unavailable energy. Statistical entropy will characterize (in respect of a macro state prepared as per a provided probability law) our uncertainty about the set of all microscopic experiments that can be conceived. Information on Shannon entropy is a measure of uncertainty or information of random events. More specifically, it is a measure of the uncertainty of the test results prior to the randomization test or the amount of information in the event subsequent to the event. Fuzzy entropy is a fuzziness measure of the system.

In the same way, identical entropy is a disorder measure. Moreover, discrepancy entropy is an order measure, and contrary entropy is a chaotic measure of uncertain systems [23]. Discrepancy includes identical and contrary in the connection degree, and therefore, discrepancy entropy is complex entropy, which comprises identical entropy and contrary entropy. The difference can be segregated with the help of the malleability of connection degree. Thus, discrepancy entropy is divided into identical entropy, discrepancy entropy and contrary entropy.

3.3. Improvement of Connection Entropy

It is quite important to improve the correlation entropy, as defined by Zhao [1], because the logarithmic function delivers no meaning, if a, b, c are likely to be 0 in the connection degree. The general expression of improvement of connection entropy is given as follows:

$$S = \sum_{i=1}^{n} a_i \ln(a_i + e) + \sum_{i=1}^{n} b_i \ln(b_i + e) \times I + \sum_{i=1}^{n} c_i \ln(c_i + e) \times J \tag{10}$$

where $S_s = \sum\limits_{i=1}^{n} a_i \ln(a_i + e)$ indicates identical entropy, $S_f = \sum\limits_{i=1}^{n} b_i \ln(b_i + e) \times I$ suggests discrepancy entropy, and $S_p = \sum\limits_{i=1}^{n} c_i \ln(c_i + e) \times J$ denotes contrary entropy.

4. Analysis Method of Water Resources Vulnerability Based on Connection Entropy

4.1. Connection Number of "Identical-Discrepancy-Contrary" Hierarchy Method

As presented in Equation (8), the connection number is established on the basis of the distinction: identical, discrepancy, and contrary, called the identical-discrepancy-contrary connection number or three dimensional connection numbers. In practice, it is not enough to divide the described object into three components and therefore it is important to extend the basic expression of contact number (i.e., Equation (8)) to include more dimensions, which are termed as the malleability of the connection number, extending the connection number, and a multi-dimensional number yield [31]:

$$u = (a_1 + a_2 + \cdots + a_r) + (b_1 I_1 + b_2 I_2 + \cdots + b_s I_s) + (c_1 J_1 + c_2 J_2 + \cdots + c_t J_t) \tag{11}$$

where a_x, b_y, c_z are connection components, $a_x, b_y, c_z \in [0, 1]$, and $\sum_{x=1}^{r} a_x + \sum_{y=1}^{s} b_y + \sum_{z=1}^{t} c_z = 1$; $I_1, I_2, \cdots, I_s$ is the discrepancy coefficients, and $I_1, I_2, \cdots, I_s \in [-1, 1]$, which sometimes perform the functions of a discrepancy mark only. Furthermore, $J_1, J_2, \cdots, J_t$ denote the coefficient of contrary degrees that have been ruled to be minus unity and sometimes only marks the contrary.

On the basis of Equation (11), the multi-connection number and the construction of the identical-discrepancy-contrary hierarchical structure of connection number can be expressed as follows:

$$u = a_1 + a_2 + b_1 I_1 + b_2 I_2 + b_3 I_3 + c_1 J_1 + c_2 J_2 \tag{12}$$

where a_1, and a_2 represent the identical degree and partial differential identical degree, respectively, and their coefficients can be assumed to be unity. Furthermore, $b_1 I_1$, $b_2 I_2$, and $b_3 I_3$ indicate partial similar, middle, and partial opposite discrepancy degrees, correspondingly and their coefficients are $I_1 \in [-1, 0]$, $I_2 \in [-0.5, 0.5]$, and $I_3 \in [0, 1]$. Lastly, $c_1 J_1$ and $c_2 J_2$ indicate partial differential contrary degree and contrary degree, respectively, and J_1, J_2 are their coefficients, which are regulated to be minus unity.

The water resources vulnerability is generally divided into five levels [14,19], and assuming that x_{ij} ($i = 1, 2, \ldots, n$; $j = 1, 2, \ldots, m$) denotes the index sample and s_{kj} ($k = 0, 1, 2, \ldots, 5$) the threshold value of standard, the connection number of the "identical-discrepancy-contrary" hierarchy method is described as follows [31]:

$$u_{ij} = \begin{cases} \frac{x_{ij}-s_{1j}}{2(s_{0j}-s_{1j})} + 0.5 + \frac{s_{0j}-x_{ij}}{2(s_{0j}-s_{1j})} I_1 + 0I_2 + 0I_3 + 0J_1 + 0J_2 & x_{ij} \in grade\ \mathrm{I} \\ 0 + \frac{x_{ij}-s_{2j}}{2(s_{1j}-s_{2j})} + 0.5I_1 + \frac{s_{1j}-x_{ij}}{2(s_{1j}-s_{2j})} I_2 + 0I_3 + 0J_1 + 0J_2 & x_{ij} \in grade\ \mathrm{II} \\ 0 + 0 + \frac{x_{ij}-s_{3j}}{2(s_{2j}-s_{3j})} I_1 + 0.5I_2 + \frac{s_{2j}-x_{ij}}{2(s_{2j}-s_{3j})} I_3 + 0J_1 + 0J_2 & x_{ij} \in grade\ \mathrm{III} \\ 0 + 0 + 0I_1 + \frac{x_{ij}-s_{4j}}{2(s_{3j}-s_{4j})} I_2 + 0.5I_3 + \frac{s_{3j}-x_{ij}}{2(s_{3j}-s_{4j})} J_1 + 0J_2 & x_{ij} \in grade\ \mathrm{IV} \\ 0 + 0 + 0I_1 + 0I_2 + \frac{x_{ij}-s_{5j}}{2(s_{4j}-s_{5j})} I_3 + 0.5J_1 + \frac{s_{4j}-x_{ij}}{2(s_{4j}-s_{5j})} J_2 & x_{ij} \in grade\ \mathrm{V} \end{cases} \tag{13}$$

where x_{ij} suggests the ith sample value in the jth index, s_{kj} ($k = 0, 1, 2, \ldots, 5$) indicates the kth standard node value in the same index, and I_1, I_2, I_3, J_1, and J_2 are the same as in Equation (12).

4.2. Connection Entropy of the Vulnerability Index

The index of water resources vulnerability is considered as set $A = \{x_{ij}$ ($i = 1, 2, \ldots, n$; $j = 1, 2, \ldots, m$)$\}$, and the thresholds of the vulnerability grade are considered as set $B_k = \{s_{kj}$ ($k = 0, 1, 2, \ldots, 5$)$\}$; thereafter, the two sets constitute a set pair $H_k = (A, B_k)$.

The connection numbers of the vulnerability index with Equation (13), and thereafter, connection entropy of the vulnerability index, are described as follows:

$$S_{ij} = a_{1ij}\ln(a_{1ij}+e) + a_{2ij}\ln(a_{2ij}+e) + b_{1ij}\ln(b_{1ij}+e)\times I_1 + b_{2ij}\ln(b_{2ij}+e)\times I_2 \\ + b_{3ij}\ln(b_{3ij}+e)\times I_3 + c_{1ij}\ln(c_{1ij}+e)\times J_1 + c_{2ij}\ln(c_{2ij}+e)\times J_2 \quad (14)$$

where $S_{s1} = a_{1ij}\ln(a_{1ij}+e)$ indicates identical entropy, $S_{s2} = a_{2ij}\ln(a_{2ij}+e)$ denotes critical identical entropy, and $S_{f1} = b_{1ij}\ln(b_{1ij}+e)\times I_1$ indicates upper discrepancy entropy, Furthermore, $S_{f2} = b_{2ij}\ln(b_{2ij}+e)\times I_2$ implies medium discrepancy entropy, $S_{f3} = b_{3ij}\ln(b_{3ij}+e)\times I_3$ represents lower discrepancy entropy, $S_{p1} = c_{1ij}\ln(c_{1ij}+e)\times J_1$ is critical contrary entropy, and $S_{p2} = c_{2ij}\ln(c_{2ij}+e)\times J_2$ suggests contrary entropy.

4.3. Synthesis Connection Entropy of Water Resources Vulnerability

In general, in accordance with the significance of the vulnerability index, respective weights for the integrated vulnerability are provided. Consequently, taking advantage of the additively weighted synthesis method (average method), multiplicatively weighted synthesis method (geometric average method) or add-multiplicatively weighted synthesis method [32], the single index connection entropy is multiplied for the purpose of generating the integrated connection entropy S_i. This brings forth the expression:

$$S_i = \prod_{j=1}^{m} w_j \times S_{ij} \quad (15)$$

where Π indicates the system synthesis method and w_j represents the weight of the jth vulnerability index.

4.4. Decision Criterion of Water Resources Vulnerability

The degree to which water resources vulnerability is subjected to a changing environment can be broken down into five grades (or levels) with 11 indices: low (I), slight (II), moderate (III), high (IV) and extreme (V). The calculation of contact entropy using Equation (14), shows an approximate value of contact entropy S with a range of $(-1.314, 1.314)$. The interval $(-1.314, 1.314)$ is divided into five parts: $[0.877, 1.314)$, $[0.292, 0.877)$, $[-0.292, 0.292)$, $[\ 0.877, -0.292)$, $(-1.314, -0.877)$, denoting low (I), slight (II), moderate (III), high (IV) and extreme (V) grades of water resources vulnerability, respectively.

5. Case Study

Anhui Province is currently considered as the development frontier in China as it absorbs the economic radiation and industrial transfer from the coastal developed areas. It is also a bridgehead between the development of Western China as well as the rise of Central China. As a consequence, the area possesses unique geographical benefits, functioning as a junction between the East and the West and connects Northern and Southern China. Various obstacles to steady development have emerged in the province, with the water environment becoming a principal factor. The per capita water resources in Anhui Province are 1125 m3, which is approximately half the national average and merely 1/8 the world average, so it goes without saying that there is a shortage of water resources throughout the province. Furthermore, regional differences between annual precipitation and the four seasons are large, giving rise to frequent droughts and flood disasters. Water pollution is another serious concern in the Yangtze River, Huaihe River, and Chaohu Lake in Anhui Province, as parts of them have lost their utility value. Thus, water conflicts have been exacerbated and the vulnerability of the regional water environment has been exposed. In accordance with the above, it is quite apparent that the increasing deterioration of regional water resources vulnerability cannot be overlooked. Therefore, it is necessary to perform an analysis of regional water resources vulnerability in Anhui Province. Here, we study used a connection entropy approach to analyze water resources vulnerability in Anhui

Province from 2001 to 2015, aiming at the provision of a theoretical basis for the improvement of the water situation in the region.

5.1. Standard Interval of the Vulnerability Index in Anhui Province

In accordance with the Section 2, and the research findings of water resources vulnerability [14,30], the standard interval of the vulnerability index in Anhui Province is presented in Table 4.

Table 4. Standard interval of the vulnerability index in Anhui Province.

Index		Standard Interval				
		Low (I)	Slight (II)	Moderate (III)	High (IV)	Extreme (V)
Sensitive	x_1 (%)	0–10	10–25	25–40	40–60	60–100
	x_2 (t/10^4 ¥)	0–10	10–25	25–40	40–55	55–100
	x_3 (t/person)	0–5	5–10	10–40	40–60	60–100
	x_4 (%)	0–1.0	1.0–2.5	2.5–4.0	4.0–4.5	4.5–5.5
Natural resilience	x_5 ($10^4 \cdot m^3/km^2$)	75–100	55–75	35–55	15–35	5–15
	x_6 (%)	90–100	80–90	70–80	50–70	0–50
	x_7 (%)	30–50	20–30	15–20	10–15	5–10
Artificial adaptation	x_8 ($m^3/10^4$ ¥)	5–24	24–140	140–610	610–1060	1060–1600
	x_9 (%)	97.5–100	92.5–97.5	85–92.5	80–85	15–80
	x_{10} (%)	80–100	60–80	40–60	20–40	0–20
	x_{11} (10^8 ¥)	400–600	200–400	100–200	60–100	0–60

5.2. Connection Entropy of Vulnerability Index in Anhui Province

The index of water resources vulnerability in Anhui Province from 2001 to 2015 is considered as set $A = \{x_{ij}\ (i = 1, 2, \ldots, 15; j = 1, 2, \ldots, 11)\}$ in Table 5, where x_{ij} suggests the sample value in 2001 if $i = 1$, the thresholds of vulnerability grade are considered as set $B_k = \{s_{kj}\ (k = 0, 1, 2, \ldots, 5)\}$ in Table 4, and the two sets constitute a set pair $H_k = (A, B_k)$.

Table 5. Index data of the vulnerability index in Anhui Province in 2001–2015.

Year	x_1	x_2	x_3	x_4	x_5	x_6	x_7	x_8	x_9	x_{10}	x_{11}
2001	45.22	17.79	11.47	3.13	34.00	41.00	27.95	603.56	95.24	7.84	128.00
2002	25.36	16.90	12.27	1.48	59.13	51.08	27.95	588.48	95.74	15.80	91.80
2003	16.83	14.88	12.13	5.72	77.65	48.69	27.95	458.79	95.88	20.14	440.67
2004	41.89	13.31	13.04	0.74	35.90	41.70	30.30	435.80	96.91	25.69	70.00
2005	28.92	10.89	14.29	3.49	51.57	44.30	26.06	386.10	97.36	31.28	209.00
2006	41.67	11.40	14.61	1.00	41.62	64.60	26.06	393.80	97.12	32.90	145.30
2007	32.60	9.99	15.24	1.98	51.80	64.70	26.06	305.90	94.77	37.64	420.50
2008	38.09	8.29	15.08	2.41	50.13	69.30	26.06	288.80	96.17	52.08	100.00
2009	39.81	7.30	15.64	1.44	52.56	72.10	26.06	290.04	96.21	72.81	86.00
2010	31.15	5.74	16.66	0.92	67.33	77.70	27.53	238.50	97.67	77.75	260.00
2011	48.94	4.62	25.07	0.61	43.17	65.50	27.53	195.00	97.47	79.04	145.00
2012	41.16	3.90	27.09	0.28	50.26	70.60	27.53	167.60	98.38	86.39	154.80
2013	50.55	3.73	28.16	0.80	41.99	72.40	27.53	155.50	98.51	88.44	187.10
2014	34.95	3.34	29.20	0.14	55.81	77.90	28.65	131.00	98.69	90.19	103.11
2015	31.58	3.25	30.07	0.34	65.54	81.80	28.65	131.20	98.70	91.80	136.89

Sources: Statistics Bureau of Anhui Province, 2002–2016, and Water Resources Bulletin of Anhui Province, 2001–2015.

Thereafter, the connection number and the connection entropy of vulnerability index in Anhui Province can be attained with the application of Equations (13) and (14), respectively. For example, the connection entropy of vulnerability index in 2015 listed in Table 6 is thus obtained.

Table 6. Connection entropy of vulnerability index in 2015.

Index	Connection Entropy	w_j
x_1	$S_{15,1} = 0.0000 + 0.0000 + 0.3083I_1 + 0.5844I_2 + 0.2363I_3 + 0.0000J_1 + 0.0000J_2$	0.120
x_2	$S_{15,2} = 0.3772 + 0.5844 + 0.1717I_1 + 0.0000I_2 + 0.0000I_3 + 0.0000J_1 + 0.0000J_2$	0.100
x_3	$S_{15,3} = 0.0000 + 0.0000 + 0.1754I_1 + 0.5844I_2 + 0.3732I_3 + 0.0000J_1 + 0.0000J_2$	0.100
x_4	$S_{15,4} = 0.3691 + 0.5844 + 0.1791I_1 + 0.0000I_2 + 0.0000I_3 + 0.0000J_1 + 0.0000J_2$	0.080
x_5	$S_{15,5} = 0.0000 + 0.2879 + 0.5844I_1 + 0.2562I_2 + 0.0000I_3 + 0.0000J_1 + 0.0000J_2$	0.150
x_6	$S_{15,6} = 0.0000 + 0.0929 + 0.5844I_1 + 0.4676I_2 + 0.0000I_3 + 0.0000J_1 + 0.0000J_2$	0.090
x_7	$S_{15,7} = 0.0000 + 0.4964 + 0.5844I_1 + 0.0692I_2 + 0.0000I_3 + 0.0000J_1 + 0.0000J_2$	0.060
x_8	$S_{15,8} = 0.0000 + 0.0385 + 0.5844I_1 + 0.5346I_2 + 0.0000I_3 + 0.0000J_1 + 0.0000J_2$	0.090
x_9	$S_{15,9} = 0.2603 + 0.5844 + 0.2838I_1 + 0.0000I_2 + 0.0000I_3 + 0.0000J_1 + 0.0000J_2$	0.075
x_{10}	$S_{15,10} = 0.3254 + 0.5844 + 0.2199I_1 + 0.0000I_2 + 0.0000I_3 + 0.0000J_1 + 0.0000J_2$	0.075
x_{11}	$S_{15,11} = 0.0000 + 0.0000 + 0.1966I_1 + 0.5844I_2 + 0.3502I_3 + 0.0000J_1 + 0.0000J_2$	0.060
$\sum$	$S_{15} = 0.1112 + 0.2776 + 0.3635I_1 + 0.2964I_2 + 0.0867I_3 + 0.0000J_1 + 0.0000J_2$	

5.3. *Weight of the Vulnerability Index in Anhui Province*

In accordance with the mechanism of water resources vulnerability in a changing environment, the weights of sensitivity, natural resilience and artificial adaptation are 0.4, 0.3, and 0.3, respectively. The weights of the sensitive index are valued at 0.3, 0.25, 0.25, 0.2, the natural resilient index are 0.5, 0.3, 0.2, and those of the artificial adaptive index are 0.3, 0.25, 0.25, 0.2. The weights w_j of the vulnerability index attained for Anhui Province are shown in Table 6.

5.4. *Synthesis of Connection Entropy of Water Resources Vulnerability in Anhui Province*

By means of a weighted average operator [32] for the synthesis of the connection entropy of water resources vulnerability, the equation for the synthesis connection entropy of water resources vulnerability in Anhui Province is presented as follows:

$$S_i = \sum_{j=1}^{11} w_j \times S_{ij} = a_{1i} + a_{2i} + b_{1i} \times I_1 + b_{2i} \times I_2 + b_{3i} \times I_3 + c_{1i} \times J_1 + c_{2i} \times J_2 \tag{16}$$

where w_j suggests the weight of vulnerability index j.

The connection entropy of the vulnerability index in Anhui Province is synthesized by Equation (16) and the results are presented in Table 7.

Table 7. Synthesize connection entropy of water resources vulnerability in Anhui Province.

Year	Synthesize Connection Entropy	Value	Grade
2001	$S_1 = 0.0000 + 0.0757 + 0.2358I_1 + 0.3786I_2 + 0.3047I_3 + 0.1166J_1 + 0.0336J_2$	−0.109	III
2002	$S_2 = 0.0000 + 0.1302 + 0.3956I_1 + 0.3372I_2 + 0.1765I_3 + 0.0996J_1 + 0.0082J_2$	0.132	III
2003	$S_3 = 0.0144 + 0.2517 + 0.3810I_1 + 0.2092I_2 + 0.1793I_3 + 0.1082J_1 + 0.0012J_2$	0.258	III
2004	$S_4 = 0.0113 + 0.1641 + 0.2431I_1 + 0.2964I_2 + 0.3117I_3 + 0.1139J_1 + 0.0077J_2$	0.019	III
2005	$S_5 = 0.0000 + 0.1185 + 0.3924I_1 + 0.3888I_2 + 0.1694I_3 + 0.0702J_1 + 0.0052J_2$	0.154	III
2006	$S_6 = 0.0002 + 0.1593 + 0.2956I_1 + 0.3815I_2 + 0.2750I_3 + 0.0320J_1 + 0.0000J_2$	0.138	III
2007	$S_7 = 0.0032 + 0.1468 + 0.3988I_1 + 0.4082I_2 + 0.1678I_3 + 0.0170J_1 + 0.0000J_2$	0.249	III
2008	$S_8 = 0.0088 + 0.1120 + 0.3518I_1 + 0.4649I_2 + 0.2066I_3 + 0.0016J_1 + 0.0000J_2$	0.192	III
2009	$S_9 = 0.0142 + 0.1683 + 0.3773I_1 + 0.3927I_2 + 0.1788I_3 + 0.0112J_1 + 0.0000J_2$	0.271	III
2010	$S_{10} = 0.0287 + 0.2735 + 0.4791I_1 + 0.3001I_2 + 0.0585I_3 + 0.0000J_1 + 0.0000J_2$	0.513	II
2011	$S_{11} = 0.0462 + 0.2157 + 0.2949I_1 + 0.3195I_2 + 0.2225I_3 + 0.0395J_1 + 0.0000J_2$	0.259	III
2012	$S_{12} = 0.0928 + 0.2184 + 0.2810I_1 + 0.3596I_2 + 0.1864I_3 + 0.0035J_1 + 0.0000J_2$	0.355	II
2013	$S_{13} = 0.0763 + 0.2184 + 0.2807I_1 + 0.3248I_2 + 0.2020I_3 + 0.0346J_1 + 0.0000J_2$	0.299	II
2014	$S_{14} = 0.1166 + 0.2293 + 0.3222I_1 + 0.3520I_2 + 0.1236I_3 + 0.0000J_1 + 0.0000J_2$	0.445	II
2015	$S_{15} = 0.1112 + 0.2776 + 0.3635I_1 + 0.2964I_2 + 0.0867I_3 + 0.0000J_1 + 0.0000J_2$	0.527	II

The value of the discrepancy entropy coefficients in Equation (16) can be measured with the application of the mean method [1]: $I_1 = 0.5$, $I_2 = 0$ and $I_3 = -0.5$. Moreover, the value of the contrary entropy coefficients, both J_1 and J_2, become equal to -1. Thus, the value of the synthesis of connection entropy of water resources vulnerability in Anhui Province is measured using Equation (16) and the findings are shown in Table 7. The curve for the synthesis of connection entropy of water resources vulnerability in Anhui Province is presented in Figure 2.

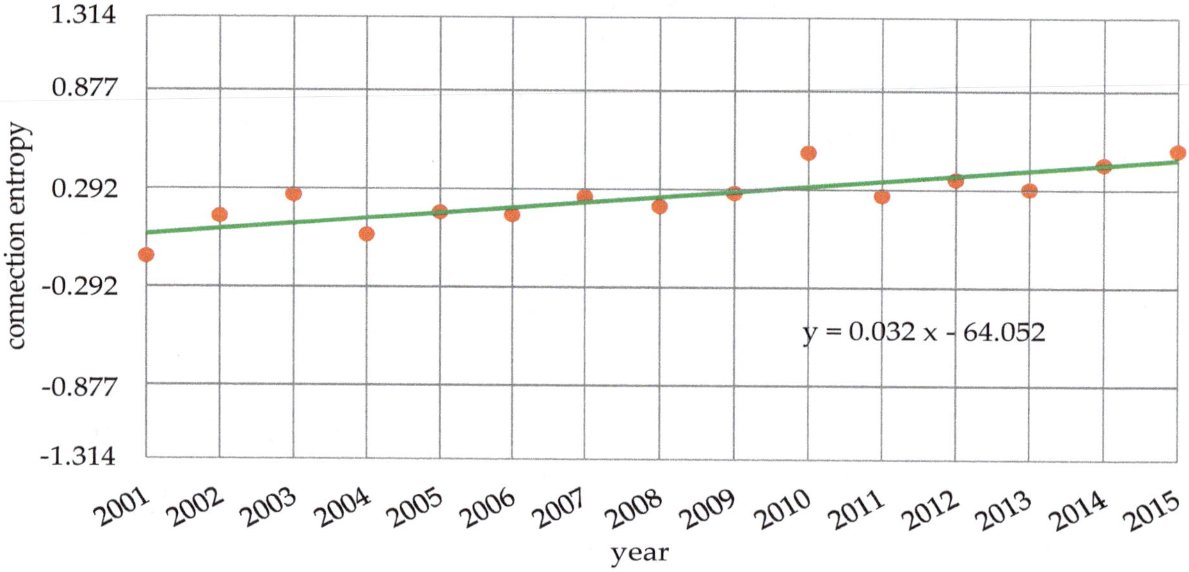

Figure 2. Synthesize connection entropy trend line of water resources vulnerability in Anhui Province.

5.5. Synthesis of the Level of Water Resources Vulnerability in Anhui Province

The synthesis of the level of water resources vulnerability in Anhui Province is attained with the help of a comparison of the value of the synthesis of connection entropy with the decision criterion in Section 4; the findings are presented in Table 7.

5.6. Analysis of Results

The synthesis of the level of water resources vulnerability in Anhui Province appeared to be moderate in 2001–2009, and 2011, and slight in 2010, and 2012–2015 (Table 7). As evident from Figure 2, the following can be observed: the vulnerability level fluctuated and exhibited an apparent improving trend from 2001 to 2015.

The analysis of vulnerability of typical years is as follows: (1) the synthesis of the level of water resources vulnerability in 2003 was highest during 2001–2005 period as it was a wet year. The average annual rainfall was 1460.9 mm, which was 24.4% higher than the multi-year average. Water resources availability was 108.301 billion m^3, which was 51.3% higher than the multi-year average. Thus, the vulnerability index water utilization rate was low, and the per unit area available amount of water resources was high. Good economic benefits are expected because of effective flood control and drought relief measures. (2) The synthesis of the level of water resources vulnerability in 2004 was the lowest during the 2002–2006 period as it was a low flow period. The average annual rainfall was 998.0 mm, 14.9% less than the multi-year average, and water resources availability was 50.065 billion m^3, being 30.1% less than the multi-year average. Thus, the value of the vulnerability index water utilization rate was high, and that of the per unit area available amount of water resources was low, together with the per unit GDP of water consumption being high. (3) The synthesis of the level of water resources vulnerability in 2010 was the highest during the 2008–2012 period as it was a partially wet period. The average annual rainfall was 1308.90 mm, being 11.6% more than the multi-year average, and water resources availability was 93.905 billion m^3, being 31.1% more than the multi-year average. Thus, the vulnerability index water utilization rate appears to be low, and the per unit area available amount of water resources is high, in addition to the reasonable economic benefits of flood control and fight drought.

6. Conclusions and Suggestions

From the finding presented in this paper, the following conclusions may be put forth:

(1) This study considered the mechanism of water resources vulnerability in a changing environment, focusing on the analysis of four key states of a water system: sensitive state, damaged state, recovery state, and equilibrium state. The state of a water system generally changes from damaged to recovery via natural factors, followed by the transition from a state of recovery to one of equilibrium primarily because of artificial factors. The former is defined as natural resilience and the latter as artificial adaptation. This mechanism results offer much-needed insights into water resources vulnerability in a changing environment, but the mechanism of transformation process between four states is the focus of future research.

(2) This paper proposed an analysis method to system uncertainty on the basis of connection entropy, extending the Zhao's [1] original concept. In addition, our analysis of water resources vulnerability in Anhui Province, China showed that the vulnerability level fluctuated, as well as exhibiting apparent improvement trends over a 15-year period. The synthesize grade of water resources vulnerability in Anhui Province appeared to be moderate in 2001–2009, and 2011, and a slight in 2010, and 2012–2015.

(3) The vulnerability level exhibited an apparent improving trend from 2001 to 2015, but still needs to be improved. There is a need to improve the perceptions of the vulnerability index, such as water utilization rate, per capita sewage discharge, surface water environment quality, per unit GDP of water consumption, and the economic benefits of flood control and fight drought.

Based on the results, we suggest the following improvements for water resources vulnerability in Anhui Province: on the one hand, there is an abated water resources sensitivity, whereby an increase in the reuse rate of industrial water can reduce water resources utilization. Moreover, sewage discharge can be reduced by the promotion of water saving policy and knowledge, encouraging water savings, and creating a greater awareness of (as well as building) a water-saving society. On the other hand, there is a need to improve natural resilience and artificial adaptation. Both can be achieved by developing sewage treatment to improve water environment quality, reduce high water consumption, eradicate backward industries, and promote new technology. Such actions will reduce the per unit GDP of water consumption and losses caused by floods and droughts, and improve the economic benefits of flood control and fight drought. Thus, for the purpose of ensuring sustainable development of economic, social and water resources in region, there is a need to improve the level of water resources vulnerability in the Anhui Province.

Acknowledgments: The authors would like to thank the support of the National Key Research and Development Program of China under Grant No. 2016YFC0401303 and No. 2016YFC0401305, the National Natural Science Foundation of China (Grant No. 51579059, No. 51509158, No. 51709071 and No. 51309004).

Author Contributions: Z.P. was responsible for the data analysis and code programming; Z.P. wrote the paper; R.Z. contributed the data; J.J. constructed the research framework and designed the study; C.L. and S.N. reviewed and edited the manuscript. All of the authors have read and approved the final manuscript.

References

1. Zhao, K.Q. *Set Pair Analysis and Preliminary Application*; China Science Technology Press: Hangzhou, China, 2000.
2. Turner, B.L., II; Kasperson, R.E.; Matson, P.A.; McCarthy, J.J.; Corell, R.W.; Christensen, L.; Eckley, N.; Kasperson, J.X.; Luers, A.; Martello, M.L.; et al. A framework for vulnerability analysis in sustainability science. *Proc. Natl. Acad. Sci. USA* **2003**, *100*, 8074–8079. [CrossRef] [PubMed]

3. Füssel, H.M. Vulnerability: A generally applicable conceptual framework for climate change research. *Glob. Environ. Chang.* **2007**, *17*, 155–167. [CrossRef]
4. Padowski, J.C.; Gorelick, S.M. Global analysis of urban surface water supply vulnerability. *Environ. Res. Lett.* **2014**, *9*, 104004. [CrossRef]
5. Allouchea, N.; Maananb, M.; Gontaraa, M.; Rollob, N.; Jmala, I.; Bouria, S. A global risk approach to assessing groundwater vulnerability. *Environ. Model. Softw.* **2017**, *88*, 168–182. [CrossRef]
6. Khakhar, M.; Ruparelia, J.P.; Vyas, A. Assessing groundwater vulnerability using GIS-based DRASTIC model for Ahmedabad district, India. *Environ. Earth Sci.* **2017**, *76*, 440. [CrossRef]
7. Rushforth, R.R.; Ruddell, B.L. The vulnerability and resilience of a city's water footprint: The case of Flagstaff, Arizona, USA. *Water Resour. Res.* **2016**, *52*, 2698–2714. [CrossRef]
8. Martinez, S.; Kralisch, S.; Escolero, O.; Perevochtchikova, M. Vulnerability of Mexico City's water supply sources in the context of climate change. *J. Water Clim. Chang.* **2015**, *6*, 518–533. [CrossRef]
9. Padowski, J.C.; Jawitz, J.W. Water availability and vulnerability of 225 large cities in the United States. *Water Resour. Res.* **2012**, *48*. [CrossRef]
10. Sullivan, C.A. Quantifying water vulnerability: A multi-dimensional approach. *Stoch. Environ. Res. Risk Assess.* **2011**, *25*, 627–640. [CrossRef]
11. McCarthy, J.J.; Canziani, O.F.; Leary, N.A.; Dokken, D.J.; White, K.S. *Climate Change 2001: Impacts, Adaptation, and Vulnerability. Contribution of Working Group II to the Third Assessment Report of the Intergovernmental Panel on Climate Change*; Cambridge University Press: Cambridge, UK, 2001.
12. Safi, A.S.; Smith, W.J.; Liu, Z. Vulnerability to climate change and the desire for mitigation. *J. Environ. Stud. Sci.* **2016**, *6*, 503–514. [CrossRef]
13. Liersch, S.; Cools, J.; Kone, B.; Koch, H.; Diallo, M.; Reinhardt, J.; Fournet, S.; Aich, V.; Hattermann, F.F. Vulnerability of rice production in the Inner Niger Delta to water resources management under climate variability and change. *Environ. Sci. Policy* **2013**, *34*, 18–33. [CrossRef]
14. Yang, X.H.; Sun, B.Y.; Zhang, J.; Li, M.S.; He, J.; Wei, Y.M.; Li, Y.Q. Hierarchy evaluation of water resources vulnerability under climate change in Beijing, China. *Nat. Hazards* **2016**, *84*, 63–76. [CrossRef]
15. Gain, A.K.; Giupponi, C.; Renaud, F.G. Climate Change Adaptation and Vulnerability Assessment of Water Resources Systems in Developing Countries: A Generalized Framework and a Feasibility Study in Bangladesh. *Water* **2012**, *4*, 345–366. [CrossRef]
16. Ionescu, C.; Klein, R.J.T.; Hinkel, J.; Kumar, K.S.K.; Klein, R. Towards a Formal Framework of Vulnerability to Climate Change. *Environ. Model. Assess.* **2009**, *14*, 1–16. [CrossRef]
17. Al-Saidi, M.; Birnbaum, D.; Buriti, R.; Diek, E.; Hasselbring, C.; Jimenez, A.; Woinowski, D. Water Resources Vulnerability Assessment of MENA Countries Considering Energy and Virtual Water. *Procedia Eng.* **2016**, *145*, 900–907. [CrossRef]
18. Shabbir, R.; Ahmad, S.S. Water resource vulnerability assessment in Rawalpindi and Islamabad, Pakistan using Analytic Hierarchy Process (AHP). *J. King Saud Univ. Sci.* **2016**, *28*, 293–299. [CrossRef]
19. Pan, Z.; Jin, J.; Wu, K.; Res, K.D. Earch on the Indexes and Decision Method of Regional Water Environmental System Vulnerability. *Resour. Environ. Yangtze Basin* **2014**, *23*, 518–525. [CrossRef]
20. Field, C.B.; Barros, V.R.; Dokken, D.J.; Mavh, K.J.; Mastrabdrea, M.D.; Bilir, T.E.; Chatterjee, M.; Ebi, K.L.; Estrada, Y.O.; Genova, R.C.; et al. (Eds.) *Climate Change 2014: Impacts, Adaptation, and Vulnerability. Summaries, Frequently Asked Questions, and Cross-Chapter Boxes. A Contribution of Working Group II to the Fifth Assessment Report of the Intergovernmental Panel on Climate Change*; World Meteorological Organization: Geneva, Switzerland, 2014.
21. Hassan, R.; Scholes, R.; Ash, N. (Eds.) Appendix D: Glossary. In *Ecosystems and Human Well-being: Current States and Trends*; Island Press: Washington, DC, USA, 2005; Volume 1, pp. 893–900.
22. Hartley, R.V. Transmission of Information. *Bell Syst. Tech. J.* **1928**, *7*, 535–563. [CrossRef]
23. Schrödinger, E. *What Is Life? The Physical Aspects of Living Cell*; Cambridge University Press: Cambridge, UK, 1944.
24. Shannon, C.E. A mathematical theory of communication. *Bell Syst. Tech. J.* **1948**, *27*, 379–423. [CrossRef]
25. De Luca, A.; Termini, S. A definition of a nonprobabilistic entropy in the setting of fuzzy sets. *Inf. Control.* **1972**, *20*, 301–312. [CrossRef]
26. Zhao, K. Study on set pair analysis and entropy. *J. Zhejiang Univ.* **1992**, *6*, 65–72.

27. Su, M.R.; Yang, Z.F.; Chen, B. Set pair analysis for urban ecosystem health assessment. *Commun. Nonlinear Sci. Numer. Simul.* **2009**, *14*, 1773–1780. [CrossRef]
28. Wang, W.; Jin, J.; Ding, J.; Li, Y. A new approach to water resources system assessment—Set pair analysis method. *Sci. China Ser. E Technol. Sci.* **2009**, *52*, 3017–3023. [CrossRef]
29. Kumar, K.; Garg, H. TOPSIS method based on the connection number of set pair analysis under interval-valued intuitionistic fuzzy set environment. *Comput. Appl. Math.* **2016**, 1–11. [CrossRef]
30. Pan, Z.; Wang, Y.; Jin, J.; Liu, X. Set pair analysis method for coordination evaluation in water resources utilizing conflict. *Phys. Chem. Earth* **2017**. [CrossRef]
31. Pan, Z.; Wu, C.; Jin, J. *Set Pair Analysis Methods for Water Resource System Evaluation and Prediction*; Science Press: Beijing, China, 2016.
32. Jin, J.; Wei, Y. *Generalized Intelligent Evaluation Method for Complex System and Its Application*; Science Press: Beijing, China, 2008.

11

Entropy Production in Stochastics

Demetris Koutsoyiannis

Department of Water Resources, School of Civil Engineering, National Technical University of Athens, 15780 Athina, Greece; dk@itia.ntua.gr

Abstract: While the modern definition of entropy is genuinely probabilistic, in entropy production the classical thermodynamic definition, as in heat transfer, is typically used. Here we explore the concept of entropy production within stochastics and, particularly, two forms of entropy production in logarithmic time, unconditionally (EPLT) or conditionally on the past and present having been observed (CEPLT). We study the theoretical properties of both forms, in general and in application to a broad set of stochastic processes. A main question investigated, related to model identification and fitting from data, is how to estimate the entropy production from a time series. It turns out that there is a link of the EPLT with the climacogram, and of the CEPLT with two additional tools introduced here, namely the differenced climacogram and the climacospectrum. In particular, EPLT and CEPLT are related to slopes of log-log plots of these tools, with the asymptotic slopes at the tails being most important as they justify the emergence of scaling laws of second-order characteristics of stochastic processes. As a real-world application, we use an extraordinary long time series of turbulent velocity and show how a parsimonious stochastic model can be identified and fitted using the tools developed.

Keywords: entropy production; conditional entropy production; stochastic processes; scaling; climacogram; turbulence

---

1. Introduction

Entropy was first recognized as a probabilistic concept in 1887 by Boltzmann [1], who established a relationship of entropy with probabilities of statistical mechanical system states, thus explaining the Second Law of Thermodynamics as the tendency of the system to run toward more probable states. In 1948 Shannon [2] used an essentially similar, albeit more general, definition of entropy as a probabilistic concept, a measure of information or, equivalently, uncertainty. In 1957 Jaynes [3] introduced the principle of maximum entropy thus equipping the entropy concept with a powerful tool for logical inference.

A decade later, probabilistic entropy and the principle of maximum entropy were used in geophysical sciences and particularly hydrology, initially for parameter estimation of models [4] and probability distributions [5]. Detailed reviews on the use of entropy in applications in hydrology, and water and environmental engineering have been provided by Singh ([6,7] and more recently [8–11]). Most applications of probabilistic entropy are static, disregarding time. In studying systems out of the equilibrium, time should necessarily be involved (e.g., [12]) and the paths of the systems should be inferred. The more recently developed, within non-equilibrium thermodynamics, extremum principles of entropy production attempt to predict the most likely path of system evolution [13–15]. Central among the latter are Prigogine's minimum entropy production principle [16] and Ziegler's maximum entropy production principle [17,18].

Entropy production in thermodynamic systems is typically defined in terms of the derivative of entropy with respect to time, while the entropy flux in open systems is also considered. Niven and Ozawa [19] provide a general definition of entropy production along with a brief review of applications of extremum principles in geophysics and particularly hydrology. It can be seen that in these applications, the entropy concept is typically used with its classical thermodynamic definition and in deterministic terms, without reference to its modern probabilistic definition.

Deterministic descriptions of the evolution of uncertain physical and geophysical systems are generally inefficient [20,21], while a much more powerful alternative is offered by stochastics. The term 'stochastics' describes the mathematics of random variables and stochastic processes, and comprises probability theory, statistics and stochastic processes. While the entropy concept is well defined within stochastics, the entropy production does not have a scientifically mature definition. By analogy to classical thermodynamic versions, we may define entropy production in stochastics as the derivative of (probabilistic) entropy with respect to time. As here we deal with a single stochastic process and not with interaction of many processes, there are no entropy fluxes and, thus, entropy production should merely rely on the time derivative of entropy. In an earlier study [22], the derivative with respect to the logarithm of time was introduced and termed 'entropy production in logarithmic time' (EPLT). Differentiation with respect to the logarithm of time has some attractive elements, such as:

- It is dimensionless as $\mathrm{d}(\ln t) = \mathrm{d}t/t$ and thus, given that entropy is a dimensionless quantity per se (see next section), the EPLT remains a dimensionless quantity.
- It is consistent with the notion of the 'arrow of time'. Usually entropy per se is related to the time asymmetry and the Second Law is regarded as the origin of the arrow. However, it may be simpler to think that it is the action of observation that creates time asymmetry. In this respect, we can set $t = 0$ for the observation time, i.e., the present, while $t < 0$ denotes the observable past and $t > 0$ denotes the unknown future. The fact that the past is observable, but the future not, generates the asymmetry. It is clarified that when the future becomes observable, it is no longer future; rather it has become present or past. Once the future ($t > 0$) is treated separately from the past, it is legitimate to differentiate with respect to the logarithm of time.
- It makes extremization of entropy production easier, particularly for asymptotic times $t \to 0$ and $t \to \infty$, as it avoids infinite or zero values of entropy production [22]; indeed, the asymptotic values of EPLT are always bounded (see Section 2.4).
- It provides the means to study or even explain the scaling behaviour often observed in geophysical processes (see Sections 2.2 and 2.4).

Here we advance the study of the EPLT concept, with particular emphasis on the conditional entropy production (CEPLT), when the past and present have been observed (Section 2.2). The theoretical properties of EPLT and CEPLT are studied in a general setting and in application to a broad set of specific types of stochastic processes (Sections 3.1 and 3.2). A main question studied is how the entropy production can be estimated from time series, i.e., realizations of stochastic processes. It turns out that there is a strong link of the EPLT with the climacogram, with the slope of a log-log plot of the latter representing the EPLT. Further, we introduce two new tools, additional to the climacogram and based on it, the differenced climacogram and the climacospectrum, where the latter has properties similar to those of the power spectrum. The slope of a log-log plot of either of the two additional tools can be used as an estimator of CEPLT. Of particular importance are the asymptotic slopes for time scale tending to zero or infinity (Section 3.3). To illustrate the theoretical and empirical properties of EPLT and CEPLT using real-world data and to test the applicability of the specific models proposed, a long time series of very fine resolution is needed in order to allow viewing the real-world behaviour at the far-left and the far-right tails. Fine resolution measurements of turbulence can provide an ideal test series and indeed a time series of length 36×10^6 of turbulent velocities is utilized for illustration and testing (Section 3.4).

To avoid bothering the reader with derivations and theoretical details, most of them have been put in a series of annexed sections organized as Appendix A.1. In addition, the details of the specific stochastic models and their behaviour have been grouped in Appendix A.2. While the two appendices are made in a stand-alone form separate from the body of the article, they are perhaps the most essential part of the study.

2. Methods

2.1. General Context

The evolution of a system state x over time t is usually represented as a trajectory $x(t)$, but such a description is often inefficient. Instead, the evolution can be represented as a stochastic process $\underline{x}(t)$, which is a collection of (infinitely many) random variables $\underline{x}$ indexed by t. A realization (sample) $x(t)$ of $\underline{x}(t)$ is a trajectory; if it is known at certain points t_i, $i = 1, 2, \ldots$, the realization is called a time series.

A random variable $\underline{x}$ is an abstract mathematical entity associated with a probability density (or mass) function $f(x)$; the realizations x of $\underline{x}$ belong to a set of possible numerical values. Notice the different notation of random variables and hence stochastic processes (underlined, according to the Dutch notation [23]) from regular ones.

Most natural processes evolve in continuous time but they are observed in discrete time, typically by averaging. Accordingly, the stochastic processes devised to represent the natural processes should evolve in continuous time and be converted into discrete time, as illustrated in Figure 1.

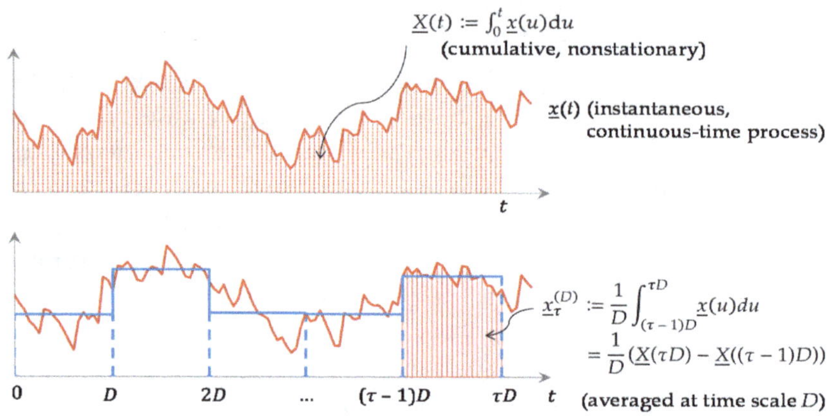

Figure 1. Explanatory sketch for a stochastic process in continuous time and in discrete time. Note that the graphs display a realization of the process (it is impossible to display the process as such) while the notation is for the process per se (adapted from [24]).

While a stochastic process denotes, by conception, change (process = a series of changes), there should be some properties that are unchanged in time. This implies the concept of stationarity [25], which is central in stochastics. According to Kolmogorov's definition [26], a stochastic process $\underline{x}(t)$ is stationary if the distributions of $(\underline{x}(t_1), \underline{x}(t_2), \ldots, \underline{x}(t_n))$ and $(\underline{x}(t_1 + \tau), \underline{x}(t_2 + \tau), \ldots, \underline{x}(t_n + \tau))$ coincide for any n, t_1, t_2, $\ldots$, t_n, τ. By negation, in a nonstationary process the probability density $f(x(t_1), x(t_2), \ldots, x(t_n))$ for some (or all) τ is not equal to $f(x(t_1 + \tau), x(t_2 + \tau), \ldots, x(t_n + \tau))$, which means that the mathematical expression of the latter should explicitly contain the time shift τ, or else that it should be a deterministic function of τ.

Stationarity is closely related to ergodicity, which in turn is a prerequisite to make inference from data, that is, induction. Within the stochastics domain ([20,27] (p. 427)) ergodicity is defined in the following manner: A stochastic process $\underline{x}(t)$ is ergodic if the time average of any function $g(\underline{x}(t))$, as time tends to infinity, equals the true (ensemble) expectation $\mathrm{E}[g(\underline{x}(t))]$. This allows the estimation (i.e., approximate calculation) of the true but unknown quantity $\mathrm{E}[g(\underline{x}(t))]$ (e.g., the true average $\mathrm{E}[\underline{x}(t)]$) from the available data, i.e., from the sample mean). Without stationarity there is no ergodicity and without ergodicity inference from data is impossible. More details about stationarity and ergodicity, and their importance, are provided by Koutsoyiannis and Montanari [25], along with highlights of the misconceptions and abuses of these concepts which abound in the literature.

For the remaining part of this article, unless otherwise stated, the processes are assumed to be stationary and ergodic, noting that nonstationary processes should be converted to stationary before their study. For example, the cumulative process $\underline{X}(t)$ in Figure 1 is nonstationary, but by differentiating it in time we obtain the stationary process $\underline{x}(t)$.

Without loss of generality we assume that the mean of the process $\underline{x}(t)$ is zero ($\mathrm{E}[\underline{x}(t)] = 0$) and we denote its variance as:

$$\gamma_0 := \mathrm{Var}[\underline{x}(t)] \tag{1}$$

Let:

$$\Gamma(t) := \mathrm{Var}[\underline{X}(t)],\ \gamma(k) := \mathrm{Var}[(1/k)(\underline{X}(t+k) - \underline{X}(t))] = \Gamma(k)/k^2 \tag{2}$$

be the variances of the cumulative and the averaged process, respectively (note that $\Gamma(0) = 0$ and $\gamma(0) = \gamma_0$). While $\Gamma(t)$, the variance of the nonstationary process $\underline{X}(t)$, is a function of t, $\gamma(k)$ is the variance of the stationary process $\underline{x}_\tau^{(k)}$, i.e., $\gamma(k) = \mathrm{Var}[\underline{x}_\tau^{(k)}]$ and thus it is not a function of time τ. We call $\Gamma(t)$ and $\gamma(k)$, as functions of time t and time scale k, the *cumulative climacogram* and the *climacogram* of the process, respectively.

The climacogram is the second central moment of the process, as a function of time scale, and thus it is a second-order characteristic of the process. Other customary second order characteristics of a stationary process are the autocovariance function $c(h)$, where h denotes time lag, the power spectrum $s(w)$, where w denotes frequency, and the structure function $v(h)$ (also known, for a stationary process, as the semivariogram or variogram). The latter has a direct analogy based on the climacogram, the climacostructure function $\xi(k)$. Definitions and notation for these second order characteristics are contained in Table 1 for the continuous time domain and in Table 2 for the discrete time case. All of these functions are transformations of one another, as shown in Table 3.

The climacogram, like the autocovariance function, is a positive definite function (see proof in Appendix A.1.2) but of the time scale k, rather than the time lag h. It is not as popular as the other tools but it has several good properties due to its simplicity, close relationship to entropy (see below), and more stable behaviour, which is an advantage in model identification and fitting from data. In particular, when estimated from data, the climacogram behaves better than all other tools, which involve high bias and statistical variation [24,28].

Table 1. Definition of main characteristics and notations for a stochastic process in continuous time; see Figure 1 for clarification.

Name of Quantity or Characteristic	Symbol and Definition	Remarks	Equation
Stochastic process of interest	$\underline{x}(t)$	Assumed stationary	
Time, continuous	t	Dimensional quantity	
Cumulative process	$\underline{X}(t) := \int_0^t \underline{x}(\xi)\mathrm{d}\xi$	Nonstationary	(3)
Variance, instantaneous	$\gamma_0 := \mathrm{Var}[\underline{x}(t)]$	Constant (not a function of t)	(4)
Cumulative climacogram	$\Gamma(t) := \mathrm{Var}[\underline{X}(t)]$	A function of t, $\Gamma(0) = 0$	(5)
Climacogram	$\gamma(k) := \mathrm{Var}[(1/k)(\underline{X}(t+k) - \underline{X}(t))]$ $= \mathrm{Var}[\underline{X}(k)/k] = \Gamma(k)/k^2$	Not a function of t, $\gamma(0) = \gamma_0$	(6)
Time scale, continuous	k	Units of time	
Autocovariance function	$c(h) := \mathrm{Cov}[\underline{x}(t), \underline{x}(t+h)]$	$c(0) = \gamma_0$	(7)
Time lag, continuous	h	Units of time	
Structure function (or semivariogram or variogram)	$v(h) := \frac{1}{2}\mathrm{Var}[\underline{x}(t) - \underline{x}(t+h)]$		(8)
Climacostructure function	$\xi(k) := \gamma_0 - \gamma(k)$		(9)
Power spectrum (or spectral density)	$s(w) := 4\int_0^\infty c(h)\cos(2\pi wh)\mathrm{d}h$		(10)
Frequency, continuous	$w = 1/k$	Units of inverse time	(11)

Table 2. Definition of main characteristics and notations for a stochastic process in discrete time and their relationship with those in continuous time; see Figure 1 for clarification.

Name of Quantity or Characteristic	Symbol and Definition	Remarks	Equation
Stochastic process, discrete time	$\underline{x}_\tau^{(D)} := \frac{1}{D}\int_{(\tau-1)D}^{\tau D} \underline{x}(u)\mathrm{d}u = \frac{1}{D}\left(\underline{X}(\tau D) - \underline{X}((\tau-1)D)\right)$		(12)
Time unit = discretization time step	D	Time step or length of time window of averaging	
Time, discrete	$\tau := t/D$	Dimensionless quantity, integer	(13)
Characteristic variance	$\mathrm{Var}[\ \underline{x}_\tau^{(D)}] = \gamma(D)$		(14)
Climacogram	$\gamma_\kappa^{(D)} := \gamma(\kappa D) = \frac{\Gamma(\kappa D)}{(\kappa D)^2}$	$\gamma_1^{(D)} = \gamma(D)$	(15)
Time scale, discrete	$\kappa := k/D$	Dimensionless quantity	(16)
Autocovariance function	$c_\eta^{(D)} := \mathrm{Cov}[\underline{x}_\tau^{(D)}, \underline{x}_{\tau+\eta}^{(D)}]$	$c_0^{(D)} = \gamma(D)$	
Time lag, discrete	$\eta := h/D$	Dimensionless quantity	(17)
Structure function	$v_\eta^{(D)} := \gamma(D) - c_\eta^{(D)}$		(18)
Power spectrum	$s_\mathrm{d}^{(D)}(\omega) := \frac{1}{D}\sum_{j=-\infty}^{\infty} s\left(\frac{\omega+j}{D}\right)\ \mathrm{sinc}^2(\pi(\omega+j))$		(19)
Frequency, discrete	$\omega := wD = 1/\kappa$	Dimensionless quantity	(20)

Note: In time-related quantities, Latin letters denote dimensional quantities and Greek letters dimensionless ones. The Latin i, j, l may also be used as integers to denote quantities τ, η, κ, depending on the context.

Table 3. Relationships between several characteristics of a process in continuous and discrete time.

Related Characteristics	Direct Relationship	Inverse Relationship	Equation
$\gamma(k) \leftrightarrow c(h)$	$\gamma(k) = 2\int_0^1 (1-\chi)c(\chi k)\mathrm{d}\chi$	$c(h) = \frac{1}{2}\frac{\mathrm{d}^2(h^2\gamma(h))}{\mathrm{d}h^2}$	(21)
$s(w) \leftrightarrow c(h)$	$s(w) := 4\int_0^\infty c(h)\cos(2\pi wh)\mathrm{d}h$	$c(h) = \int_0^\infty s(w)\cos(2\pi wh)\mathrm{d}w$	(22)
$\gamma(k) \leftrightarrow s(w)$	$\gamma(k) = \int_0^\infty s(w)\ \mathrm{sinc}^2(\pi wk)\mathrm{d}w$	$s(w) := 2\int_0^\infty \frac{\mathrm{d}^2(h^2\gamma(h))}{\mathrm{d}h^2}\cos(2\pi wh)\mathrm{d}h$	(23)
$v(h) \leftrightarrow c(h)$	$v(h) = \gamma_0 - c(h)$	$c(h) = v(\infty) - v(h) \quad (v(\infty) = \gamma_0)$	(24)
$\xi(k) \leftrightarrow \gamma(k)$	$\xi(k) := \gamma_0 - \gamma(k)$	$\gamma(k) = \xi(\infty) - \xi(k) \quad (\xi(\infty) = \gamma_0)$	(25)
$\xi(k) \leftrightarrow v(h)$	$\xi(k) = 2\int_0^1 (1-\chi)v(\chi k)\mathrm{d}\chi$	$v(h) = \frac{1}{2}\frac{\mathrm{d}^2(h^2\xi(h))}{\mathrm{d}h^2}$	(26)
$\gamma_\kappa^{(D)} \equiv \gamma(\kappa D) \leftrightarrow c_\eta^{(D)}$	$\gamma_\kappa^{(D)} = \frac{1}{\kappa}\left(c_0^{(D)} + 2\sum_{\eta=1}^{\kappa-1}(1-\frac{\eta}{\kappa})c_\eta^{(D)}\right)$ Alternatively, $\gamma_\kappa^{(D)} = \frac{\Gamma(\kappa D)}{(\kappa D)^2}$ where, in recursive mode, $\Gamma(\kappa D) = 2\Gamma((\kappa-1)D) - \Gamma((\kappa-2)D) + 2c_{j-1}^{(D)}D^2$ with $\Gamma(0) = 0$, $\Gamma(D) = c_0^{(D)}D^2$	$c_\eta^{(D)} = \frac{1}{D^2}\left(\frac{\Gamma(\lvert\eta+1\rvert D)+\Gamma\left((\lvert\eta-1\rvert D\right)}{2} - \Gamma(\lvert\eta\rvert D)\right)$	(27)
$c_\eta^{(D)} \leftrightarrow s_\mathrm{d}^{(D)}(\omega)$	$s_\mathrm{d}^{(D)}(\omega) = 2c_0^{(D)} + 4\sum_{\eta=1}^{\infty} c_\eta^{(D)}\cos(2\pi\eta\omega)$	$c_\eta^{(D)} = \int_0^{1/2} s_\mathrm{d}^{(D)}(\omega)\cos(2\pi\omega\eta)\mathrm{d}\omega$	(28)
$v_\eta^{(D)} \leftrightarrow c_\eta^{(D)}$	$v_\eta^{(D)} = \gamma(D) - c_\eta^{(D)}$	$c_\eta^{(D)}\gamma(D) - v_\eta^{(D)}$	(29)

The climacogram involves bias too, but this can be determined analytically and included in the estimation. Specifically, assuming that we have n observations of the averaged process $\underline{x}_i^{(k)}$, at scale $k = \kappa D$, so that $T = nk$ is the observation period (rounded off to an integer multiple of k), the standard statistical estimator of variance has expected value:

$$\mathrm{E}\left[\underline{\hat{\gamma}}(k)\right] = \theta(k,T)\gamma(k) \tag{30}$$

where θ is the bias correction coefficient, given by [22,24]:

$$\theta(k,T) = \frac{1-\gamma(T)/\gamma(k)}{1-k/T} = \frac{1-(k/T)^2\Gamma(T)/\Gamma(k)}{1-k/T} \tag{31}$$

It becomes clear from the above equations, which are valid for any process, that direct estimation of the variance $\gamma(k)$ for any time scale k, let alone that of the instantaneous process γ_0, is not possible merely from the data. We need to know the ratio $\gamma(T)/\gamma(k)$ and thus we should assume a stochastic model which evidently influences the estimation of $\gamma(k)$. Once the model is assumed and its parameters estimated based on the data, we can expand our calculations to estimate the variance for any time scale, including that of the instantaneous scale γ_0.

2.2. Entropy and Entropy Production

Historically entropy was introduced in thermodynamics but its rigorous definition was given within probability theory. Thermodynamic and probabilistic entropy are regarded by many as different concepts having in common only their name. However, here we embrace the view that they are essentially the same thing, as has articulated elsewhere [29,30]. According to the latter interpretation, the mathematical description of thermodynamic systems could be produced by the probabilistic definition of entropy. As an example indicating how powerful this interpretation could be, the law of phase change transition of water (Clausius-Clapeyron equation) has been produced in [30], with impressive agreement with reality, by maximizing, in a formal probabilistic frame, the entropy, i.e., the total uncertainty about the state of a single molecule of water.

In this frame, entropy (often called Boltzmann-Gibbs-Shannon entropy to give credit to its pioneers and distinguish it from other forms of generalized entropies) is a *dimensionless* measure of uncertainty defined as follows:

- For a discrete random variable $\underline{x}$ with probability mass function $P_j := P\{\underline{x} = x_j\}, j = 1, \ldots, w$:

$$\Phi[\underline{x}] := \mathrm{E}[-\ln\ P\underline{x}] = -\textstyle\sum_{j=1}^{w} P_j \ln P_j \tag{32}$$

- For a *continuous random variable* $\underline{z}$ with probability density function $f(z)$:

$$\Phi[\underline{x}] := \mathrm{E}\left[-\ln\frac{f(\underline{x})}{m(\underline{x})}\right] = -\int_{-\infty}^{\infty} \ln\frac{f(x)}{m(x)} f(x)dx \tag{33}$$

where $m(x)$ is the density of a background measure (usually Lebesgue, i.e., $m(x) = 1$, with dimensions $[x^{-1}]$).

Entropy acquires its importance from the *principle of maximum entropy* [3], which postulates that the entropy of a random variable should be at maximum, under some conditions, formulated as constraints, which incorporate the information that is given about this variable. Its physical counterpart, the tendency of entropy to become maximal (Second Law of thermodynamics) is the driving force of natural change.

The definition of the entropy of a stochastic process directly follows that of a random variable. Thus, the entropy of the cumulative process $\underline{X}(t)$ with probability density function $f(X; t)$ is a dimensionless quantity defined as:

$$\Phi\ [\underline{X}(t)] := \mathrm{E}\left[-\ln\frac{f(\underline{X};t)}{m(\underline{X})}\right] = -\int_{-\infty}^{\infty} \ln\frac{f(X;\ t)}{m(X)} f(X;\ t)dX \tag{34}$$

In a stochastic process, which by definition involves time, it is natural to consider the change of entropy in time. Intuitively, for the stochastic process $\underline{X}(t)$ we can define entropy production as the time derivative, i.e., $\Phi'[\underline{X}(t)] := \mathrm{d}\Phi[\underline{X}(t)]/\mathrm{d}t$. This has dimensions of inverse time. Koutsoyiannis [22] defined the entropy production in logarithmic time (EPLT) as the derivative of entropy in logarithmic time:

$$\phi(t) \equiv \phi[\underline{X}(t)] := \Phi'[\underline{X}(t)]\, t \equiv \mathrm{d}\Phi[\underline{X}(t)]/\mathrm{d}(\ln t) \tag{35}$$

and showed that it has some advantages over $\Phi'[\underline{x}(t)]$ as described above.

Assuming that the background density is constant, $m(X) \equiv m$, and using simple constraints related to the preservation of the mean and variance, the principle of maximum entropy yields that the probability distribution of $\underline{X}$ will be Gaussian. For a Gaussian process the entropy depends on its variance $\Gamma(t)$ and is:

$$\Phi[\underline{X}(t)] = (1/2)\ln(2\pi\mathrm{e}\,\Gamma(t)/m^2) \tag{36}$$

Hence:

$$\phi(t) = \Gamma'(t)\, t/2\Gamma(t) = 1 + \gamma'(t)\, t/2\gamma(t) \tag{37}$$

Because of the simplicity of the Gaussian case, we are basing further analyses on that assumption, noting that for large times, due to the Central Limit Theorem, the cumulative process $\underline{X}(t)$ will tend to Gaussian, even if for the smallest scales it deviates from Gaussian.

A question arises whether $\phi(t)$ is a totally abstract concept useful for theoretical analyses only or it can be related to data, and visualized or estimated from them. The reply is fortunate if, in addition, we consider another concept, the log-log derivative (LLD) of a function $f(x)$, formally expressed by:

$$f^{\#}(x) := \frac{\mathrm{d}(\ln f(x))}{\mathrm{d}(\ln x)} = \frac{x f'(x)}{f(x)} \tag{38}$$

This is visualized as the (local) slope on the popular log-log plot (plot of $\log f(x)$ vs. $\log x$). It is then seen that (37) can be written as:

$$\phi(t) = ½\, \Gamma^{\#}(t) = 1 + ½\, \gamma^{\#}(t) \tag{39}$$

In other words, EPLT is visualized and estimated by the slope of a log-log plot of the climacogram.

The quantity $\phi(t)$ represents the entropy production when nothing is known (observed) about the process realization. More interesting is the conditional EPLT (CEPLT, $\phi_{\mathrm{C}}(t)$) when the past ($t < 0$) and the present ($t = 0$) are observed. In this case, instead of the unconditional variance $\Gamma(t)$ we should use a variance $\Gamma_{\mathrm{C}}(t)$ conditional on the past and present. It has been shown [22,24] that for a Markov process (see definition of the Markov process in Section 3.1 and its details in Appendix A.2.1), the conditional variance is:

$$\Gamma_{\mathrm{C}}(t) \approx 2\Gamma(t) - \frac{\Gamma(2t)}{2}, \qquad \gamma_{\mathrm{C}}(k) \approx 2(\gamma(k) - \gamma(2k)) \tag{40}$$

We note that the difference $\gamma(k) - \gamma(2k)$ is always nonnegative (this is shown in Appendix A.1.3, Equation (A21)) and thus it can indeed represent a (conditional) variance.

For a non-Markov process the following adaptation is necessary:

$$\gamma_{\mathrm{C}}(k) = \varepsilon(\gamma(k) - \gamma(2k)) \tag{41}$$

where ε is a constant, ensuring that as $k \to \infty$, $\gamma_{\mathrm{C}}(k)/\gamma(k) \to 1$; the rationale of this limit is that the information about the past and present will tend not to affect the far distant future. As shown in the Appendix A.1.4, its value is:

$$\varepsilon = \frac{1}{1 - 2^{\gamma^{\#}(\infty)}} \tag{42}$$

It is easily verified that for a Markov process in which $\gamma^{\#}(\infty) = -1$ (see Appendix A.2.1), $\varepsilon = 2$, while for a Hurst-Kolmogorov process (original or filtered; see details in [24], in next session and in Appendix A.2) for which $\gamma^{\#}(\infty) = 2H - 2$, $\varepsilon = 1/(1 - 2^{2H-2})$, where H is the Hurst parameter.

Due to its mathematical form, we will refer to $\gamma_C(k)$ as the differenced climacogram. The inverse transformation, i.e., that giving the climacogram $\gamma(k)$ once the differenced climacogram $\gamma_C(k)$ is known, can be easily deduced by iterative evaluations also observing that $\gamma(\infty) = \gamma_C(\infty) = 0$ (this is a necessary condition in order for a stochastic process to be ergodic [27] (p. 429)) and takes both the following forms:

$$\gamma(k) = \frac{1}{\varepsilon}\sum_{i=0}^{\infty}\gamma_C\left(2^i k\right) = \gamma(0) - \frac{1}{\varepsilon}\sum_{i=1}^{\infty}\gamma_C\left(2^{-i} k\right) \tag{43}$$

whereas for numerical evaluations it should be kept in mind that for large time scales k, $\gamma(k) \approx \gamma_C(k)$. Note that (43) entails:

$$\gamma(0) = \frac{1}{\varepsilon}\sum_{i=-\infty}^{\infty}\gamma_C\left(2^i k\right) \tag{44}$$

and this is valid for any k. The importance of the differenced climacogram stems from the fact that it is directly related to the CEPLT, through its obvious relationship (similar to (39)):

$$\varphi_C(k) = 1 + {}^1\!/_2\gamma_C^{\#}(k) \tag{45}$$

As the differenced climacogram can be easily visualized from data, because it only needs calculation of variances, it directly provides a means for estimation and visualization of the CEPLT in terms of the slope in a log-log plot of the differenced climacogram.

In Appendix A.1.5 it is shown that the asymptotic properties of $\gamma_C(k)$ are:

$$\gamma_C(0) = \gamma_C(\infty) = 0, \quad \gamma_C^{\#}(0) = \xi^{\#}(0), \quad \gamma_C^{\#}(\infty) = \gamma^{\#}(\infty) \tag{46}$$

In other words, we need both the climacogram $\gamma(k)$ and the climacostructure function $\xi(k)$ to express the asymptotic behaviour of $\gamma_C(k)$. Once this is known, the asymptotic behaviour of CEPLT is also known from (45).

As shown in Appendix A.1.6, the differenced climacogram for persistent processes (see their definition in next subsection) is not integrable in $(0, \infty)$. However, by slightly changing it multiplying with k, we can obtain an additional tool which is integrable and has some additional good properties. Specifically, we introduce the following tool, which, as we will see next, resembles the power spectrum and thus we refer to it as the climacospectrum:

$$\zeta(k) := \frac{k(\gamma(k) - \gamma(2k))}{\ln 2} \tag{47}$$

The climacospectrum is also written in an alternative manner in terms of frequency $w = 1/k$:

$$\tilde{\zeta}(w) := \zeta(1/w) = \frac{\gamma(1/w) - \gamma(2/w)}{(\ln 2)w} \tag{48}$$

The inverse transformation, i.e., that giving the climacogram $\gamma(k)$ once the climacospectrum $\zeta(k)$ is known, can be easily deduced with the same method as before, and takes the form:

$$\gamma(k) = \ln 2\sum_{i=0}^{\infty}\frac{\zeta(2^i k)}{2^i k} = \gamma(0) - \ln 2\sum_{i=1}^{\infty}\frac{\zeta(2^{-i}k)}{2^{-i}k} \tag{49}$$

A very good approximation in which the sum is replaced by an integral, is:

$$\gamma(k) \approx \frac{(\varepsilon \ln 2)k + 2}{2k + 2} I(k), \quad I(k) := \int_k^{\infty}\frac{\zeta(x)}{x^2}\mathrm{d}x = \int_0^{1/k}\tilde{\zeta}(w)\mathrm{d}w \tag{50}$$

This holds for all k and its details are given in Appendix A.1.7. For small k, it is easy to see that the multiplier of the integral $I(k)$ is 1. It follows that for small time scales k or large frequencies $w = 1/k$ the climacospectrum, expressed in terms of frequency, has an area $I(k)$ on the left of w equal to the variance $\gamma(1/w)$. As shown in Appendix A.1.6, the entire area under the curve $\tilde{\zeta}(w)$ (i.e., $I(0)$) is precisely equal to the variance $\gamma(0)$ of the instantaneous process, a property also shared by the power spectrum $s(w)$. This is not the only connection with the power spectrum. The climacospectrum has also the same asymptotic behaviour with the power spectrum, i.e.,

$$\tilde{\zeta}^{\#}(0) = -\zeta^{\#}(\infty) = s^{\#}(0), \quad \tilde{\zeta}^{\#}(\infty) = -\zeta^{\#}(0) = s^{\#}(\infty) \tag{51}$$

(see Appendix A.1.9). This property holds almost always, with the exception of the cases where $\zeta^{\#}(0)$ is a specific integer ($\zeta^{\#}(\infty) = -1$ or $\zeta^{\#}(0) = 3$, as explained in Appendix A.1.9).

Again the connection of the climacospectrum with the CEPLT emerges through log-log derivatives. Specifically, combining (41), (45) and (47) we find that $\zeta^{\#}(k) = 1 + \gamma_C^{\#}(k) = 1 + 2\varphi_C(k) - 2$; hence:

$$\varphi_C(k) = 1/2\left(1 + \zeta^{\#}(k)\right) = 1/2\left(1 - \tilde{\zeta}^{\#}(1/k)\right) \tag{52}$$

By virtue of (51) and with the exceptions mentioned, the asymptotic CEPLT can also be inferred from the slopes of the power spectrum and, as the climacospectrum slopes do not differ substantially from those of the power spectrum even for intermediate time scales, the power spectrum slopes could also be used for a qualitative estimation if the CEPLT.

Obviously, however, the climacospectrum is more advantageous than the power spectrum in this respect, because the connection with conditional entropy production is more precise and without exceptions at all. In addition, intuitively the variance, on which the definition of the climacospectrum is based, is more closely related to uncertainty, and hence entropy of the process, than the power spectrum and the autocovariance. Other advantages are the easy calculation only using the concept of variance, without any need to perform laborious transformations (such as Fourier) and the fact that, like the climacogram, it is not affected by discretization (while autocovariance and power spectrum are). The empirical climacogram and climacospectrum are easily determined from data using nothing more than the standard statistical estimator of variance and they have a smooth shape, much smoother than those of the empirical autocovariance and power spectrum, thus enabling better model identification and fitting.

When the objective is the model fitting, we should be aware that any of the statistical tools entail estimation bias, which should be accounted for in the fitting. Again the climacogram and the climacospectrum are preferable than other tools as the bias is explicitly calculated from the assumed model (Equations (30) and (31)). In particular, the bias of the climacospectrum is usually very small (see the relevant graphs in Appendix A.2) because of its definition as a difference of two variances, in which the biases tend to cancel out. In comparison with another version of climacogram-based spectrum [24], the climacospectrum introduced here has advantages. Namely, these are its simpler expression, as it is a linear transformation (difference) of the climacogram, and the absence of the variance $\gamma_0 \equiv \gamma(0)$ from the definition, which makes possible the calculation of the empirical climacospectrum prior to specifying the model (note that γ_0 is not known before a model is specified and its parameters are estimated).

2.3. Scaling

In the previous subsection we have shown that the EPLT and the CEPLT are related to LLDs or slopes of log-log plots of second order tools such as climacogram, climacospectrum, power spectrum, etc. With a few exceptions, these slopes are nonzero asymptotically, hence entailing asymptotic scaling or asymptotic power laws with the LLDs being the scaling exponents. It is intuitive to expect that an emerging asymptotic scaling law would provide a good approximation of the true law for a range of scales. If the scaling law was appropriate for the entire range of scales, then we would have a simple scaling law. Such simple scaling sounds attractive from a mathematical point of view, but, for reasons that will be explained in the next subsection, it turns out to be impossible to make it appropriate for physical processes.

It is thus physically more realistic to expect two different types of asymptotic scaling laws, one in each of the ends of the continuum of scales. We call these two behaviours local scaling or fractal behaviour, when $k \to 0$ or $w \to \infty$, and global scaling, persistence or Hurst-Kolmogorov behaviour, when $k \to \infty$ or $w \to 0$. The respective scaling exponents are (see Appendix A.1.8):

- For local scaling: $\gamma_C^{\#}(0) = \xi^{\#}(0) = v^{\#}(0) = \zeta^{\#}(0) - 1 = 2\varphi_C(0) - 2 = -s^{\#}(\infty) - 1$
- For global scaling: $\gamma_C^{\#}(\infty) = \gamma^{\#}(\infty) = c^{\#}(\infty) = \zeta^{\#}(\infty) - 1 = 2\varphi_C(\infty) - 2 = -s^{\#}(0) - 1$

Here, the emergence of scaling has been related to maximum entropy considerations, and this may provide the theoretical background in modelling complex natural processes by such scaling laws. However, as shown in Appendix A.1.8, scaling laws are a mathematical necessity and could be constructed for virtually any continuous function defined in $(0, \infty)$. In other words, there is no magic in power laws, except that they are, logically and mathematically, a necessity. (No assumption of criticality, self-organization, fractal or multi-fractal generating mechanisms, is necessary to justify their emergence). Here we have focused on second order characteristics of a stochastic process. If we also considered higher order (raw) moments, it is natural to expect that some other (different) power laws would emerge (or could be constructed using the procedure described in Appendix A.1.8), and then one would speak about multifractals everywhere, etc.

2.4. Bounds of Scaling and Entropy Production

It is not well known that the asymptotic scaling exponents cannot take on any arbitrary values but they should be constrained in rather narrow ranges. Instead, misconceptions and incorrect applications resulting in inconsistent values abound in the literature. One of the most typical examples is the asymptotic scaling of the power spectrum for $w \to 0$, for which often steep slopes are reported for low frequencies (down to 0), based on data analyses. Usually these reports are accompanied with a claim that the process with $-s^{\#}(0) > 1$ is nonstationary, going further to characterize the power spectrum as a tool to identify whether a process is stationary or nonstationary (with values $-s^{\#}(0)$ below or above 1 suggesting stationarity and nonstationarity, respectively). As thoroughly studied elsewhere [31], this entire line of thought is theoretically inconsistent and flawed, while such results often reported are usually artefacts due to insufficient data or inadequate estimation algorithms. In fact, it has been shown [31] that a process with $-s^{\#}(0) > 1$ is nonergodic. Inference from data is only possible when the process is ergodic and thus, claiming that $-s^{\#}(0) > 1$ based on data is self-contradictory. Furthermore, claiming nonstationarity using the power spectrum as a function of frequency is also self-contradictory as in a nonstationary process both the autocovariance and the spectral density, i.e., the Fourier transform of the autocovariance, are functions of two variables, one being related to "absolute" time (see e.g., [32]).

An example of the conditions leading to misinterpreting a stationary process as nonstationary is discussed in Section 3.3 and Appendix A.2.2. We must clarify though that steep slopes ($-s^{\#}(w) > 1$) are mathematically and physically possible for medium and large w (see, e.g., Figure A7) and indeed they are quite frequent in geophysical and other processes. In this respect, the estimation from data of steep slopes is possible and not problematic, if we are conscious that such slopes are for intermediate or large frequencies (actually, the estimation algorithms are bandpass filtered with the bounds of the filter determined by the resolution and length of the series). The problem arises when such slopes are interpreted as asymptotic ones and particularly when they are used for inference resulting in self-contradicting conclusions.

Because of the equality of slopes of power spectrum and climacospectrum, the ergodicity limitation holds also for the slope of the climacospectrum, i.e., $\zeta^{\#}(\infty) = -\tilde{\zeta}^{\#}(0) < 1$. On the other hand, too steep negative asymptotic slopes of the climacospectrum are also impossible. Indeed, $\zeta^{\#}(k) = -\tilde{\zeta}^{\#}(1/k) < -1$ would entail (because of (52)) $\varphi_C(k) < 0$ and (because of (37) and (45)),

$\Gamma'_C(k) < 0$. This means that the variance of the cumulative process would be a decreasing function of time, which is absurd. This holds both for the global case ($k \to \infty$, in which the conditional variance $\Gamma_C(\infty)$ equals the unconditional $\Gamma(\infty)$) and the local case ($k \to 0$, for the conditional variance $\Gamma_C(0)$).

However, in the local case there is a more severe limitation imposed by physical reasoning. As demonstrated in Appendix A.1.10, the case $\zeta^{\#}(0) = -s^{\#}(\infty) < 1$ would entail infinite variance. Infinite variance would require infinite energy to emerge, which is physically inconsistent. Therefore, the physical lower limit for $\zeta^{\#}(0) = -s^{\#}(\infty)$ is 1. Interestingly, the inconsistency if this constraint is violated stems from the behaviour at high frequencies (see Appendix A.1.10), even though the power spectrum at high frequencies in such cases tends to zero rather diverging to infinity. Therefore, the related "catastrophe" is of "ultraviolet" type, while the infinite value of the power spectrum at low frequencies, which was highlighted in several texts as "infrared catastrophe" [33] is not actually a problem at all, nor does it impose any limitation on the scaling law additional to the limitation for ergodicity discussed above.

A final—and quite severe—limitation is an upper bound of the local scaling exponent, which is 3 for $\zeta^{\#}(0) = -s^{\#}(\infty)$. Proof is provided in Appendix A.1.11. The problem if this limitation is violated is that the resulting autocovariance function is not positive definite or, equivalently, that the resulting power spectrum is not always (for any frequency w) positive but takes on negative values for some w. Likewise, the Fourier transform of the climacogram takes on negative values for some w.

Because of the relationship of the entropy production with the scaling exponents, the bounds on the spectrum slopes translate directly into bounds of the CEPLT, i.e., $1 \le \varphi_C(0) \le 2$ and $0 \le \varphi_C(\infty) \le 1$. These define the "green square" of admissible values of φ_C in Figure 2, which is also depicted in terms of admissible values of slopes $\zeta^{\#}$ and $s^{\#}$ (noting that, as discussed above, $s^{\#}$ can, by exception, take on values out of the square when $\varphi_C(0) = 2$ or $\varphi_C(\infty) = 0$. The reasons why a process out of the square would be impossible or inconsistent, as discussed above, are also marked in Figure 2.

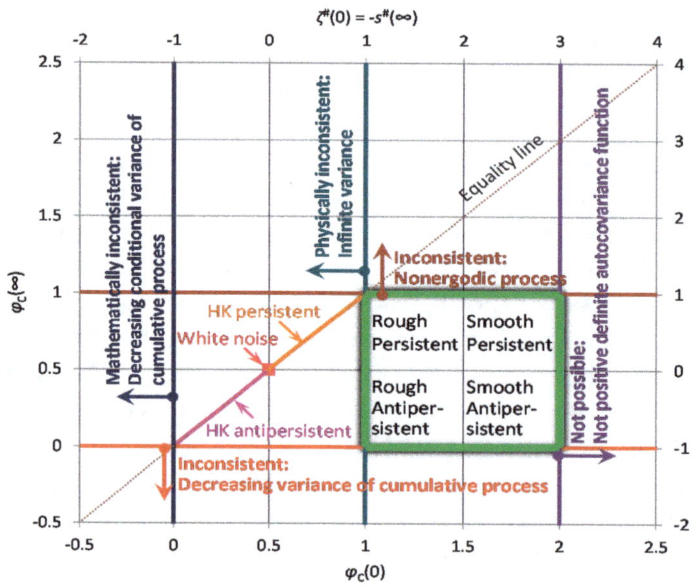

Figure 2. Bounds of asymptotic values of CEPLT, $\varphi_C(0)$ and $\varphi_C(\infty)$, and corresponding bounds of the log-log slopes of power spectrum and climacospectrum The "green square" represents the admissible region (note that $s^{\#}$ can, by exception, take on values out of the square when $\varphi_C(0) = 2$ or $\varphi_C(\infty) = 0$). The reasons why a process out of the square would be impossible or inconsistent are also marked. The lines $\varphi_C(0) = 3/2$ and $\varphi_C(\infty) = 1/2$ represent neutrality and support the classification of stochastic processes into the indicated four categories (smaller squares within the "green square").

Large values of $\varphi_C(0)$ indicate a smooth process and small ones a rough process. Also, large values of $\varphi_C(\infty)$ indicate a persistent process and small ones an antipersistent process. The centre of the square, with coordinates $\varphi_C(0) = 3/2$, $\varphi_C(\infty) = 1/2$ represents a neutral process. As will be seen in next section, neutrality corresponds to a Markov process, which is neither rough nor smooth, and neither antipersistent nor persistent. Even though sometimes it is said to reflect short-term persistence, we must have in mind that a discretized Markov process at time scale k tends to be uncorrelated in time as k increases and therefore we may not have it in mind as persistent. In other words, here persistence is used as synonymous to long-term persistence, while high autocorrelations at small lags or scales, which die off at large lags or scales, are better described with the term smoothness. Both smoothness and persistence reflect high entropy production, locally and globally, respectively.

A useful observation in Figure 2 is that the entire "green square" lies below the equality line, which means that the same scaling exponent is not possible for both local and global behaviour, or else, it is impossible to have a physically realistic simple scaling process. There is one exception, the upper-left corner of the "green square", which corresponds to the so-called "pink noise" or "$1/f$ noise" and will be discussed further in Section 3.3.

Since we have pointed out several inconsistent or contradictory results that are reported in the literature, it may be useful to mention another case, which is frequently met in geostatistics. The so-called *intrinsic* models are common in geostatistics but susceptible to inconsistent use, even though the literature presents them as nonstationary to avoid some of the emerging problems. Such models are defined in terms of their structure function (variogram), $v(h)$, which tends to infinity as lag h tends to infinity (e.g., [34]). An example is $v_I(h) = a\, h^b$ with $b > 0$, in which we note that its formulation with respect to a single scalar argument, h, does not reveal the nonstationarity. Thus, it may be treated as if it was stationary. In that case, from (24) we have $\gamma_0 = v_I(\infty) = \infty$ and thus the intrinsic process violates the constraint for physical consistency. Moreover, since $v_I(h) = \gamma_0 - c(h) < \infty$ for $h < \infty$, it turns out that $c(h) = \infty$ for any h and, hence, (21) entails that $\gamma(k) = \infty$ for any k. In other words, the infinity of $v_I(\infty)$ is transferred to the entire autocorrelation function and the entire climacogram. In particular, the property $\gamma(k) = \infty$ means that the process is non-ergodic (in ergodic processes $\gamma(k)$ should tend to zero as k tends to infinity; see [27] (p. 429)). Thus, in addition to being physically inconsistent such a treatment of the process is mathematically inconsistent and logically self-contradictory. A consistent way of treatment is to identify the intrinsic process with the (nonstationary) cumulative process $\underline{X}(t)$ [35], derive the stationary process $\underline{x}(t)$ and treat the latter further regularly. In that case the variogram is no longer related to the structure function of $\underline{x}(t)$ but to its climacogram; namely, it is identical to the cumulative climacogram $v_I(h) = \Gamma(h) = \gamma(h)\, h^2$.

Another peculiarity in geostatisical analyses is the so-called "nugget effect", which is also problematic or enigmatic [36]. Namely, this is a discontinuity of the structure function at the origin. An example is $v_E(h) = a\,(h + c)^b$, $h > 0$, with $c, b > 0$, while $v_E(0) = 0$. Investigation shows that the "nugget effect" does not necessarily create inconsistency. It is obviously associated with an infinite derivative of the structure function at the origin, i.e., $v'(0) = \infty$. However, the LLD, $v^{\#}(h) = v'(h)h/v(h)$ can be finite at the limit as $h \to 0$ (because of the multiplication by h). Therefore, the resulting $\varphi_C(0) = 1 + v^{\#}(0)/2$ does not necessarily lie out of the "green square".

Figure 3 illustrates both the "intrinsic" case and the "nugget effect" and provides hints how to avoid both, adopting much better modelling alternatives. The above example of $v_E(h)$ was used with parameters indicated in the figure caption. It is easy to find that a FHK-C model which is both persistent ($H = 0.75 > 1/2$) and rough ($M = 0.2 < 1/2$) can replace the $v_E(h)$ and be used further.

Conversely, we can imagine that in the model identification and fitting of $v_E(h)$, persistence was regarded as an "intrinsic" characteristic and roughness was interpreted as "nugget effect". Note that because $M < 1/2$, $v'(h) = -c'(h) = \infty$. Thus, a rough process seems like if it exhibited the "nuggest effect". However, if instead of the standard plot of the variogram, a logarithmic plot was made (also shown in Figure 3 with respect to the upper horizontal axis), then the "nugget" would disappear.

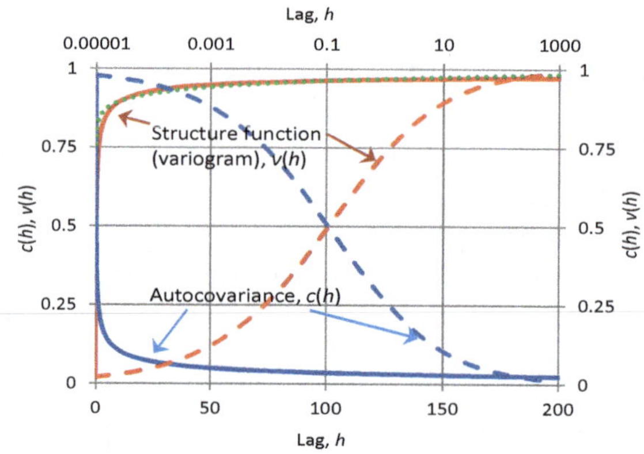

Figure 3. A power type variogram model with "nugget effect" (green dotted line), with equation $v_{\mathrm{E}}(h) = 0.83\ (h + 0.05)^{0.03}$ (corresponding to a "nugget" value $v(0) = 0.83\ (0.05)^{0.03} = 0.76$) and its replacement with a FHK-C model with $H = 0.75$ and $M = 0.2$.

3. Results

3.1. Revisiting Earlier Results

In an earlier study (already mentioned, [22]) on subjects related to the present one, it was suggested that extremization of EPLT in a continuous time representation could determine the entire dependence structure of the process of interest based on simple constraints. The specific premises for EPLT extremization were:

(a) Lebesgue background measure;
(b) constrained mean μ and variance $\gamma(1)$ at a specified (observation) time scale;
(c) constrained lag-one autocorrelation ρ at the specified time scale;
(d) an inequality constraint $\phi_{\mathrm{C}}(t) \geq \phi(t)$ to ensure physical realism as, naturally, by observing the present and past state of a process, the future entropy is reduced, whereas as $t \to \infty$ conditional and unconditional entropies should tend to be equal, which, however, cannot happen if the entropy production is consistently lower in the conditional than in the unconditional case;
(e) extremization of entropy production at asymptotic time scales, i.e., $t \to 0$ and $t \to \infty$.

These premises after systematic analyses resulted in two processes extremizing entropy production:

- A Markov process:

$$c(h) = \lambda \mathrm{e}^{-h/\alpha},\ \gamma(k) = \frac{2\lambda}{k/\alpha}\left(1 - \frac{1 - \mathrm{e}^{-k/\alpha}}{k/\alpha}\right) \tag{53}$$

maximizes entropy production for small times ($t \to 0$) but minimizes it for large times ($t \to \infty$).

- A *Hurst-Kolmogorov* (HK) process:

$$\gamma(k) = \lambda(\alpha/k)^{2-2H} \tag{54}$$

maximizes entropy production for large times ($t \to \infty$) but minimizes it for small times ($t \to 0$).

In these definitions α and λ are scale parameters with dimensions of $[t]$ and $[x^2]$, respectively. The parameter H, known as the Hurst parameter, determines the global properties of the process with the notable property:

$$H = \varphi(\infty) = \varphi_{\mathrm{C}}(\infty) \tag{55}$$

Here we revisit these results as well as the premises, in light of the theoretical analyses of the previous section and incorporating empirical experience based on various data sets. These allow the following notes:

1. The minimal entropy production of the HK process at small times may not be plausible for complex processes; rather we should generally expect that complex processes maximize entropy production at both small and large times (asymptotically as $t \to 0$ and ∞).
2. The HK process is characterized by infinite variance ($\gamma(0) = \infty$), which, as discussed in the previous section, makes it physically inconsistent.
3. The premise $\phi_C(t) \geq \phi(t)$ should certainly hold asymptotically (for $t \to 0$ and $t \to \infty$) but not necessarily at intermediate times. A specific example is the process shown in Figure 4h which includes a periodic component where the curve of $\phi_C(t)$ intersects that of $\phi(t)$ (the latter is not shown in the figure), even though asymptotically the inequality holds. It can thus be conjectured that if a process incorporates a deterministic component (e.g., periodic) while it is treated stochastically without separating the deterministic component, the inequality for EPLT and CEPLT could be violated at intermediate times.
4. While the Markov process represents the maximal EPLT as $t \to 0$ ($\phi(0) = 1$), this property is shared by many other processes (see next subsection). Furthermore, in terms of CEPLT, while the Markov process corresponds to $\phi_C(0) = 3/2$, there are processes which have higher CEPLT than the Markov, up to $\phi_C(0) = 2$. These are smoother than the Markov process and have structure function with $v^{\#}(0) = \xi^{\#}(0) > 1$, where the value 1 corresponds to the Markov process.

Based on these remarks, the Markov process remains a valid candidate of physically consistent processes maximizing entropy production, if only the local asymptotic behaviour is of interest. However, additional constraints to the above premises (a)–(e) (from [22]) are needed to confirm its appropriateness, both on small scales (to enable determination of $\phi_C(0)$) and large scales (to enable determination of $\phi_C(\infty) = \phi(\infty)$).

On the contrary, the HK process should not be regarded as a candidate of general validity, even though it is valid for large scales. Indeed, it is still quite useful as it results in high entropy production at large scales, a property that cannot hold for processes with exponential decrease of autocovariance such as the Markov process. We can thus regard the HK process as a useful mathematical construct which does not appear in nature but can be filtered appropriately to make a physically consistent process. This is the same with white noise (WN), which again is not physically consistent as it involves infinite variance in continuous time (see the specific location and the relationship of the WN and HK processes in the diagram of Figure 2).

It is well known that if a WN process $\underline{y}(t)$ is the input to a moving average filter $\underline{x}(t) = \frac{1}{2T}\int_{t-T}^{t+T} \underline{y}(a)da$, then it produces a physically consistent process $\underline{x}(t)$ with finite variance and autocovariance linearly varying in the interval [0, 2T] ([27] (p. 325)). Also, if $\underline{y}(t)$ is the input to a linear system corresponding to the linear differential equation $\underline{x}'(t) + \alpha\underline{x}(t) = \underline{y}(t)$, then the output $\underline{x}(t)$ is a Markov process ([27] (p. 326)), physically consistent.

Likewise if in these two cases the input $\underline{y}(t)$ is an HK process, then it is easy to see that the filtered output is a physically realistic process with finite variance $\gamma(0)$, practically unaffected climacogram $\gamma(k)$ at large scales, with $\gamma^{\#}(\infty) = 2H - 2$ (as in the original HK process) but highly modified climacogram at small scales, thus having a valid structure (and climacostructure) function with $v^{\#}(0) = \xi^{\#}(0) = 2H$.

3.2. Specific Processes

Following the theoretical analyses of Section 2 and the remarks of the Section 3.1, we describe here a number of stochastic processes specified through any of their second order functions. All of these processes respect the limitations discussed in Section 2.4 and thus they are consistent mathematically and physically. Some of them correspond to high entropy production, but we also discuss processes with low entropy production. The complete details of these processes, along with graphs illustrating their properties and results of their application, are contained in Appendix A.2.

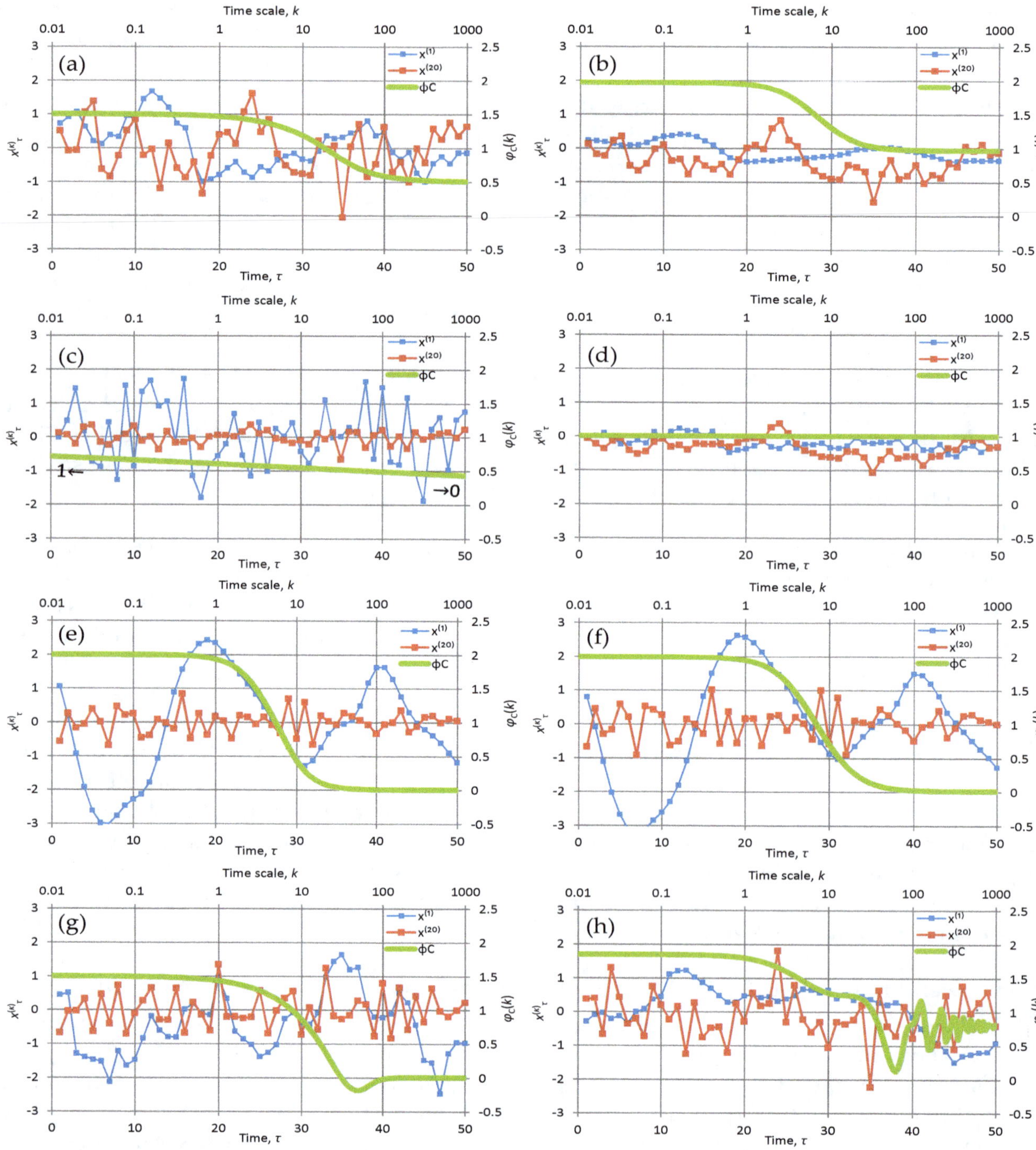

Figure 4. The first fifty terms at time scales $k = 1$ and 20 of time series produced by various models, along with "stamps" of the models (green lines plotted with respect to the secondary axes) represented by the CEPLT, $\phi_C(k)$. The different models are (**a**) Markov; (**b**) FHK, with CEPLT close to the absolute maximum ($H = M = 0.97$); (**c**) FHK, with CEPLT close to the absolute minimum ($H = M = 0.05$); (**d**) FHK, with CEPLT close to the absolute maximum for large scales ($H = 0.99$) and close to the absolute minimum for small scales ($M = 0.01$); (**e**) GP with $n = 1$ (blackbody spectrum) with CEPLT equal to the absolute minimum (0) for large scales and to the absolute maximum (2) for small scales; (**f**) Approximation of (e) by FHK ($H = 0.01$, $M = 0.99$, also using the same initial sequence of random numbers as in (e) to visualize similarity); (**g**) AE with $n = 2$; (**h**) average of a FHK with $H = M = 0.8$ and a harmonic oscillation with $T = 5\alpha = 50$. In all cases the discretization time scale is $D = 1$, the characteristic time scale $a = 10$, and the characteristic variance scale λ is chosen so that for time scale D, $\gamma(D) = 1$. The mean is 0 in all cases and the marginal distribution is normal. The FHK is implemented using the Cauchy-type climacogram.

We start with a general Filtered Hurst-Kolmogorov (FHK) process (which in some cases, e.g., [24], has also been called the Hybrid Hurst-Kolmogorov—HHK—process). In Section 3.1 we briefly discussed two types of linear filters to the HK process, whose asymptotic properties are determined based on a single parameter, H. Here we generalize this result also making the asymptotic behaviour on the left independent from that on the right (cf. [37]) through introducing another parameter denoted as M (in honour of Mandelbrot, following [31]) which is called the smoothness (or fractal) parameter. We avoid specifying the linear filter needed to convert the HK process into the types of the FHK process given as this is not necessary (and in some cases it would be too involved). Rather we specify these types in terms of a convenient expression of the climacogram. The first case is defined through a generalized Cauchy-type climacogram (FHK-C) (see also [24]):

$$\gamma(k) = \lambda\left(1 + (k/\alpha)^{2M}\right)^{\frac{H-1}{M}} \tag{56}$$

Both M and H are dimensionless parameters, and M and $1 - H$ vary in the interval $(0, 1]$ with $M < 1/2$ or $> 1/2$ indicating a rough or a smooth process, respectively, and with $H < 1/2$ or $> 1/2$ indicating an antipersistent or a persistent process, respectively.

The second case is defined through a Dagum-type (FHK-D) climacogram:

$$\gamma(k) = \lambda\left(1 - \left(1 + (k/\alpha)^{2(H-1)}\right)^{\frac{M}{H-1}}\right) \tag{57}$$

and, despite its different expression, its behaviour, as well as the meaning and ranges of parameters are the same as in the Cauchy-type climacogram. For $M = 1 - H$ both result in precisely the same special case:

$$\gamma(k) = \frac{\lambda}{1 + (k/\alpha)^{2(1-H)}} \tag{58}$$

The third case is obtained by summing a Cauchy-type climacogram with $M = 1$ and a Dagum-type climacogram with $H = 0$. The climacogram of the thus produced model (FHK-CD) is:

$$\gamma(k) = \lambda_1(1 + (k/\alpha_1)^2)^{H-1} + \lambda_2(1 - (1 + (k/\alpha_2)^{-2})^{-M}) \tag{59}$$

This is a convenient form, in which the first additive term determines merely the persistence of the process and the second one the smoothness of the process. In addition, it is more flexible and richer than its constituents, as it contains two couples of scale parameters; however, if parsimony is sought, then it can take the same number of parameters as each of the constituents by setting $\alpha_1 = \alpha_2$ and $\lambda_1 = \lambda_2$ (note that, for dimensional consistency, one λ and one α are minimal parameter requirements).

Even in its most parsimonious form, the FHK in any of the above three variants can cover the entire admissible range, i.e., the entire "green square" in Figure 2. The different patterns in time series generated by different M and H (specifically for the Cauchy-type climacogram) are illustrated in some plots of Figure 4 also in comparison with other models, first of which is the most customary Markov model (Figure 4a). The Markov model is good as a benchmark for comparisons because, as already discussed in Section 2, it is fully neutral (neither rough nor smooth as $\phi_C(0) = 3/2$, and neither antipersistent nor persistent as $\phi_C(\infty) = 1/2$).

Each of the panels shows the first fifty terms of time series produced by each of the model implementations at time scales $k = 1$ and 20. In addition, each panel contains a "stamp" of the specific model represented by the plot of CEPLT, $\phi_C(k)$. Additional plots for all second order tools of each of the models are given in Appendix A.2. The time series for these models were generated quite easily using the generic model proposed in [24]; their length is 1024 and this is also the length of the series of coefficients a used for the generating symmetric moving average (SMA) scheme. All calculations are of algebraic type (based on the equations of Section 2 and Appendix A.2) and were performed in Excel spreadsheets without difficulties.

In Figure 4b the CEPLT is close to the absolute maximum both for small and large scales ($H = M = 0.97$ so as to obtain $\phi_C(0) = 1.97 \approx 2$ and $\phi_C(\infty) = 0.97 \approx 1$); notable is the very smooth shape at scale 1 and the large departures from the mean (which is 0) at scale 20. On the contrary, in Figure 4c the CEPLT is close to the absolute minimum for all scales ($H = M = 0.05$, so as to obtain $\phi_C(0) = 1.05 \approx 1$ and $\phi_C(\infty) = 0.05 \approx 0$—for better visualization it was preferred not to use values of H and $M < 0.05$). Furthermore, in Figure 4d the CEPLT is close to the absolute maximum for large scales ($H = \phi_C(\infty) = 0.99 \approx 1$) and close to the absolute minimum for small scales ($M = 0.01$ resulting in $\phi_C(0) = 1.01 \approx 1$). Finally, in Figure 4f the conditions are opposite to those in 4d, i.e., the CEPLT is equal to the absolute minimum for large scales ($H = \phi_C(\infty) = 0.01 \approx 0$) and to the absolute maximum (2) for small scales ($M = 0.99$ resulting in $\phi_C(0) = 1.99 \approx 2$).

The particular case of Figure 4d is close to what is usually called "pink noise" or "$1/f$ noise", as the power spectrum has almost constant slope -1 for the entire frequency domain (which is the same in the climacospectrum). This means that using the FHK model we can theoretically represent and practically produce even "pink noise" in a consistent stationary setting without linking it to a nonstationary process [38,39], which actually involves several theoretical inconsistencies [31]. Indeed, as can be seen more thoroughly in the detailed graphs of Figure A5 in Appendix A.2.2, which are for the same application with that of Figure 4d, the small change of slope of from 0.99 to 1.01 is not actually visible, especially in view of the very rough shape of the empirical periodogram, which certainly cannot support differentiation between 0.99 and 1. The FHK model can be used also in other ways to produce "pink noise", that is, by selecting a very large (small) parameter α so as to expel from our field of vision the asymptotic behaviour on large (small) scales. And we can imagine that in several cases of empirical explorations using observations of natural processes, the observation resolution and length, compared to characteristic scale(s) of the process, are such as to hide the asymptotic behaviour of the process.

We can use this as a trick to obtain virtually constant power spectrum slopes much steeper than -1. This is illustrated Figure A7 of Appendix A.2.2 where the FHK was used with $H = M = 0.75$ and $\alpha = 100$. These yield theoretical slopes $\widetilde{\zeta}^{\#}(\infty) = s^{\#}(\infty) = -(1+2M) = -2.5$ and $\widetilde{\zeta}^{\#}(0) = s^{\#}(0) = -(2H-1) = -0.75$. However, the large α does not allow viewing the asymptotic behaviour at low frequencies or large scales and the slope -2.5 dominates everywhere. Actually the empirical periodogram estimates an even steeper constant slope, $s^{\#} = -2.6$. This should not mislead us to conclude that the process is nonstationary because the slope is steeper than -1 (as happens in many studies in the literature). Likewise, in the case that the time series represented observations of a natural process, such a result must not be misinterpreted as evidence that natural processes can lead to slopes constant for all scales and steeper than -1. Clearly here, the process is stationary, the slope -2.5 refers to large frequencies or small scales but, because of the large characteristic scale and the limited observation period (1024 in this example), we cannot see what happens at larger scales. If we saw, then the slope would be -0.75, in accord with the theory.

As discussed, the Markov process is neutral but, by modifying its power spectrum, we can obtain processes which are smooth or antipersistent. Two types of modifications are studied in Appendix A.2, introduced in terms of their power spectrum, i.e.,

$$s(w) = \frac{4^{n+1}(n!)^2}{(2n)!} \frac{\lambda\alpha}{\left(1+(2\pi\alpha w)^2\right)^{n+1}} \tag{60}$$

and:

$$s(w) = \frac{4^{n+1}(n!)^2}{(2n)!} \frac{\lambda\alpha(2\pi\alpha w)^{2n}}{\left(1+(2\pi\alpha w)^2\right)^{n+1}} \tag{61}$$

where $n = 0, 1, 2 \ldots$, and the value $n = 0$ results precisely in a Markov process. For $n > 0$, (60) corresponds to a smooth process with $\phi_C(0) = 2$ and (61) corresponds to an antipersistent process with $\phi_C(\infty) = 0$. The resulting autocovariance function retains the characteristic exponential form of the Markov process but is multiplied by a polynomial function of lag (see Appendix A.2). Therefore, we call these two processes the smooth exponential (SE) and the antipersistent exponential (AE), respectively. In each of the cases, a most extreme value of CEPLT is achieved (maximal for small scales in SE and minimal for large scales in AE). Figure 4g, shows an example of AE for $n = 2$; the hole in the function $\phi_C(k)$ for moderate k is a characteristic of this process.

An interesting process that is simultaneously extremely smooth ($\phi_C(0) = 2$) and extremely antipersistent ($\phi_C(\infty) = 0$) is defined by the following power spectrum:

$$s(w) = \frac{c_n \lambda \alpha (\alpha w)^{2n+1}}{e^{2\pi\alpha w} - 1} \tag{62}$$

where c_n is a normalizing constant making the area of the power spectrum equal to $\gamma_0 = \lambda$. This process, which we call generalized Planck (GP) is obtained by a generalization of Planck's law of black-body radiation, whereas it is identical to this law if $n = 1$. The detailed equations of this process are also given in Appendix A.2. A characteristic plot of time series generated from this process is shown in Figure 4e, and is almost indistinguishable from Figure 4f which, as already discussed, was produced as an approximation by FHK.

A final case examined in the Appendix A.2 is a harmonic oscillation, which is very easily modelled as a deterministic process but here it is treated as a stochastic process. Of course stochastic treatment is not advisable in this case, but often such periodic behaviours appear as components of stochastic processes. Obviously the second-order characteristics of such processes are affected by periodic components and therefore we need to know which equations should be superimposed in those of the pure stochastic process (see also [40,41]). As an example of such a case, Figure 4h shows the behaviour of a process defined as the average of a FHK with $H = M = 0.8$ and a harmonic oscillation. It is difficult to identify the presence of the oscillation from visual inspection of the time series, but the detailed graph of any of the second order characteristics, plotted in Appendix A.2., captures the periodic behaviour.

3.3. Comparison of Asymptotic Properties

Most of the models examined in the previous subsection can only reproduce specific values of asymptotic properties. An exception is the FHK model which is the most powerful as it can perform in the entire admissible domain of asymptotic properties. A visual comparison of all models examined, it terms of the asymptotic values of the different second order characteristics is made in Figure 5, while Figure 6 provides an envelope for all models, with classification of the ranges in terms of smoothness and persistence.

Further comparative information about which model can perform for the most extremal cases is provided in Table 4. These figures and the table can be useful for the model selection based on data.

Table 4. Models that can perform in each of the specified extremal or neutral cases of smoothness and persistence.

Persistence	Smoothness		
	Maximal Roughness $\varphi_C(0) = 1$	**Neutral Smoothness** $\varphi_C(0) = 3/2$	**Maximal Smoothness** $\varphi_C(0) = 2$
Maximal Persistence, $\varphi_C(\infty) = 1$	~FHK	~FHK	~FHK
Neutral Persistence $\varphi_C(\infty) = 1/2$	~FHK	Markov, FHK	SE, GP, FHK
Maximal Antipersistence $\varphi_C(\infty) = 0$	~FHK	AE, FHK	GP, FHK

Note: the symbol '~' indicates that the model can perform in close approximation but not precisely.

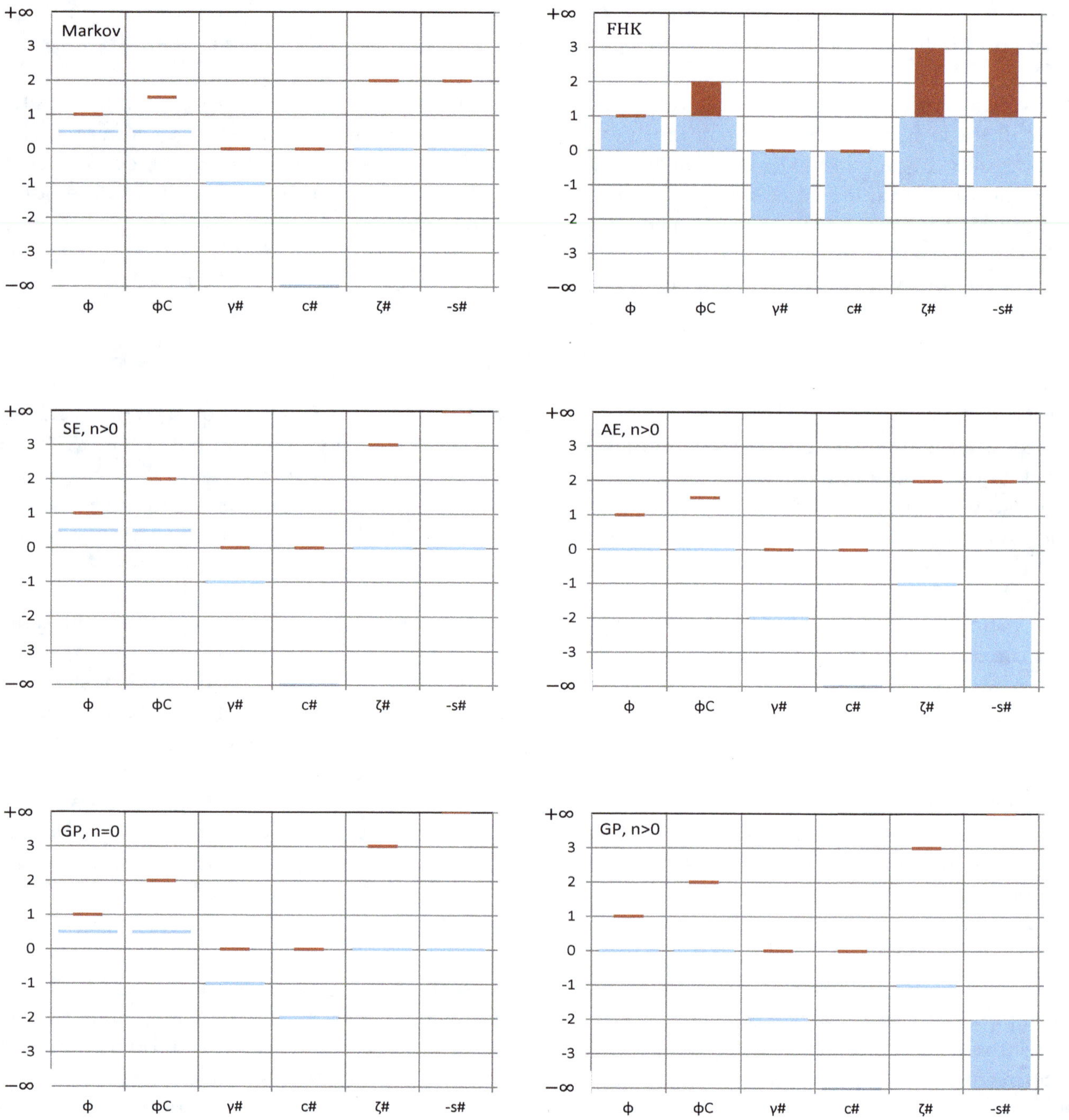

Figure 5. Ranges of asymptotic values for each of the most important second order functions of the examined processes. Wide light blue lines/boxes represent global properties (k, $h \to \infty$, $w \to 0$) while narrow dark red lines/boxes represent local properties (k, $h \to 0$, $w \to \infty$). Markov: Markov process; FHK: filtered Hurst-Kolmogorov process in any of the three forms; SE: smooth process with generalized exponential-type autocovariance; AE: antipersistent process with generalized exponential-type autocovariance; GP: generalized Planck. Note that the Markov process is identical with the SE and PE for $n = 0$.

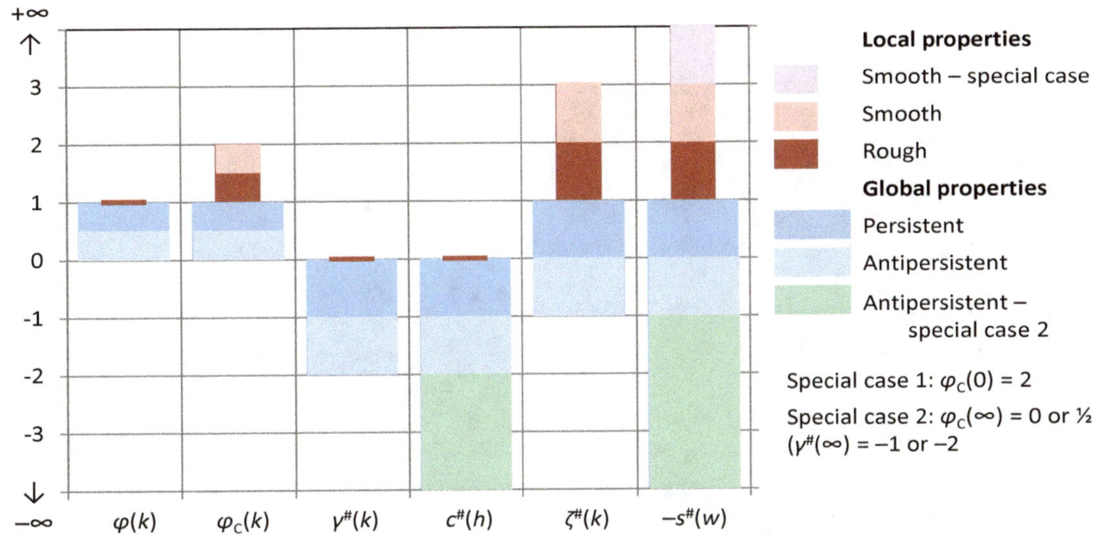

Figure 6. Envelopes of ranges of asymptotic values for each of the most important second order functions for all examined processes.

3.4. Real World Case Study

Empirical testing of the theoretical analyses presented here needs fine resolution of measurements and long time series, in order to reliably determine the asymptotic properties from data. Geophysical time series usually do not satisfy either of these two conditions. However, laboratory measurements of turbulent velocity can meet the conditions.

Indeed, in laboratory experiments at sampling intervals of μs, very large samples can be formed which can enable viewing the asymptotic behaviour. Here we use grid data of nearly isotropic turbulence from the Corrsin Wind Tunnel at a high-Reynolds-number [42], which were made available on the Internet by the Johns Hopkins University.

This dataset consists of 40 time series with $n = 3\ 6 \quad \times 10^6$ data points of wind velocity along the flow direction and an equal number of time series of cross-stream velocity, all measured at a sampling time interval $D = 2\ 5$ μs by X-wire probes placed downstream of the grid. Here we use part of the data, namely the series of wind velocity along the flow direction at the first of the probes (first column of files in http://pages.jh.edu/~cmeneve1/datasets/Activegrid/M20/H1/). More data and analyses have been used by Dimitriadis and Koutsoyiannis [43] at a somewhat different context.

Characteristic plots of the turbulent velocity time series are shown in the Figure 7a for various time scales. The large persistence of turbulence is evident from the large variability at the largest time scales. Impressively, an FHK-CD model with five parameters describes well the data at the entire range of available time scales which spans almost seven orders of magnitude.

The fitted model parameters are: $H = 5/6 = 0.83$, $M = 1/3 = 0.33$, $a_1 = a_2 = 2170\ D = 54$ ms, $\lambda_1 = 1.13$ m^2/s^2, $\lambda_2 = 2.39$ m^2/s^2. These confirm the intense persistence and a small roughness, which is reflected in a climacospectrum slope of $-5/3$ for large frequencies (the neutral would be -2), in accord with Kolmogorov's theory.

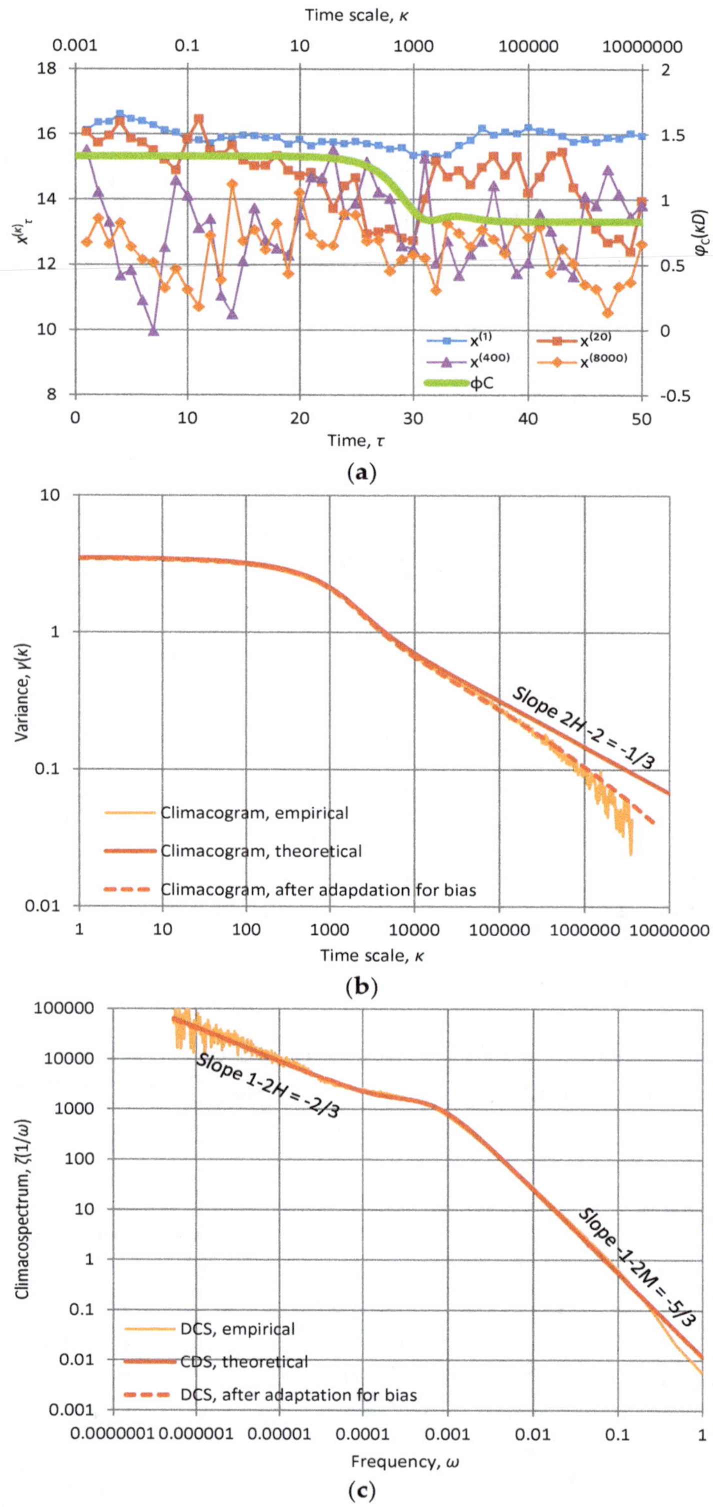

Figure 7. Characteristic graphs of the turbulent velocity time series: (**a**) plot of the first fifty terms of the time series at time scales $\kappa = 1, 20, 400$ and 8000, along with the "stamp" $\phi_C(k)$ of the fitted model (green line plotted with respect to the secondary axes); (**b**) empirical and fitted theoretical climacogram; (**c**) empirical and fitted theoretical climacospectrum.

4. Discussion and Conclusions

The question whether the probabilistic (information) entropy incorporates, as a special case, thermodynamic entropy or the two are different concepts, is still debated [44,45]. If we adopt the former proposition of the dilemma, as we have done here, the requirement emerges that entropy production should be formulated in a stochastic context, while current attempts mostly use the classical thermodynamic entropy definition as in heat transfer. This is not an easy task and certainly requires exploration of a broad spectrum of alternatives for a definition of the concept of entropy production. Exploration of extremum principles for entropy production in a stochastic framework is also crucial. The two versions proposed here, entropy production in logarithmic time, unconditionally (EPLT) or conditionally on the past and present having been observed (CEPLT), may have some usefulness and potential as explanatory and modelling tools. The findings of this study include the following:

- Definition of entropy production with respect to logarithmic time has some attractive characteristics and is consistent with the notion of the time arrow.
- The asymptotic values of entropy production for time zero and infinity seem to be crucial for its extremization and define important features of a process.
- In particular, the asymptotic value of CEPLT for zero time defines the roughness (fractality) of the stochastic process while the common asymptotic value of EPLT and CEPLT for infinite time defines the (long-term) persistence of the process and equals the Hurst parameter.
- Both ELPT and CEPLT are finite and constrained within specific theoretically justified bounds. The scaling laws of all second order characteristics of stochastic processes are a result of these finite values and are also subject to bounds (with few exceptions fully described). The proofs of existence of these bounds offer the means to fight common misunderstanding, misuse, false reporting and inconsistent estimation of characteristics of geophysical processes, as well as to develop consistent estimation algorithms.
- The climacogram and its derivative tools introduced here, namely, the differenced climacogram and the climacospectrum are very powerful and support the estimation of entropy production, in either of its forms, from data series.
- The broad family of models for stochastic processes investigated here covers the entire admissible range of asymptotic behaviours and offers the means for modelling any type of processes, either using a single model or a combination of more than one model.
- The real-world application presented with the extraordinary long time series of turbulent velocity shows how a parsimonious stochastic model can be identified and fitted using the tools developed.

Measurements of diverse turbulent phenomena can potentially provide equally long time series with fine resolution, shaping a broader empirical basis for future work on identifying natural behaviours, specifically on larger spatial and temporal scales, and further testing the adequacy of the models. Characteristics of order higher than second, along with the effects of non-Gaussianity on entropy production, are also interesting to examine in light of the developments for second-order ones. A recent study [43] has already provided interesting results with respect to high-order moments. Connections of microscale turbulence and macroscale atmospheric phenomena are also a broad field for future studies. Further applications, also extending similar recent works in hydrology, geophysics and ecosystems [30,40,41], would enrich our knowledge on natural behaviours.

Acknowledgments: I thank Huijuan Cui, Bellie Sivakumar and Vijay P. Singh for the invitation to write an article for the Special Issue, and the two anonymous reviewers for the encouraging and constructive comments. I gratefully acknowledge discussions about several issues on stochastics covered here with members of the Itia research team, colleagues and friends, and particularly with Panayiotis Dimitriadis, Federico Lombardo, Spencer Stevens, Any Iliopoulou, Hristos Tyralis, Andreas Efstratiadis, Katerina Tzouka, Georgia Papacharalambous, Panagiotis Kossieris, Yannis Tsoukalas, Andreas Langousis, Yannis Markonis, Christoforos Pappas, Simon Papalexiou, John Halley, Elena Volpi, Yannis Dialynas, Ilias Deligiannis, Harry Lins, Christian Onof, Alin Carsteanu, Amilcare Porporato and Alberto Montanari, and earlier with Tim Cohn and Vit Klemes who are no longer with us.

Appendix A

Appendix A.1. Derivations and Theoretical Details

Appendix A.1.1. Properties of the LLD

Here we examine some asymptotic properties of the log-log derivative of general validity, which are useful in other derivations herein. First we examine the product of two functions, i.e.,

$$g(t) := f_1(t) f_2(t) \tag{A1}$$

By applying the definition of LLD (Equation (38)) and using standard calculus, it is easy to show that:

$$g^{\#}(t) = f_1^{\#}(t) + f_2^{\#}(t) \tag{A2}$$

which we can conveniently denote as:

$$(f_1(t) f_2(t))^{\#} = f_1^{\#}(t) + f_2^{\#}(t) \tag{A3}$$

Likewise, it is easily shown that:

$$(f_1(t) / f_2(t))^{\#} = f_1^{\#}(t) - f_2^{\#}(t) \tag{A4}$$

By setting in (A3) $f_1(t) = \lambda$ (constant), $f_2(t) = f(t)$ and observing that $f_1^{\#}(t) = 0$, it is easily verified that:

$$(\lambda\, f(t))^{\#} = f^{\#}(t) \tag{A5}$$

By setting in (A3) $f_1(t) = f_2(t) = f(t)$ it is easily verified that:

$$\left(f(t)^2\right)^{\#} = 2 f^{\#}(t) \tag{A6}$$

and likewise, extending for any exponent λ:

$$\left(f(t)^{\lambda}\right)^{\#} = \lambda f^{\#}(t) \tag{A7}$$

We now examine the sum of two functions, i.e.,

$$g(t) := f_1(t) + f_2(t) \tag{A8}$$

By applying the definition of LLD and using standard calculus, it is easy to show that:

$$g^{\#}(t) := \frac{f_1(t) f_1^{\#}(t) + f_2(t) f_2^{\#}(t)}{f_1(t) + f_2(t)} \tag{A9}$$

In other words:

$$(f_1(t) + f_2(t))^{\#} = \frac{f_1(t) f_1^{\#}(t) + f_2(t) f_2^{\#}(t)}{f_1(t) + f_2(t)} \tag{A10}$$

Finally, we examine the composition of two functions, i.e.,

$$(f \circ g)(t) = f(g(t)) \tag{A11}$$

Again by applying the definition of LLD and using standard calculus, we easily find:

$$((f \circ g)(t))^{\#} = (f(g(t))^{\#} = f^{\#}(g(t))\, g^{\#}(t) \tag{A12}$$

Appendix A.1.2. Proof of the Positive Definiteness of the Climacogram

We will demonstrate that $\gamma(k)$ is a positive definite function based on the well-known fact that the autocovariance function $c(h)$ is positive definite. It suffices to show that the Fourier transform of $\gamma(k)$, i.e.,

$$s_\gamma(w) = 4\int_0^\infty \gamma(k)\cos(2\pi w k)\mathrm{d}k \tag{A13}$$

is positive. Expressing $\gamma(k)$ in terms of $c(k)$ from (21) we have:

$$s_\gamma(w) = 8\int_0^\infty\int_0^1 (1-\chi)c(\chi k)\mathrm{d}\chi\ \cos(2\pi w k)\mathrm{d}k = 8\int_0^1 (1-\chi)\int_0^\infty c(\chi k)\ \cos(2\pi w k)\mathrm{d}k\,\mathrm{d}\chi \tag{A14}$$

Setting $\chi k = \psi$, $\chi \mathrm{d}k = \mathrm{d}\psi$, we find:

$$s_\gamma(w) = 8\int_0^1 (1-\chi)\int_0^\infty c(\psi)\ \cos(2\pi w\psi/\chi)(1/\chi)\mathrm{d}\psi\,\mathrm{d}\chi \tag{A15}$$

and invoking the definition of the power spectrum as the Fourier transform of $c(h)$ (Equation (10)), we get:

$$s_\gamma(w) = 2\int_0^1 \frac{1-\chi}{\chi} s(w/\chi)\mathrm{d}\chi \tag{A16}$$

Both quantities $s(w/\chi)$ and $(1-\chi)/\chi$ are positive for $0 < \chi < 1$ and thus $s_\gamma(w) \geq 0$.

Appendix A.1.3. Proof that the Differenced Climacogram Is Nonnegative

First we will prove a more general inequality that will be used for the proof, i.e.,

$$(h+r)\sqrt{\gamma(h+r)} \leq h\sqrt{\gamma(h)} + r\sqrt{\gamma(r)},\ \forall h,r > 0 \tag{A17}$$

With reference to the cumulative process, as seen in Figure 1, we have:

$$\begin{aligned}\Gamma(h+r) &= \mathrm{Var}[\underline{X}(h+r)] = \mathrm{Var}[(\underline{X}(h+r) - \underline{X}(h)) + \underline{X}(h)] \\ &= \mathrm{Var}[\underline{X}(h+r) - \underline{X}(h)] + \ \mathrm{Var}[\underline{X}(h)] + 2\mathrm{Cov}[(\underline{X}(h+r) - \underline{X}(h)),\ \underline{X}(h)] \\ &\leq \mathrm{Var}[\underline{X}(h+r) - \underline{X}(h)] + \ \mathrm{Var}[\underline{X}(h)] + 2\sqrt{\mathrm{Var}[\underline{X}(h+r) - \underline{X}(h)]\ \mathrm{Var}[\underline{X}(h)]}\end{aligned} \tag{A18}$$

and since $\mathrm{Var}[\underline{X}(h+r) - \underline{X}(h)] = \mathrm{Var}[\underline{X}(r)]$, we find:

$$\Gamma(h+r) \leq \ \Gamma(r) + \Gamma(h) + 2\sqrt{\Gamma(r)\Gamma(h)} = \left(\sqrt{\Gamma(r)} + \sqrt{\Gamma(h)}\right)^2 \tag{A19}$$

or, taking the square roots as the quantities are positive:

$$\sqrt{\Gamma(h+r)} \leq \sqrt{\Gamma(h)} + \sqrt{\Gamma(r)} \tag{A20}$$

By substituting $\Gamma(h) = h^2\ \gamma(h)$ we get (A17). As a corollary, by setting $r = h$ we derive:

$$\gamma(2h) \leq \gamma(h) \tag{A21}$$

which ensures that the differenced climacogram is nonnegative.

Appendix A.1.4. Determination of the Characteristic Constant ε of CEPLT

From (41) and the condition put as desideratum (i.e., that, as $k \to \infty$, $\gamma_C(k)/\gamma(k) \to 1$), it follows that:

$$\frac{1}{\varepsilon} = 1 - \lim_{t\to\infty} \frac{\gamma(2t)}{\gamma(t)} \tag{A22}$$

Now assuming that $b = \gamma^{\#}(\infty)$, it follows that the limit:

$$\lim_{t\to\infty} t^{-b}\gamma(t) = c \tag{A23}$$

is finite. Likewise:

$$\lim_{t\to\infty} (2t)^{-b}\gamma(2t) = c \tag{A24}$$

By dividing the last two equations we find:

$$\lim_{t\to\infty} \frac{\gamma(2t)}{\gamma(t)} \frac{(2t)^{-b}}{(t)^{-b}} = 1 \tag{A25}$$

and finally:

$$\lim_{t\to\infty} \frac{\gamma(2t)}{\gamma(t)} = 2^b \tag{A26}$$

which proves (42).

Appendix A.1.5. Asymptotic Properties of CEPLT

The fact that $\gamma_C(0) = \gamma_C(\infty) = 0$ follows directly from the definition. For the LLDs, starting from the definition of the differenced climacogram we proceed as follows:

$$\gamma_C(k) = \varepsilon(\gamma(k) - \gamma(2k)) = \varepsilon\gamma(k)\left(1 - \frac{\gamma(2k)}{\gamma(k)}\right) \tag{A27}$$

Hence, by virtue of (A3):

$$\gamma_C^{\#}(k) = \gamma^{\#}(k) + \left(1 - \frac{\gamma(2k)}{\gamma(k)}\right)^{\#} \tag{A28}$$

According to the derivations in Appendix A.1.4, the quantity in parentheses has a finite limit as $k \to \infty$, which specifically is $1 - 2^{\gamma^{\#}(\infty)} = 1/\varepsilon$. Thus, its LLD tends to zero as $k \to \infty$ (see explanation in the last paragraph in Appendix A.1.8) and hence $\gamma_C^{\#}(\infty) = \gamma^{\#}(\infty)$. Likewise, $\gamma^{\#}(0) = 0$ because $\gamma(0) = \gamma_0$ (finite). However the term in parenthesis is nonzero for $k \to 0$, so to determine the asymptotic behaviour of $\gamma_C^{\#}(k)$ in that case we follow a different path.

Specifically, we can express $\gamma_C(k)$ in an alternative manner in terms of the climacostructure function. Using the definition of the latter, we can write:

$$\gamma_C(k) = \varepsilon(\xi(2k) - \xi(k)) = \varepsilon\xi(k)\left(\frac{\xi(2k)}{\xi(k)} - 1\right) \tag{A29}$$

Hence:

$$\gamma_C^{\#}(k) = \xi^{\#}(k) + \left(\frac{\xi(2k)}{\xi(k)} - 1\right)^{\#} \tag{A30}$$

Proceeding as above we derive:

$$\lim_{k\to 0} \frac{\xi(2k)}{\xi(k)} = 2^{\xi^{\#}(0)} \tag{A31}$$

and hence the quantity in parentheses in (A30) has a finite limit as $k \to 0$, which specifically is $2^{\xi^{\#}(0)} - 1$. Thus, its log-log derivative tends to zero as $k \to 0$ and hence $\gamma_C^{\#}(0) = \xi^{\#}(0)$.

We further note that from the definition of the climacostructure function and using Equation (A10) we get:

$$\xi^{\#}(k) = \frac{-\gamma^{\#}(k)\gamma(k)}{\gamma_0 - \gamma(k)} \quad \text{(A32)}$$

As $k \to \infty$, the numerator is zero, while the denominator is γ_0, so that $\xi^{\#}(\infty) = 0$. The concluding remark is that, because $\gamma^{\#}(0) = \xi^{\#}(\infty) = 0$, none of the two tools, climacogram and climacostructure function suffices alone to express both the local and global asymptotic properties of $\gamma_C(k)$. On the other hand, the combination of the two suffices.

Appendix A.1.6. Area of Climacospectrum

From the definition of climacospectrum (48), the entire area under it is:

$$A := \int_0^\infty \tilde{\zeta}(w)\mathrm{d}w = \frac{1}{\ln 2}\int_0^\infty \frac{\gamma(1/w) - \gamma(2/w)}{w}\mathrm{d}w \quad \text{(A33)}$$

By changing variable, setting $k = 1/w$, $\mathrm{d}k = -(1/w^2)\,\mathrm{d}w$, we get:

$$A = \frac{1}{\ln 2}\int_0^\infty \frac{\gamma(k) - \gamma(2k)}{k}\mathrm{d}k \quad \text{(A34)}$$

We will evaluate A by taking the limit as $x \to 0$ of:

$$A(x) = A_1(x) - A_2(x),\ A_1(x) := \frac{1}{\ln 2}\int_x^\infty \frac{\gamma(k)}{k}\mathrm{d}k,\ A_2(x) := \frac{1}{\ln 2}\int_x^\infty \frac{\gamma(2k)}{k}\mathrm{d}k \quad \text{(A35)}$$

The latter can be written as:

$$A_2(x) = \frac{1}{\ln 2}\int_{2x}^\infty \frac{\gamma(k)}{k}\mathrm{d}k = A_1(2x) \quad \text{(A36)}$$

so that:

$$A(x) = \frac{1}{\ln 2}\int_x^{2x} \frac{\gamma(k)}{k}\mathrm{d}k \quad \text{(A37)}$$

For small x, because $\gamma(k)$ is finite for any k and we are interested about the limit of $A(k)$ as $x \to 0$, at the limit we can replace $\gamma(k)$ with $\gamma(0)$ and get:

$$\lim_{x\to 0} A(x) = \frac{\gamma(0)}{\ln 2}\int_x^{2x} \frac{1}{k}\mathrm{d}k = \frac{\gamma(0)}{\ln 2}(\ln 2x - \ln x) = \frac{\gamma(0)\ln 2}{\ln 2} = \gamma(0) \quad \text{(A38)}$$

For completeness we note that, as can be easily seen, the integral:

$$B := \int_0^\infty (\gamma(k) - \gamma(2k))\mathrm{d}k \quad \text{(A39)}$$

evaluates to:

$$B = \frac{1}{2}\int_0^\infty \gamma(k)\mathrm{d}k \quad \text{(A40)}$$

However, this is finite only for antipersistent processes, while it diverges to $+\infty$ for persistent processes.

Appendix A.1.7. Details and Derivation of the Inverse Transformation of Climacospectrum

Combining both forms of (49) we find:

$$\gamma(0) = \ln 2 \sum_{i=-\infty}^{\infty} \frac{\zeta(2^i k)}{2^i k} \tag{A41}$$

or by setting $k = 2^x$:

$$\gamma(0) = \ln 2 \sum_{i=-\infty}^{\infty} \frac{\zeta(2^{x+i})}{2^{x+i}} \tag{A42}$$

and this is valid for any x. This means that the function $\zeta(x)$ cannot take any arbitrary form but its form should be such that (A42) is satisfied for any x. At the same time, as shown in Appendix A.1.6:

$$\gamma(0) = \int_0^{\infty} \tilde{\zeta}(w)\mathrm{d}w = \int_0^{\infty} \frac{\zeta(k)}{k^2}\mathrm{d}k = \ln 2 \int_{-\infty}^{\infty} \frac{\zeta(2^x)}{2^x}\mathrm{d}x \tag{A43}$$

where for the very last term we used the transformation $k = 2^x$, $\mathrm{d}k = (\ln 2)\, 2^x\, \mathrm{d}x$. Combining (A42) and (A43) we find that, for any k:

$$\sum_{i=-\infty}^{\infty} \frac{\zeta(2^i k)}{2^i k} = \int_{-\infty}^{\infty} \frac{\zeta(2^x)}{2^x}\mathrm{d}x = \frac{\gamma(0)}{\ln 2} \tag{A44}$$

Note, however, that the equality of the sum with the integral holds only if both limits for summation and integration are infinite. It can be used also as an approximation for small k, by setting the lower integration limit to k; thus the following approximations hold:

$$\gamma(k) = \begin{cases} I(k) & k \to 0 \\ \varepsilon \ln 2 \frac{\zeta(k)}{k} = \frac{\varepsilon \ln 2}{2} I(k) & k \to \infty \end{cases} \tag{A45}$$

where $I(k)$ has been defined in Equation (50). Combining these two cases we heuristically derive the left part of Equation (50), which was shown by an extensive numerical investigation to be almost indistinguishable from the exact solution (49). The advantage of (50) over (49) is that, to calculate a sufficient approximation by (49), we need to go to very large time scales ($2^i\, k$), which may not be feasible when working with data series, while (50) can perform well in reasonably smaller time scales. In numerical applications, the evaluation of the integral $I(t)$ goes sequentially from large to small time scales. To start the calculation, for the largest time scale $k = q$ we exploit the second case of (A45) and get $I(q) = 2\, \zeta(q)/q$. It should be kept in mind that both $I(k)$ and $\zeta(k)/k$ are decreasing functions of k for large scale. Even if we assumed that $\zeta(k)/k$ is constant, rather than decreasing, for a range of scales close to q, we would derive from the second case of (A45) that $I(q - iD) = (I(q)/q)\, (2 + 1/(q/D - 1) + \ldots + 1/(q/D - i))$, which clearly shows that $I(q)$ is a decreasing function of k.

Appendix A.1.8. Remarks on Asymptotic Scaling

We maintain that every nonzero continuous function $f(x)$ defined in $(0, \infty)$, whose limits as $x \to 0$ and $x \to \infty$ exist, is associated with two asymptotic power laws, or else two asymptotic scaling behaviours, one on each of the two ends, 0 (local scaling behaviour) and ∞ (global scaling behaviour). To prove this, first we examine the asymptotic behaviour of $f(x)$ as $x \to \infty$ and we try to identify the first asymptotic law. We perform the following steps:

1. We find $\beta_1 := \lim_{x \to \infty} f(x)$. If $\beta_1 \neq \pm\infty$ then we replace $f(x)$ with $1/(f(x) - \beta_1)$, otherwise we keep it as it is. Clearly then, $|\lim_{x \to \infty} f(x)| = \infty$.

2. We find $\beta_2 := \lim_{c \to \infty}\left(\lim_{x\to\infty} x^{-c} f(x)\right)$. If $\beta_2 \neq 0$ then we replace $f(x)$ with $\ln|f(x)|$ (which preserves the property $\lim_{x \to \infty} f(x) = \infty$); if necessary, we make iterations so that eventually $\lim_{x \to \infty} f(x) = \infty$ and $\lim_{c \to \infty}\left(\lim_{x\to\infty} x^{-c} f(x)\right) = 0$.
3. Given the properties in point 2, there exists a unique b, $0 < b < \infty$ satisfying:

$$|\lim_{x\to\infty} x^{-b} f(x)| < \infty \tag{A46}$$

so that for any $b' \neq b$:

$$|\lim_{x\to\infty} x^{-b'} f(x)| = \begin{cases} 0, & \forall b' < b \\ \infty, & \forall b' > b \end{cases} \tag{A47}$$

4. The constant b defines an asymptotic power law with exponent b (cf. Hausdorff dimension; the case $b = 0$ signifies an improper scaling).

To identify the second asymptotic law as $x \to 0$, we define $\widetilde{f}(x) := f(1/x)$ and we proceed in the same manner to construct a function $\widetilde{f}(x)$ satisfying the conditions in point 2 above. Proceeding likewise, we determine the unique a for which $|\lim_{x\to\infty} x^{-a}\widetilde{f}(x)| < \infty$. This determines an asymptotic power law with exponent a for $f(x)$ as $x \to 0$.

We remark that the two power laws refer to the same function $f(x)$ but, due to the replacement steps of the procedure, may eventually correspond to different functions, say, $f_a(x)$ and $f_b(x)$ for the asymptotic behaviours as $x \to 0$ (local behaviour) and $x \to \infty$ (global behaviour), respectively. However, it is easy to construct a single function that combines both, making a final replacement, e.g., $f(x) = f_a(x)f_b(x)$, but many different $f(x)$ can actually be constructed. Thus, as well as any object has a dimension, any continuous function entails asymptotic power laws; generally not one but two, which in the special case of simple scaling can be identical.

Now, coming to the asymptotic values of $f^{\#}(x)$ for $x \to 0$ and ∞, symbolically $f^{\#}(0)$ and $f^{\#}(\infty)$, these are:

$$f^{\#}(\infty) = b, \; f^{\#}(0) = a \tag{A48}$$

To prove the former case, we proceed as follows:

$$\lim_{x\to\infty} x^{-b} f(x) = \lim_{x\to\infty} \frac{f(x)}{x^b} = \lim_{x\to\infty} \frac{f'(x)}{b\,x^{b-1}} = \lim_{x\to\infty}\left(\frac{xf'(x)}{b\,f(x)}\,\frac{f(x)}{x^b}\right) = \lim_{x\to\infty}\left(\frac{f^{\#}(x)}{b}\,\frac{f(x)}{x^b}\right) = \lim_{x\to\infty}\frac{f^{\#}(x)}{b}\lim_{x\to\infty}\frac{f(x)}{x^b} \tag{A49}$$

This implies that $\lim_{x\to\infty} f^{\#}(x)/b = 1$ and thus $f^{\#}(\infty) = b$. The latter case ($f^{\#}(0) = a$) can be proved in the same manner.

A useful property is that if $\lim_{x\to\infty} f(x)$ is a finite value ($<\infty$), then $b = f^{\#}(\infty) = 0$, because for any $b' > 0$, $\lim_{x\to\infty} x^{-b'} f(x) = \lim_{x\to\infty} f(x)/x^{b'} = 0$. Likewise, if $\lim_{x\to 0} f(x)$ is a finite value, then $a = f^{\#}(0) = 0$. Thus, to have asymptotic (for $x \to 0$ and ∞) values of $f^{\#}$ different for 0 and ∞, the asymptotic values of f should be either 0 or ∞. These properties are useful and are utilized in other proofs herein.

Appendix A.1.9. Asymptotic Scaling of the Different Second Order Tools

We assume that the asymptotic behaviour of the climacogram-based second order tools is given as:

$$\begin{aligned} \gamma_C^{\#}(0) = \xi^{\#}(0) = \zeta^{\#}(0) - 1 = 2\varphi_C(0) - 2 = a > 0 \\ \gamma_C^{\#}(\infty) = \gamma^{\#}(\infty) = \zeta^{\#}(\infty) - 1 = 2\varphi_C(\infty) - 2 = b < 0 \end{aligned} \tag{A50}$$

and we will determine that of the autocovariance-based tools. Note that the equality of the different function limits in (A50) has been proved in Appendix A.1.5.

We will fist examine the asymptotic properties of the autocovariance function. Initially we note that, because $\gamma(0) = c(0) = \gamma_0$ (finite), it follows that $\gamma^{\#}(0) = c^{\#}(0) = 0$. Now, since $b = \gamma^{\#}(\infty) < 0$, it follows that:

$$\lim_{k\to\infty} k^{-b}\gamma(k) = c \tag{A51}$$

where c is finite. We use the relationship between $\gamma(k)$ and $c(h)$ (Equation (21)) to find:

$$\lim_{h\to\infty} h^{-b}c(h) = \lim_{h\to\infty}\frac{c(h)}{h^b} = \lim_{h\to\infty}\frac{1}{2}\frac{\mathrm{d}^2(h^2\gamma(h))/\mathrm{d}h^2}{h^b} = \lim_{h\to\infty}\left(\frac{\gamma(h)}{h^b} + 2\frac{\gamma'(h)}{h^{b-1}} + \frac{1}{2}\frac{\gamma''(h)}{h^{b-2}}\right) \tag{A52}$$

Because both $\gamma(h)$ and $h^b \to 0$ as $h \to \infty$, applying l'Hôpital's rule we find:

$$\lim_{h\to\infty}\frac{\gamma(h)}{h^b} = \lim_{h\to\infty}\frac{\gamma'(h)}{bh^{b-1}} = \lim_{h\to\infty}\frac{\gamma''(h)}{b(b-1)h^{b-2}} \tag{A53}$$

Since $b < 0$, $\gamma'(h)$, $\gamma''(h)$, h^{b-1} and h^{b-2} will also $\to 0$ as $h \to \infty$. Thus:

$$\lim_{h\to\infty} h^{-b}c(h) = \lim_{h\to\infty}\left(\tfrac{\gamma(h)}{h^b} + 2b\tfrac{\gamma(h)}{h^b} + \tfrac{b(b-1)}{2}\tfrac{\gamma(h)}{h^b}\right) = \tfrac{1}{2}(b+1)(b+2)\lim_{h\to\infty}\tfrac{\gamma(h)}{h^b} = \tfrac{1}{2}(b+1)(b+2)c \tag{A54}$$

Unless $b = -1$ or $b = -2$, the limit $\lim\limits_{h\to\infty} h^{-b}c(h)$ is 0, finite or ∞, if and only if $\lim\limits_{h\to\infty} h^{-b}\gamma(h)$ is 0, finite or ∞, respectively. Therefore, if $b \neq -1, -2$, then $c^{\#}(\infty) = \gamma^{\#}(\infty) = b$, otherwise $\lim\limits_{h\to\infty} h^{-b}c(h) = 0$, which does not guarantee (nor does it exclude) that $c^{\#}(\infty) = \gamma^{\#}(\infty) = b$.

Next we will examine the asymptotic properties of the structure function. Initially we note that, because $\xi(\infty) = v(\infty) = \gamma_0$ (finite), it follows that $\xi^{\#}(\infty) = v^{\#}(\infty) = 0$. Now, since $a = \xi^{\#}(0) > 0$, it follows that:

$$\lim_{k\to 0} k^{-\alpha}\xi(k) = d \tag{A55}$$

where d is finite. We use the relationship between $\xi(k)$ and $v(h)$ (Equation (26)) to find:

$$\lim_{h\to 0} h^{-a}v(h) = \lim_{h\to 0}\frac{v(h)}{h^a} = \lim_{h\to 0}\frac{1}{2}\frac{\mathrm{d}^2(h^2\xi(h))/\mathrm{d}h^2}{h^\alpha} = \lim_{h\to 0}\left(\frac{\xi(h)}{h^a} + 2\frac{\xi'(h)}{h^{a-1}} + \frac{1}{2}\frac{\xi''(h)}{h^{a-2}}\right) \tag{A56}$$

Because both $\xi(h)$ and $h^a \to 0$ as $h \to 0$, applying l'Hôpital's rule we find:

$$\lim_{h\to 0}\frac{\xi(h)}{h^a} = \lim_{h\to 0}\frac{\xi'(h)}{ah^{a-1}} \tag{A57}$$

If $a > 1$, then both $\gamma'(h)$ and h^{a-1} will also $\to 0$ as $h \to 0$. In this case, again applying l'Hôpital's rule:

$$\lim_{h\to 0}\frac{\xi'(h)}{ah^{a-1}} = \lim_{h\to 0}\frac{\xi''(h)}{a(a-1)h^{a-2}} \tag{A58}$$

If $a < 1$, then $\gamma'(h)$ and h^{a-1} will both $\to \infty$ as $h \to 0$. Thus, again applying l'Hôpital's rule, we will conclude with (A58) again. Thus, in both cases:

$$\lim_{h\to 0} h^{-a}v(h) = \lim_{h\to 0}\left(\tfrac{\xi(h)}{h^a} + 2a\tfrac{\xi(h)}{h^a} + \tfrac{a(a-1)}{2}\tfrac{\xi(h)}{h^a}\right) = \tfrac{1}{2}(a+1)(a+2)\lim_{h\to 0}\tfrac{\xi(h)}{h^a} = \tfrac{1}{2}(a+1)(a+2)d \tag{A59}$$

Since $a > 0$ the rightmost expression cannot be zero unless $d = 0$. Therefore, the limit $\lim\limits_{h\to 0} h^{-a}v(h)$ is 0, finite or ∞, if and only if $\lim\limits_{h\to 0} h^{-a}\xi(h)$ is 0, finite or ∞, respectively. Therefore, $v^{\#}(0) = \xi^{\#}(0) = a$.

If $a = 1$ precisely, then $h^{a-1} = 1$ (it does not tend to 0 or ∞), while $\lim\limits_{h\to 0} h\xi''(h) = 0$. Thus, from (A56) we get $\lim\limits_{h\to 0} h^{-1}v(h) = 3d$ (finite), which again means that $v^{\#}(0) = \xi^{\#}(0) = a$. Also, if $a = 2$ precisely, then $h^{a-2} = 1$ and from (A56) we get $\lim_{h\to 0} h^{-1}v(h) = 6d$ (finite). In conclusion, in all cases it holds that $v^{\#}(\infty) = \xi^{\#}(\infty) = 0$ and $v^{\#}(0) = \xi^{\#}(0) = a$, which means that $v(k)$ has the same asymptotic behaviour with $\xi(k)$ in both ends. We can thus rewrite (A50) as:

$$\begin{gathered}\gamma_C^{\#}(0) = \xi^{\#}(0) = v^{\#}(0) = \zeta^{\#}(0) - 1 = 2\varphi_C(0) - 2 = a > 0 \\ \gamma_C^{\#}(\infty) = \gamma^{\#}(\infty) = c^{\#}(\infty) = \zeta^{\#}(\infty) - 1 = 2\varphi_C(\infty) - 2 = b < 0 \\ \text{Possible exceptions for } c^{\#}(\infty):\ b = \gamma_C^{\#}(\infty) = \gamma^{\#}(\infty) = \{-1, -2\}, \\ \zeta^{\#}(\infty) = \{0, -1\},\ \varphi_C(\infty) = \{1/2, 0\}\end{gathered} \tag{A60}$$

Now coming to the asymptotic behaviour of the power spectrum, according to Gneiting and Schlather [37] (who also cite [46–48]), the power spectrum is "typically" associated with the asymptotic relationship $s(w) \sim w^{-a-1}$ as $w \to \infty$ and $s(w) \sim w^{-b-1}$ as $w \to 0$, where a (>0) and b (<0) are the LLDs of the structure function, $v^{\#}(0)$ and autocorrelation function $c^{\#}(\infty)$, respectively, same to the above discourse. By inspection, it is seen that the slopes $-(a+1)$ and $-(b+1)$ are precisely the log-log slopes of the climacospectrum expressed in terms of frequency, $\tilde{\zeta}(w)$, or equivalently $-\zeta^{\#}(0)$ and $-\zeta^{\#}(\infty)$, respectively. It can be conjectured that, since this is the "typical" behaviour, the non-typical cases correspond to integral values of a and b.

In plain language, the above results can be expressed as follows:

1. The asymptotic behaviour of the autocovariance function is the same with that of the climacogram.
2. The asymptotic behaviour of the structure function is the same with that of the climacostructure function.
3. The asymptotic behaviour of the power spectrum is the same with that of the climacospectrum.

Exceptions to the rule 1 may occur when $b = -1$ or $b = -2$ and this was indeed verified by thorough analytical work using the series of models examined in Appendix A.2. Exception to rule 2 can also appear and indeed they were found when $b = -2$ or $a = 2$. The entire list of exceptions found in this analytical work follows, also mentioning the specific models in which they appeared:

1. $b = \gamma^{\#}(\infty) = -1$ and $c^{\#}(\infty) = -\infty$: Markov, Smooth Exponential models
2. $b = \gamma^{\#}(\infty) = -1$ and $c^{\#}(\infty) = -2$: Generalized Planck model for $n = 0$
3. $b = \gamma^{\#}(\infty) = -2$ and $c^{\#}(\infty) = -\infty$, also $-s^{\#}(0) = -2n$ while $\zeta^{\#}(\infty) = -1$: Antipersistent Exponential model for $n > 0$
4. $b = \gamma^{\#}(\infty) = -2$ and $c^{\#}(\infty) = -2(n+1)$, also $-s^{\#}(0) = -2n$ while $\zeta^{\#}(\infty) = -1$: Generalized Planck model for $n > 0$
5. $a = \xi^{\#}(0) = 2$ and $-s^{\#}(\infty) = \infty$ while $\zeta^{\#}(0) = 3$: Generalized Planck model
6. $a = \xi^{\#}(0) = 2$ and $-s^{\#}(\infty) = 2(n+1)$ while $\zeta^{\#}(0) = 3$: Smooth Exponential for $n > 0$

Appendix A.1.10. Proof of Physical Inconsistency of too Mild Slopes at High Frequencies in Power Spectrum

Let us assume that for small scales $k < \epsilon$ or high frequencies $w > 1/\epsilon$, with ϵ however small, the log-log derivative is $s^{\#}(w) \approx \beta$, or else $s(w) \sim w^{\beta}$ where β is a constant satisfying $\beta > -1$. As the area of the power spectrum equals the variance of the continuous time process, we will have:

$$\gamma(0) = \int_0^{\infty} s(w)\mathrm{d}w = \int_0^{1/\epsilon} s(w)\mathrm{d}w + \int_{1/\epsilon}^{\infty} s(w)\mathrm{d}w \tag{A61}$$

Because $s(w) \sim w^{\beta}$, there exist $\alpha > 0$ so that $s(w) \geq \alpha\, w^{\beta}$ for $w > 1/\epsilon$. Thus the, rightmost of the above integrals can be evaluated as:

$$\int_{1/\epsilon}^{\infty} s(w)\mathrm{d}w \geq \int_{1/\epsilon}^{\infty} \alpha w^{\beta}\,\mathrm{d}w = \left.\frac{\alpha w^{\beta+1}}{\beta+1}\right|_{1/\epsilon}^{\infty} \tag{A62}$$

Clearly if $\beta > -1$, $w^{\beta+1} \to \infty$ as $w \to \infty$ and thus $\gamma(0) = \infty$. The same result could be obtained if the climacospectrum $\widetilde{\zeta}(w)$ was used instead of the power spectrum. Because infinite variance means infinite energy, which is physically inconsistent, it turns out that the assumption $\beta > -1$ is physically absurd.

Appendix A.1.11. Proof of the Inconsistency of too Steep Slopes at High Frequencies in Power Spectrum

We prove that $\varphi_C(0) \leq 2$ or, equivalently, $\xi^{\#}(0) \leq 2$, by contradiction, i.e., assuming $\xi^{\#}(0) = 2M$ with $M > 1$. In the latter case for small k we may approximate $\xi(k)$ as:

$$\xi(k) \approx \lambda \frac{k^{2M}}{1 + k^{2M}} \tag{A63}$$

which indeed yields $\xi^{\#}(0) = 2M$. The climacogram for small k is thus:

$$\gamma(k) = \frac{\lambda}{1 + k^{2M}} \tag{A64}$$

The positive definiteness of $\gamma(k)$ entails that any matrix A with elements $a_{i,j} = \gamma(k_i - k_j)$ should be positive definite. Choosing three points, $k_i = k_0$, $k_0 + k$ and $k_0 + 2k$, we form the 3×3 matrix:

$$A = \begin{bmatrix} \gamma(0) & \gamma(k) & \gamma(2k) \\ \gamma(k) & \gamma(0) & \gamma(k) \\ \gamma(2k) & \gamma(k) & \gamma(0) \end{bmatrix} \tag{A65}$$

whose determinant can be easily evaluated. Specifically, after algebraic manipulations we obtain:

$$B(k, M) := \mathrm{Det}(A)/\lambda^3 = 1 - \frac{2}{(1 + k^{2M})^2} - \frac{1}{\left(1 + (2k)^{2M}\right)^2} + \frac{2}{(1 + k^{2M})^2\left(1 + (2k)^{2M}\right)} \tag{A66}$$

This can also be written as:

$$\begin{gathered} B(k, M) = \frac{2^{2M} k^{4M} G(k,M)}{(1+k^{2M})^2 \left(1+(2k)^{2M}\right)^2}, \\ G(k, M) := 4 - 2^{2M} + 2(1 + 2^{2M}) k^{2M} + 2^{2M} \left(k^{2M}\right)^2 \end{gathered} \tag{A67}$$

We observe that $G(k, M)$ is a second-order polynomial in terms of $y := k^{2M}$ with discriminant $D(M) = 1 + 2^{2M+1}(2^{2M} - 1)$, which is positive for any $M > 0$ and thus the equation $G(k, M) = 0$ has two real solutions:

$$y_{1,2} = 2^{-2M}\left(-1 - 2^{2M} \mp \sqrt{1 + 2^{2M+1}(2^{2M} - 1)}\right) \tag{A68}$$

Clearly, $y_1 < 0$ for any M, while $y_2 < 0$ only when $M < 1$ (for $M = 1$, $y_2 = 0$ precisely). Therefore, if $M > 1$, then for $0 < k^{2M} < y_2$, $G(k, M) < 0$ and hence $B(k, M) < 0$, which proves that A is not positive definite as its principal determinant is negative.

Solved for k, the interval where $B(k, M) < 0$ is:

$$0 < k < \frac{1}{2}\left(-1 - 2^{2M} + \sqrt{1 + 2^{2M+1}(2^{2M} - 1)}\right)^{1/2M} \tag{A69}$$

This means that, if $M > 1$, $B(k, M)$ is negative for arbitrarily small k (in the neighbourhood of 0), which justifies the approximation we have used for $\xi(k)$. Therefore the assumption $M > 1$ should be rejected.

As an illustration by a different approach, Figure A1 shows for the FHK-C model, a single value of the power spectrum, namely $s(\alpha)$, where α is the time scale parameter of the model, as a function of the parameter M. As seen in the plot, for $M > 1$, $s(\alpha) < 0$ which confirms absence of positive definiteness, now of the autocorrelation function. (We note that for $M > 1$ but very close to 1, the plot shows $s(\alpha)$ slightly positive but in fact it is negative again for some $w > \alpha$—not shown in the figure).

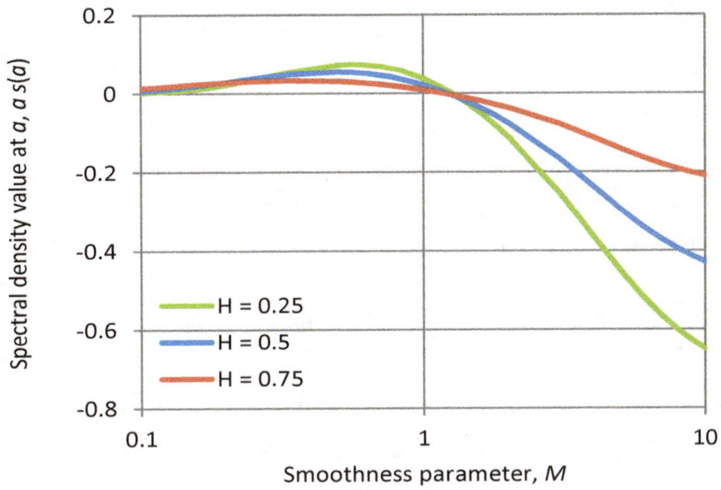

Figure A1. Plot of the spectral density value at $w = \alpha$, standardized by α, for the FHK-C model with smoothness parameter M varying in [0.1, 10] and for the indicated values of the Hurst parameter H.

Appendix A.2. Specific Stochastic Models and Their Behaviour

In Appendix A.2 we provide detailed tables with mathematical expressions for all the models used in the paper and detailed graphs with applications of these models. All models include two scale parameters, α (scale parameter of time with dimensions of $[t]$) and λ (scale parameter of variance, with dimensions of $[x^2]$). These are absolutely necessary for dimensional consistency. Some models contain additional dimensionless shape parameters, such as the smoothness parameter M, the persistence (Hurst) parameter H, or other as specified in each of the models. The theoretical characteristics depicted in the graphs are determined from the expressions given in the tables. The time series also depicted in the graphs, along with and their empirical characteristics, were produced by the SMA model as described in the text; their length is 1024 in all cases. For facilitating comparisons, in all applications the parameter α is the same, $\alpha = 10$ (except if stated otherwise in the caption), while λ is chosen so that $\gamma(1) = 1$.

Appendix A.2.1. Markov Process

The Markov (also known as Ornstein–Uhlenbeck) process is the most convenient as all its second-order characteristics have simple expressions (see Table A1) and, simultaneously, the most parsimonious as it includes no other parameter additional to the two necessary scale parameters α and λ. For these reasons it is the most common in applications. On the other hand, its neutrality in terms of smoothness and persistence, and more specifically the low entropy production for large time scales, does not make it a good candidate to model natural behaviours.

The application depicted in Figure A2 corresponds to an unusually high lag-one autocorrelation ($\rho = 0.94$). Note that the model does not allow control of autocorrelation once the parameter α is specified.

Table A1. Mathematical expressions of the second order characteristics of the Markov process and their asymptotic properties.

Characteristic	Mathematical Expression	Local Asymptotic Properties	Global Asymptotic Properties	Equation
Climacogram	$\gamma(k) = \frac{2\lambda}{(k/\alpha)^2}\left(-1 + k/\alpha + \mathrm{e}^{-k/\alpha}\right)$	$\gamma^{\#}(0) = 0$	$\gamma^{\#}(\infty) = -1$	(A70)
Climacospectrum	$\zeta(k) := \frac{k(\gamma(k)-\gamma(2k))}{\ln 2}$ $= \frac{\lambda\left(-3+2k/\alpha+4\mathrm{e}^{-k/\alpha}-\mathrm{e}^{-2k/\alpha}\right)}{2(\ln 2)k/\alpha}$	$\zeta^{\#}(0) = 2$	$\zeta^{\#}(\infty) = 0$	(A71)
Entropy production	$\varphi(k) = \frac{2\lambda\left(1-\mathrm{e}^{-k/\alpha}\right)}{\gamma(k)(k/\alpha)}$ $= \frac{(k/\alpha)\left(1-\mathrm{e}^{-k/\alpha}\right)}{2\left(-1+k/\alpha+\mathrm{e}^{-k/\alpha}\right)}$ $\varphi_{\mathrm{C}}(k) = \frac{\varphi(k)\gamma(k)-\varphi(2k)\gamma(2k)}{\gamma(k)-\gamma(2k)}$	$\varphi(0) = 1$ $\varphi_{\mathrm{C}}(0) = 3/2$	$\varphi(\infty) = \varphi_{\mathrm{C}}(\infty) = 1/2$	(A72)
Autocovariance function	$c(h) = \lambda \mathrm{e}^{-h/\alpha}$	$c^{\#}(0) = 0$	$c^{\#}(\infty) = -\infty$	(A73)
Power spectrum	$s(w) = \frac{4\lambda\alpha}{1+(2\pi\alpha w)^2}$	$-s^{\#}(\infty) = 2$	$-s^{\#}(0) = 0$	(A74)

Appendix A.2.2. FHK Process with Cauchy-Type Climacogram (FHK-C)

General description of the FHK process with Cauchy-type climacogram (FHK-C) is given in Section 3.2 and its detailed mathematical expressions and asymptotic processes are given in Table A2.

Table A2. Mathematical expressions of the second order characteristics of the FHK process with Cauchy-type climacogram (FHK-C) and their asymptotic properties.

Characteristic	Mathematical Expression	Local Asymptotic Properties	Global Asymptotic Properties	Equation
Climacogram	$\gamma(k) = \lambda\left(1 + (k/\alpha)^{2M}\right)^{\frac{H-1}{M}}$	$\gamma^{\#}(0) = 0$	$\gamma^{\#}(\infty) = 2H - 2$	(A75)
Climacospectrum	$\zeta(k) := \frac{k(\gamma(k)-\gamma(2k))}{\ln 2}$	$\zeta^{\#}(0) = 1 + 2M$	$\zeta^{\#}(\infty) = 2H - 1$	(A76)
Entropy production	$\varphi(k) = H + \frac{1-H}{1+(k/\alpha)^{2M}}$ $\varphi_{\mathrm{C}}(k) = \frac{\varphi(k)\gamma(k)-\varphi(2k)\gamma(2k)}{\gamma(k)-\gamma(2k)}$	$\varphi(0) = 1$ $\varphi_{\mathrm{C}}(0) = 1 + M$	$\varphi(\infty) = \varphi_{\mathrm{C}}(\infty) = H$	(A77)
Autocovariance function	$c(h) = \gamma(h) \times \frac{1+(3H+2MH-2M-1)(h/\alpha)^{2M}+H(2H-1)(h/\alpha)^{4M}}{1+2(h/\alpha)^{2M}+(h/\alpha)^{4M}}$	$c^{\#}(0) = 0$	$c^{\#}(\infty) = 2H - 2$	(A78)
Power spectrum	Analytical expression not possible except in special cases; numerical solution easily derived from (10)	$-s^{\#}(\infty) = 1 + 2M$	$-s^{\#}(0) = 2H - 1$	(A79)

This model is very rich as it covers the entire admissible range, by appropriate choice of the smoothness parameter ($0 < M \leq 1$) and persistence parameter ($0 \leq H < 1$). It represents smooth processes (if $M > ½$), rough processes (if $M < ½$), persistent processes (if $H > ½$) and antipersistent processes (if $H < ½$). For $M = H = \frac{1}{2}$ it becomes a neutral process, practically identical to Markov.

The FHK process can even produce near pink ($1/f$) noise if $M = 1 - H \approx 0$, which is associated with maximal entropy production for large scales and minimal for small scales. In this case the quantity $(k/\alpha)^{2M}$ is close to 1 and the following approximation holds:

$$\gamma(k) = \frac{\lambda}{1 + (k/\alpha)^{2(1-H)}} \approx \frac{\lambda(1 + (H-1)\ln(k/\alpha))}{2} \tag{A80}$$

This can be verified by observing that for x close to 1 the quantities $\ln x$ and $\left(x^2 - 1\right) / \left(x^2 + 1\right)$ are close to each other, differing only by a factor of $\mathrm{O}[x - 1]^3$.

For $M = H \approx 0$, we have minimal entropy production for all scales. In this case:

$$\gamma(k) = \lambda\left(1 + (k/\alpha)^{2H}\right)^{\frac{H-1}{H}} \approx \lambda(k/\alpha)^{H-1} \tag{A81}$$

This can be verified by observing that for x close to 1 the logarithms of the two quantities differ only by a factor of $H\,\mathrm{O}[x-1]^2$. We note that both (A80) and (A81) violate the condition for physical consistency (finite variance).

A set of applications of the FHK-C are depicted in detail in Figures A3–A7, which are also discussed in Section 3.2. It is observed that the SMA generation scheme performs almost perfectly even in the most extreme cases and that the climacogram-based characteristics perform better than the autocorrelation-based ones.

Appendix A.2.3. FHK Process with Dagum-Type Climacogram (FHK-D)

The FHK process with Dagum-type climacogram (FHK-D) has expressions for its second order characteristics different from those of the Cauchy-type climacogram (see Table A3). However, the behaviour of the two models, as well as the meaning and ranges of parameters, are the same. For $M = 1 - H$ the two models become precisely identical. The model is particularly useful in combination with the FHK-C model as shown below.

Table A3. Mathematical expressions of the second order characteristics of the FHK process with Dagum-type climacogram (FHK-D) and their asymptotic properties.

Characteristic	Mathematical Expression	Local Asymptotic Properties	Global Asymptotic Properties	Equation
Climacogram	$\gamma(k) = \lambda\left(1 - \left(1 + (k/\alpha)^{2(H-1)}\right)^{\frac{M}{H-1}}\right)$	$\gamma^{\#}(0) = 0$	$\gamma^{\#}(\infty) = 2H - 2$	(A82)
Climacospectrum	$\zeta(k) := \frac{k(\gamma(k)-\gamma(2k))}{\ln 2}$	$\zeta^{\#}(0) = 1 + 2M$	$\zeta^{\#}(\infty) = 2H - 1$	(A83)
Entropy production	$\varphi(k) = 1 - \frac{\lambda-\gamma(k)}{\gamma(k)} \frac{M}{(1+(k/\alpha)^{2-2H})}$ $\varphi_C(k) = \frac{\varphi(k)\gamma(k)-\varphi(2k)\gamma(2k)}{\gamma(k)-\gamma(2k)}$	$\varphi(0) = 1$ $\varphi_C(0) = 1 + M$	$\varphi(\infty) = \varphi_C(\infty) = H$	(A84)
Autocovariance function	$c(h) = \lambda - (\lambda - \gamma(h))\times$ $\frac{1+(2+M+2MH)(h/\alpha)^{2H-2}+(1+M)(1+2M)(h/\alpha)^{4H-4}}{1+2(h/\alpha)^{2H-2}+(h/\alpha)^{4H-4}}$	$c^{\#}(0) = 0$	$c^{\#}(\infty) = 2H - 2$	(A85)
Power spectrum	Analytical expression not possible except in special cases; numerical solution easily derived from (10)	$-s^{\#}(\infty) = 1 + 2M$	$-s^{\#}(0) = 2H - 1$	(A86)

Appendix A.2.4. FHK Process with Climacogram Equal to the Sum of a Cauchy and a Dagum Climacograms (FHK-CD)

By summing a Cauchy-type climacogram with $M = 1$ and a Dagum-type climacogram with $H = 0$ we get the FHK-CD model whose second-order characteristics are shown in Table A4. This is a convenient model, in which the first additive term determines merely the persistence of the process and the second one the smoothness of the process. It is more convenient that its constituents as most of mathematical expressions are simpler; in particular, it results in an explicit relationship of the power spectrum, which however is not very simple. In addition, it is more flexible and richer than its constituents, as it contains two couples of scale parameters; however, if parsimony is sought, then it can take the same number of parameters as each of the constituents by setting $\alpha_1 = \alpha_2$ and $\lambda_1 = \lambda_2$ (it is reminded that one λ and one α are absolutely minimal requirements of parameters).

An application of the model based on real measurements of turbulent velocities is contained in Section 3.4.

Another option, of the same type, not studied in detail here, would be to sum a Cauchy-type climacogram with $M = ½$ and a Dagum-type climacogram with $H = ½$. The climacogram of the thus formed model is:

$$\gamma(k) = \lambda_1(1 + k/\alpha_1)^{2H-2} + \lambda_2(1 - (1 + \alpha_2/k)^{-2M}) \tag{A87}$$

The model (A87), however, can only produce rough persistent processes with $0 < \varphi_C(0) \leq 3/2$ and $1/2 \leq \varphi_C(\infty) < 1$. Indeed, it is easily shown that in this case $\varphi_C(0) = \min(1 + M, 3/2)$ and

$\varphi(\infty) = \varphi_C(\infty) = \max(H, 1/2)$. The turbulent velocity case, studied in Section 3.4, belongs to this domain and thus the model (A87) could also be used as an alternative.

Table A4. Mathematical expressions of the second order characteristics of the FHK process formed as the sum of a Cauchy and a Dagum process (FHK-CD) dealing separately with persistence and smoothness, respectively, and their asymptotic properties.

Characteristic	Mathematical Expression	Local Asymptotic Properties	Global Asymptotic Properties	Equation
Climacogram	$\gamma(k) = \gamma_1(k) + \gamma_2(k)$ $\gamma_1(k) = \lambda_1(1 + (k/\alpha_1)^2)^{H-1}$ $\gamma_2(k) = \lambda_2(1 - (1 + (k/\alpha_2)^{-2})^{-M})$	$\gamma^\#(0) = 0$	$\gamma^\#(\infty) = 2H - 2$	(A88)
Climacospectrum	$\zeta(k) := \frac{k(\gamma(k) - \gamma(2k))}{\ln 2}$	$\zeta^\#(0) = 1 + 2M$	$\zeta^\#(\infty) = 2H - 1$	(A89)
Entropy production	$\varphi(k) = \frac{\varphi_1(k)\gamma_1(k) + \varphi_2(k)\gamma_2(k)}{\gamma(k)}$ $\varphi_1(k) = \frac{1 + H(k/\alpha_1)^2}{1 + (k/\alpha_1)^2}$ $\varphi_2(k) = 1 - \frac{M}{(1 + (k/\alpha_2)^2)((1 + (k/\alpha_2)^{-2})^M - 1)}$ $\varphi_C(k) = \frac{\varphi(k)\gamma(k) - \varphi(2k)\gamma(2k)}{\gamma(k) - \gamma(2k)}$	$\varphi(0) = 1$ $\varphi_C(0) = 1 + M$	$\varphi(\infty) = \varphi_C(\infty) = H$	(A90)
Autocovariance function	$c(h) = c_1(h) + c_2(h)$ $c_1(h) = \gamma_1(h) \times$ $\frac{1 + (5H-3)(h/\alpha_1)^2 + H(2H-1)(h/\alpha_1)^4}{1 + 2(h/\alpha_1)^2 + (h/\alpha_1)^4}$ $c_2(h) = \lambda_2 - (\lambda_2 - \gamma_2(h)) \times$ $\frac{(1+M)(1+2M) + (2+M)(h/\alpha)^2 + (h/\alpha)^4}{1 + 2(h/\alpha)^2 + (h/\alpha)^4}$	$c^\#(0) = 0$	$c^\#(\infty) = 2H - 2$	(A91)
Power spectrum	Analytical expression too complex, derived from (10)	$-s^\#(\infty) = 1 + 2M$	$-s^\#(0) = 2H - 1$	(A92)

Appendix A.2.5. Smooth Process with Generalized Exponential-Type Autocovariance (SE)

By generalizing the power spectrum of a Markov process, as seen in Table A5, we can obtain a smooth processes (the smooth exponential process—SE). Generalization is made by introducing an integral parameter $n = 0, 1, 2, \ldots$, where the value $n = 0$ results precisely in a Markov process. A notable characteristic of the process is that it corresponds to the absolute maximum of entropy production for small scales.

Appendix A.2.6. Antipersistent Process with Generalized Exponential-Type Autocovariance (AE)

This is again a generalization of the Markov process by introducing an integral parameter $n = 0, 1, 2, \ldots$, where the value $n = 0$ results precisely in a Markov process. Its characteristics are shown in Table A6. The process is antipersistent and its most notable characteristic is that it corresponds to the absolute minimum of entropy production for large scales. An application of the process is shown in Figure A8 for $n = 2$; the hole in the function $\phi_C(k)$ for moderate k is a characteristic of this process.

Appendix A.2.7. Generalized Planck Model (GP)

The generalized Planck (GP) model is obtained by a generalization of Planck's law of black-body radiation, introducing an integral parameter $n = 0, 1, 2, \ldots$, where the value $n = 1$ corresponds to Planck's law precisely. The process, whose detailed equations are given in Table A7, is simultaneously extremely smooth ($\phi_C(0) = 2$) and extremely antipersistent ($\phi_C(\infty) = 0$). An application for $n = 1$ is depicted in Figure A9.

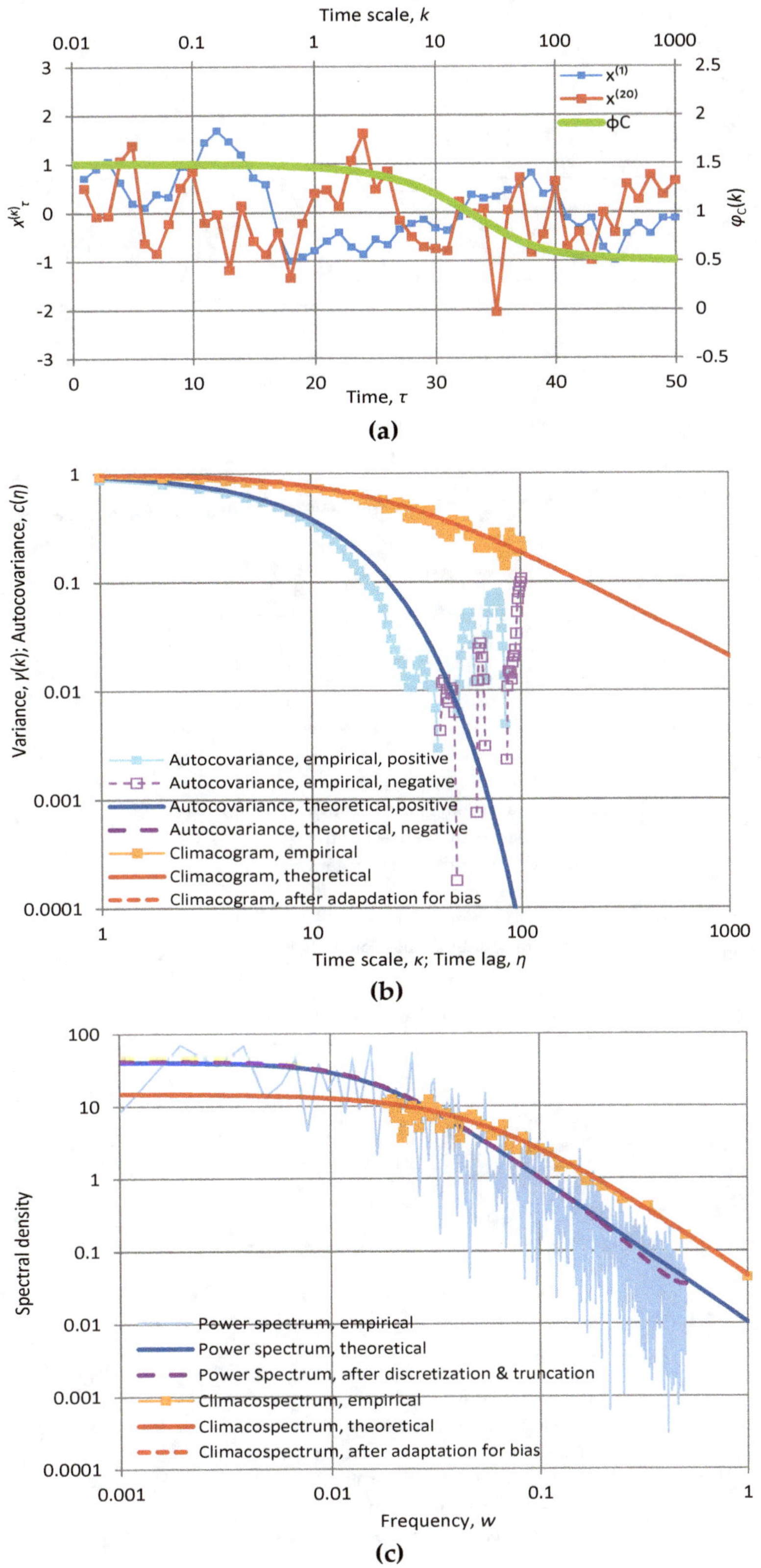

Figure A2. Characteristic graphs of a time series of length 1024 generated by the Markov model with $\alpha = 10$ and $\lambda = 1.03$ (so that $\gamma(1) = 1$): (**a**) plot of the first fifty terms of the at time scales $\kappa = 1$ and 20, along with the "stamp" $\phi_C(k)$ of the fitted model (green line plotted with respect to the secondary axes); (**b**) empirical and theoretical climacogram and autocovariance function; (**c**) empirical and fitted theoretical climacospectrum and power spectrum.

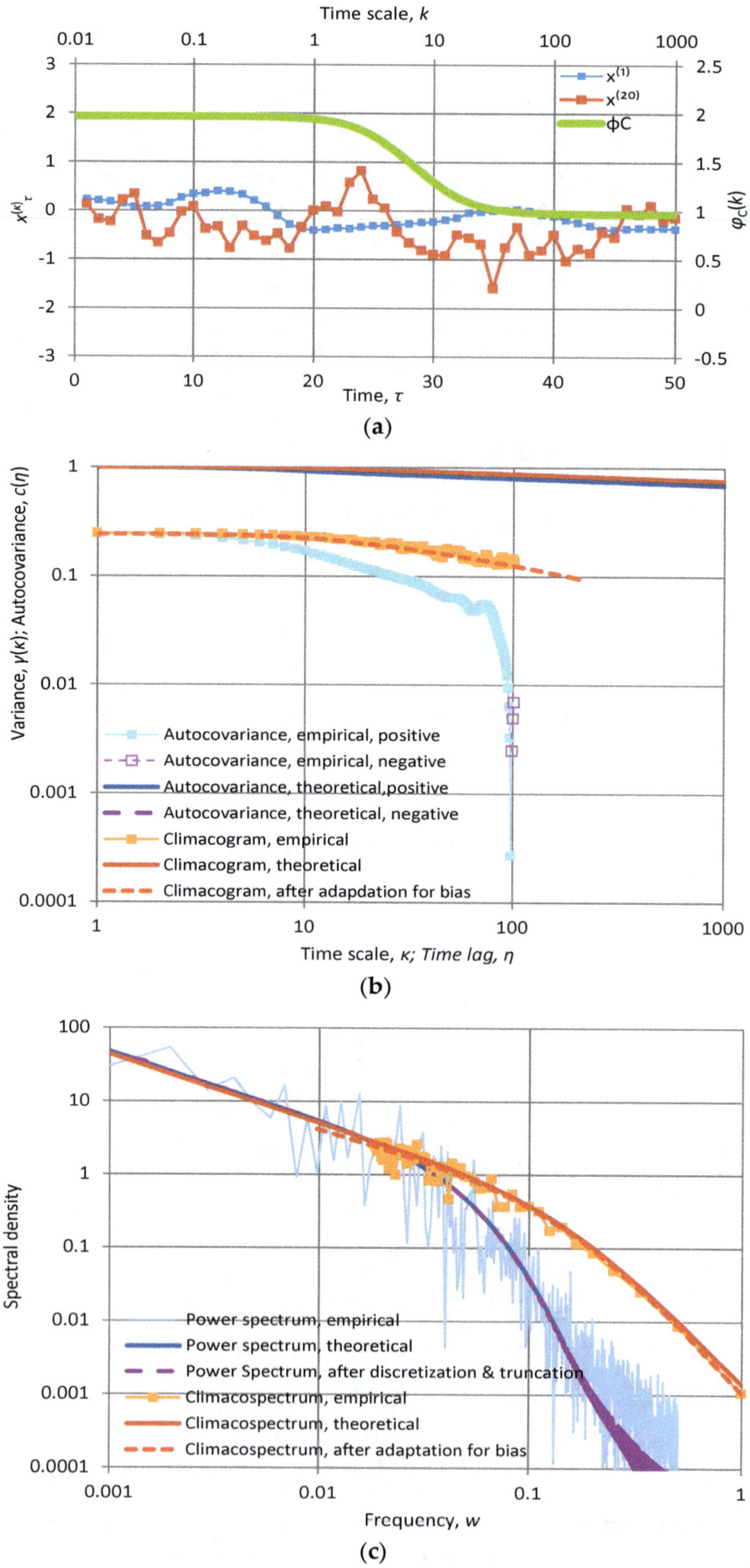

Figure A3. Characteristic graphs as in Figure A2 but for a time series generated by the FHK-C model with $\gamma(1) = 1$, $\alpha = 10$ and $H = M = 0.97$.

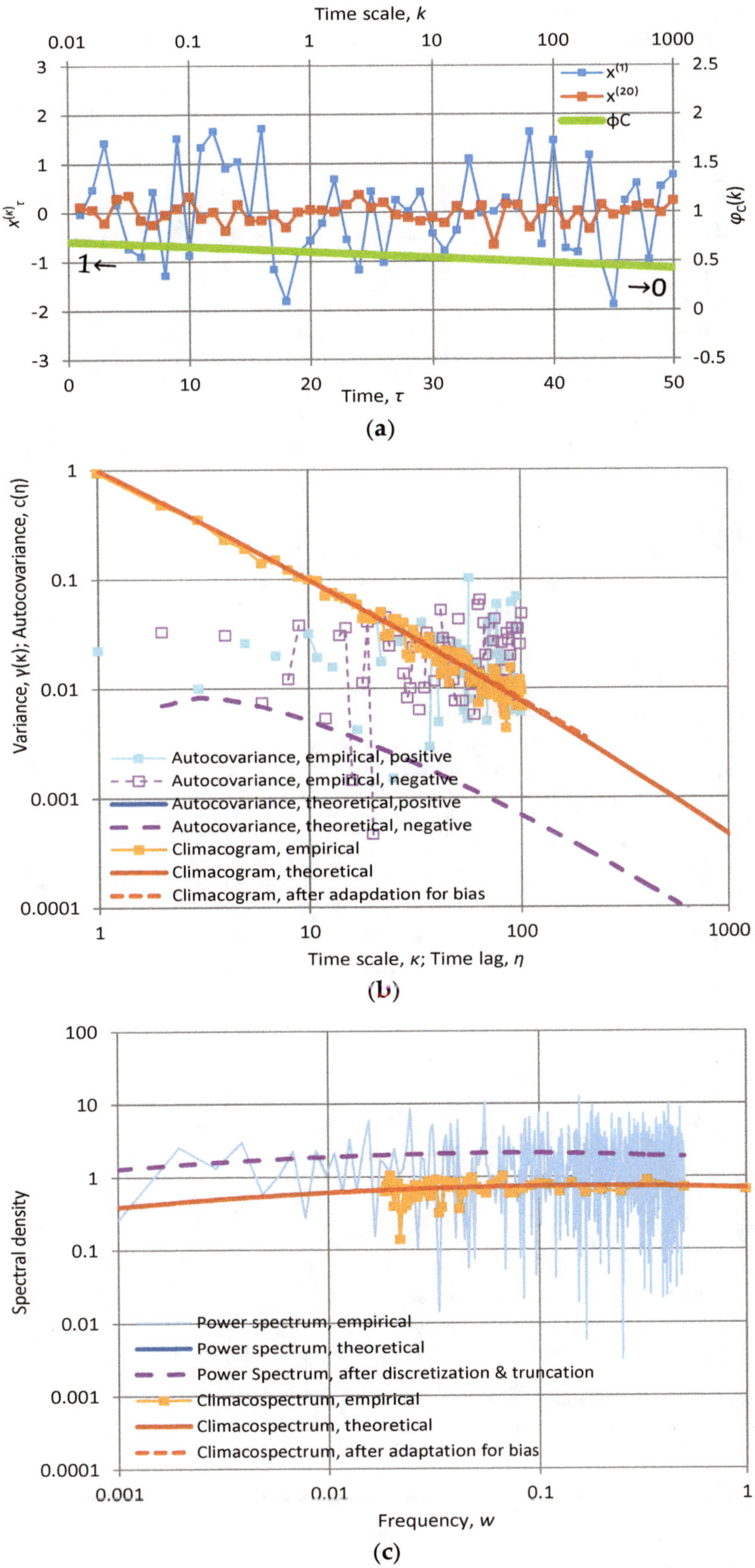

Figure A4. Characteristic graphs as in Figure A2 but for a time series generated by the FHK-C model with $\gamma(1) = 1$, $\alpha = 10$, and $H = M = 0.05$.

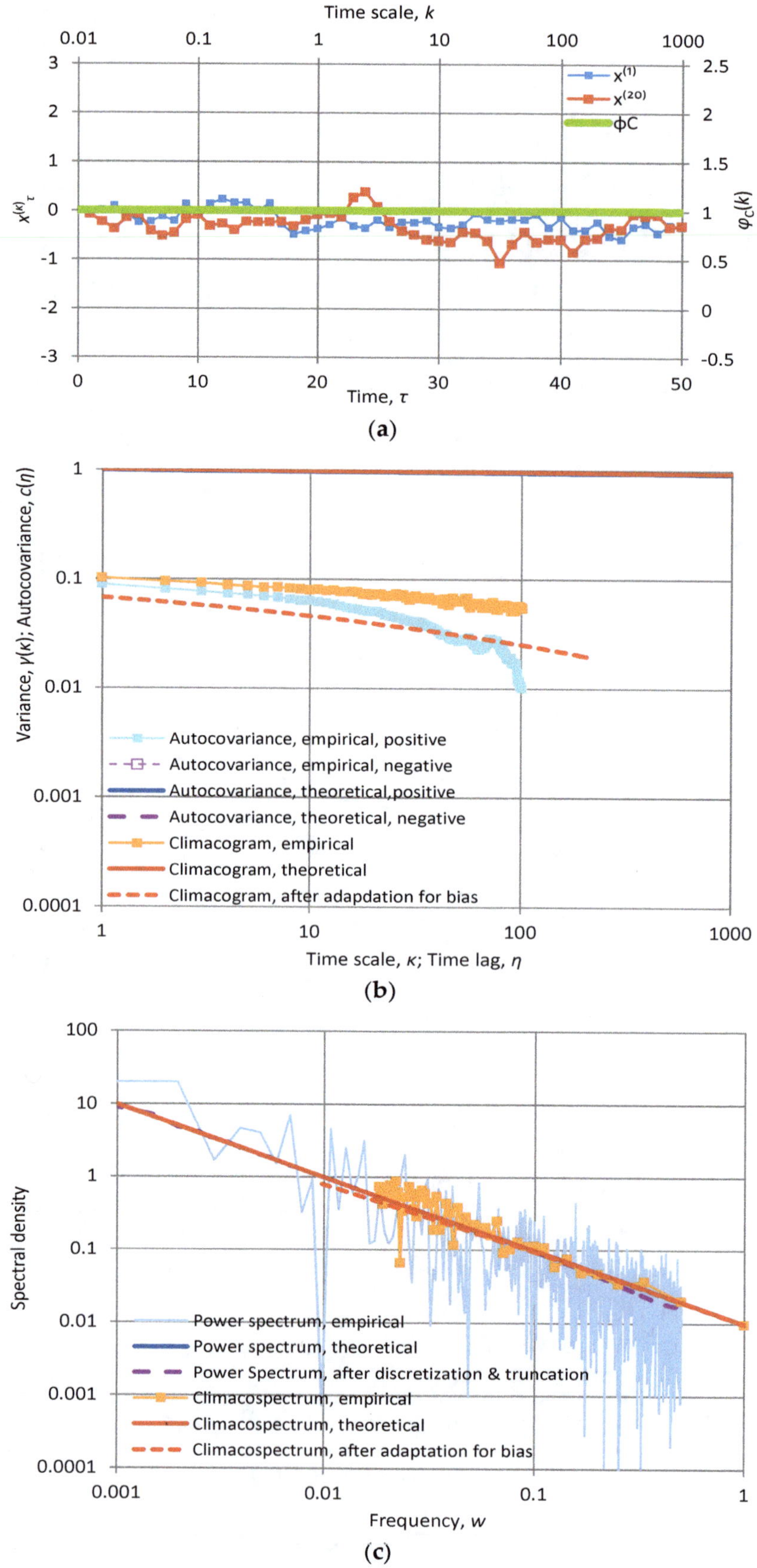

Figure A5. Characteristic graphs as in Figure A2 but for a time series generated by the FHK-C model with $\gamma(1) = 1$, $\alpha = 10$, $H = 0.99$ and $M = 0.01$ (close to "pink noise").

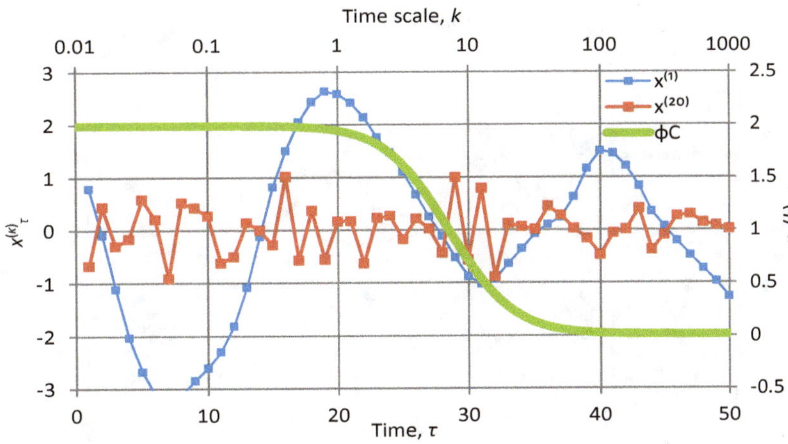

(a)

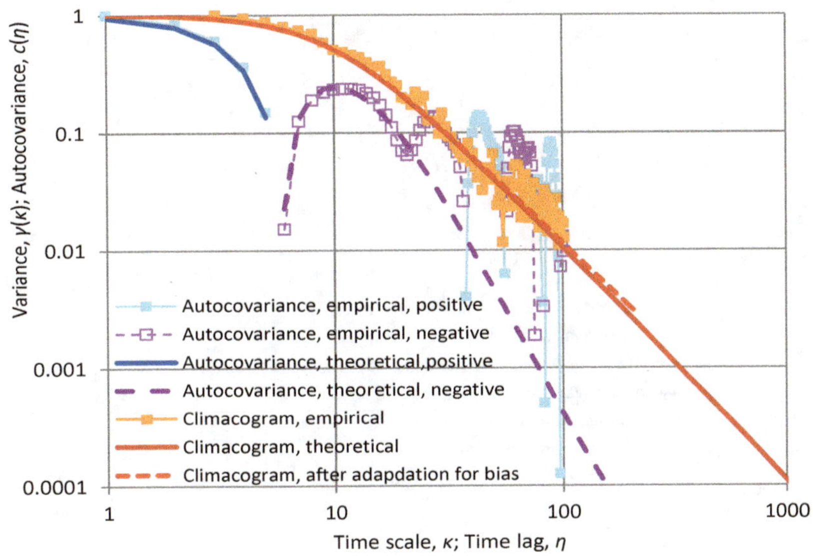

(b)

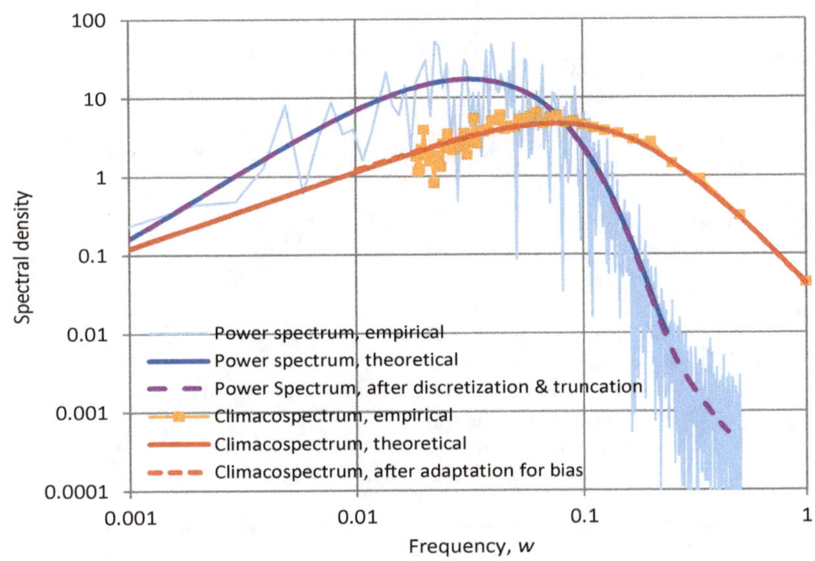

(c)

Figure A6. Characteristic graphs as in Figure A2 but for a time series generated by the FHK-C model with $\gamma(1) = 1$, $\alpha = 10$, $H = 0.01$ and $M = 0.99$.

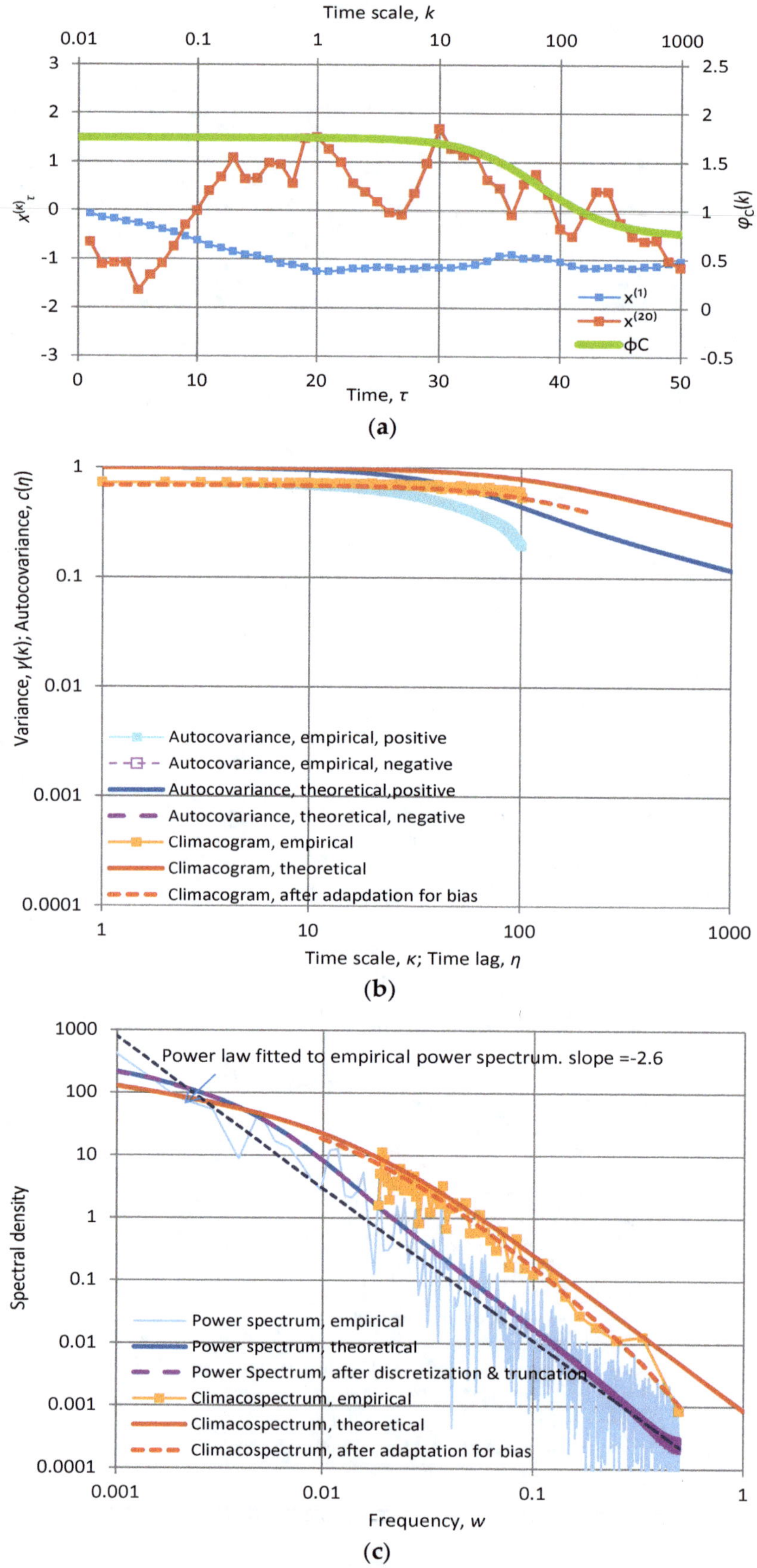

Figure A7. Characteristic graphs as in Figure A2 but for a time series generated by the FHK-C model with $\gamma(1) = 1$, $\alpha = 100$ and $H = M = 0.75$.

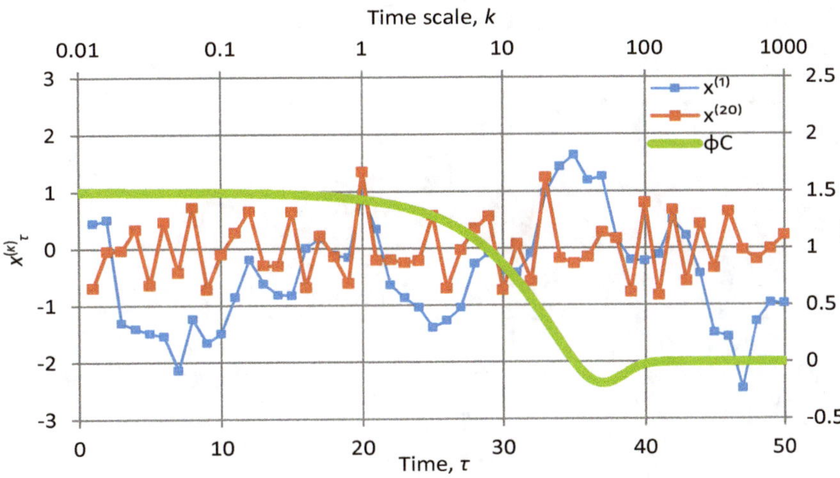

(a)

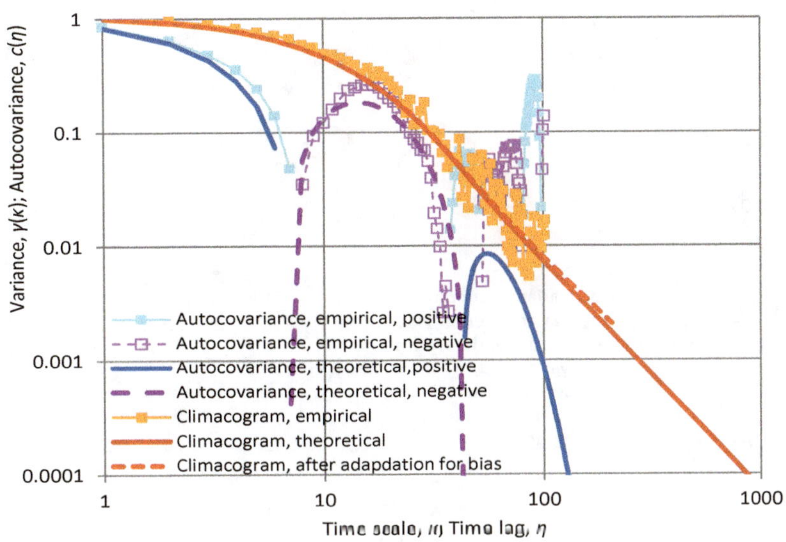

(b)

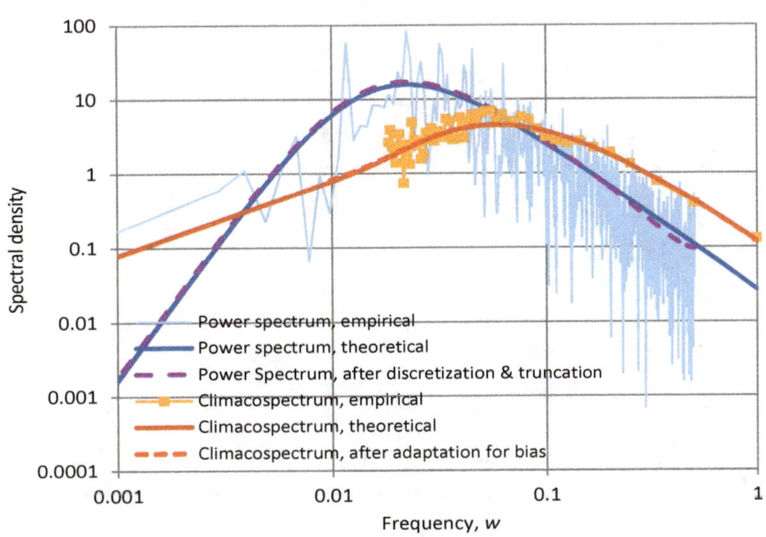

(c)

Figure A8. Characteristic graphs as in Figure A2 but for a time series generated by the AE model with $\gamma(1) = 1$, $\alpha = 10$ and $n = 2$.

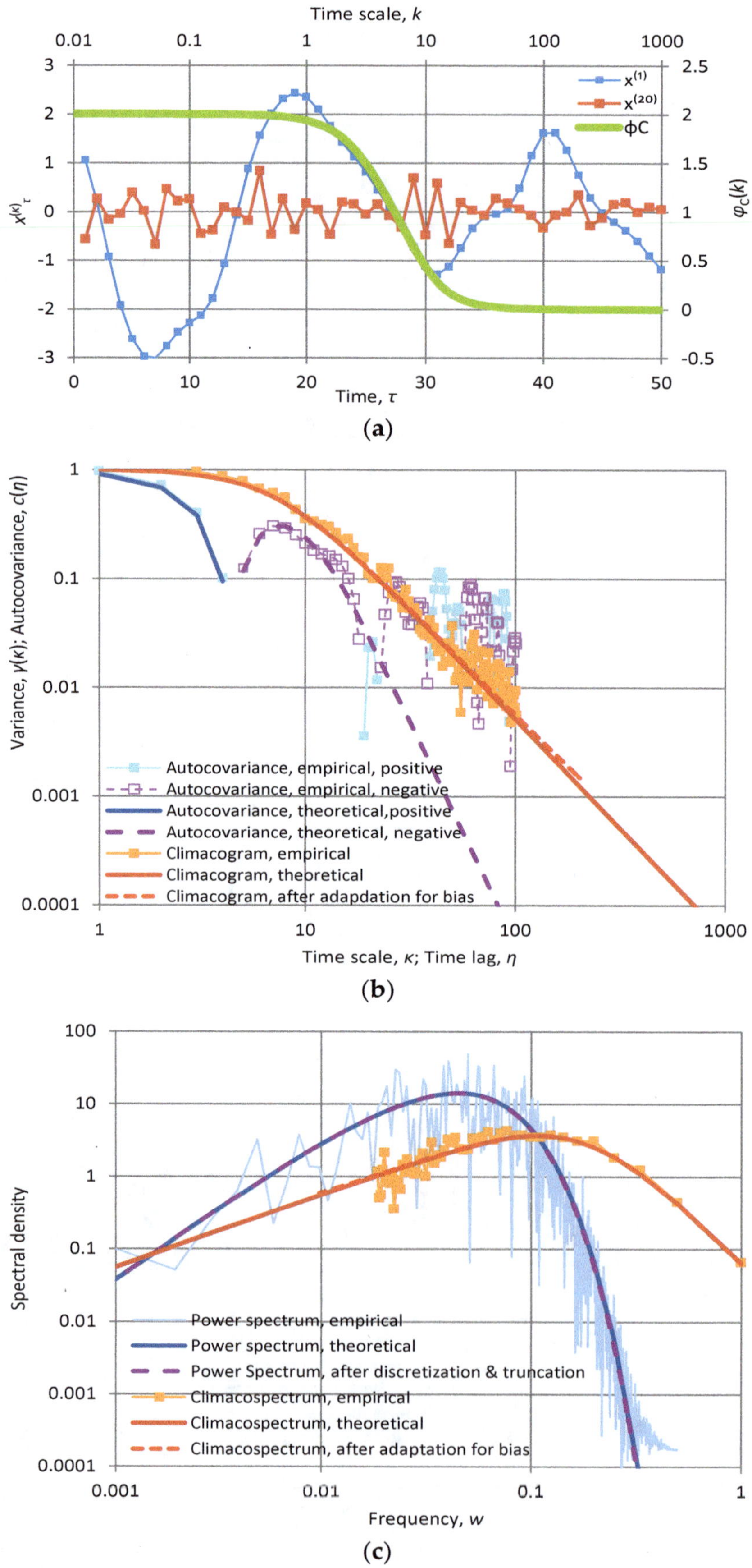

Figure A9. Characteristic graphs as in Figure A2 but for a time series generated by the GP model with $\gamma(1) = 1$, $\alpha = 10$ and $n = 1$.

Table A5. Mathematical expressions of the second order characteristics of the SE process and their asymptotic properties.

Characteristic	Mathematical Expression	Local Asymptotic Properties	Global Asymptotic Properties	Equation
Climacogram	$\gamma(k) = \frac{2\lambda}{(k/\alpha)^2} \times \left(-2n - 1 + a_n^{\mathrm{PE}} k/\alpha + \mathrm{e}^{-k/\alpha} P_n^{\mathrm{PE}}(k/\alpha)\right)$ where $n = 0, 1, 2, \ldots$, $P_n^{\mathrm{PE}}(x)$ is a polynomial and a_n^{PE} a coefficient, both determined using (23); the first five are: $a_0^{\mathrm{PE}} = 1,\ P_0^{\mathrm{PE}}(x) = 1$ $a_1^{\mathrm{PE}} = 2, P_1^{\mathrm{PE}}(x) = 3 + x$ $a_2^{\mathrm{PE}} = \frac{8}{3}, P_2^{\mathrm{PE}}(x) = 5 + \frac{7}{3}x + \frac{1}{3}x^2$ $a_3^{\mathrm{PE}} = \frac{16}{5}, P_3^{\mathrm{PE}}(x) = 7 + \frac{57}{15}x + \frac{12}{15}x^2 + \frac{1}{15}x^3$ $a_4^{\mathrm{PE}} = \frac{128}{35}, P_4^{\mathrm{PE}}(x) = 9 + \frac{561}{105}x + \frac{141}{105}x^2 + \frac{18}{105}x^3 + \frac{1}{105}x^4$	$\gamma^{\#}(0) = 0$	$\gamma^{\#}(\infty) = -1$	(A93)
Climacospectrum	$\zeta(k) := \frac{k(\gamma(k) - \gamma(2k))}{\ln 2}$	$\zeta^{\#}(0) = 2,\ n = 0$ $\zeta^{\#}(0) = 3,\ n > 0$	$\zeta^{\#}(\infty) = 0$	(A94)
Entropy production	$\varphi(k) = \frac{2\lambda}{\gamma(k)(k/\alpha)}\left(a_n^{\mathrm{PE}} - \mathrm{e}^{-k/\alpha} R_n^{\mathrm{PE}}(k/\alpha)\right)$ where a_n^{PE} are coefficients as above and $R_n^{\mathrm{PE}}(x)$ are polynomials determined using (39); the first five are: $R_0^{\mathrm{PE}}(x) = 1$ $R_1^{\mathrm{PE}}(x) = 2 + x$ $R_2^{\mathrm{PE}}(x) = \frac{8}{3} + \frac{5}{3}x + \frac{1}{3}x^2$ $R_3^{\mathrm{PE}}(x) = \frac{48}{15} + \frac{33}{15}x + \frac{9}{15}x^2 + \frac{1}{15}x^3$ $R_4^{\mathrm{PE}}(x) = \frac{384}{105} + \frac{279}{105}x + \frac{87}{105}x^2 + \frac{14}{105}x^3 + \frac{1}{105}x^4$ $\varphi_{\mathrm{C}}(k) = \frac{\varphi(k)\gamma(k) - \varphi(2k)\gamma(2k)}{\gamma(k) - \gamma(2k)}$	$\varphi(0) = 1$ $\varphi_{\mathrm{C}}(0) = 3/2,\ n = 0$ $\varphi_{\mathrm{C}}(0) = 2,\ n > 0$	$\varphi(\infty) = \varphi_{\mathrm{C}}(\infty) = 1/2$	(A95)
Autocovariance function	$c(h) = \lambda \mathrm{e}^{-h/\alpha} Q_n^{\mathrm{PE}}(h/\alpha)$ where $Q_n^{\mathrm{A}}(x)$ is a polynomial determined by the power spectrum (see below) using (22); the first five are: $Q_0^{\mathrm{PE}}(x) = 1$ $Q_1^{\mathrm{A}}(x) = 1 + x$ $Q_2^{\mathrm{A}}(x) = 1 + x + \frac{1}{3}x^2$ $Q_3^{\mathrm{A}}(x) = 1 + x + \frac{6}{15}x^2 + \frac{1}{15}x^3$ $Q_4^{\mathrm{A}}(x) = 1 + x + \frac{45}{105}x^2 + \frac{10}{105}x^3 + \frac{1}{105}x^4$	$c^{\#}(0) = 0$	$c^{\#}(\infty) = -\infty$	(A96)
Power spectrum	$s(w) = \frac{4^{n+1}(n!)^2}{(2n)!} \frac{\lambda\alpha}{\left(1 + (2\pi\alpha w)^2\right)^{n+1}}$	$-s^{\#}(\infty) = 2(n+1)$	$-s^{\#}(0) = 0$	(A97)

Table A6. Mathematical expressions of the second order characteristics of the AE process and their asymptotic properties.

Characteristic	Mathematical Expression	Local Asymptotic Properties	Global Asymptotic Properties	Equation
Climacogram	$\gamma(k)=\frac{2\lambda}{(k/\alpha)^2}\left(\frac{1}{2n-1}-\mathrm{e}^{-k/\alpha}P_n^{\mathrm{AE}}(k/\alpha)\right)$ where $n=0,1,2\ldots$, $P_0^{\mathrm{AE}}(x)=-1-x\mathrm{e}^x$ while for $n>0$ $P_n^{\mathrm{AE}}(x)$ is a polynomial determined using (23); the first four are: $P_1^{\mathrm{AE}}(x)=1+x$ $P_2^{\mathrm{AE}}(x)=\frac{1}{3}+\frac{1}{3}x-\frac{1}{3}x^2$ $P_3^{\mathrm{AE}}(x)=\frac{3}{15}+\frac{3}{15}x-\frac{6}{15}x^2+\frac{1}{15}x^3$ $P_4^{\mathrm{AE}}(x)=\frac{15}{105}+\frac{15}{105}x-\frac{45}{105}x^2+\frac{14}{105}x^3-\frac{1}{105}x^4$	$\gamma^{\#}(0)=0$	$\gamma^{\#}(\infty)=-1,\ n=0$ $\gamma^{\#}(\infty)=-2,\ n>0$	(A98)
Climacospectrum	$\zeta(k):=\frac{k(\gamma(k)-\gamma(2k))}{\ln 2}$	$\zeta^{\#}(0)=2$	$\zeta^{\#}(\infty)=0,\ n=0$ $\zeta^{\#}(\infty)=-1,\ n>0$	(A99)
Entropy production	$\varphi(k)=\frac{2\lambda\mathrm{e}^{-k/\alpha}}{\gamma(k)}R_n^{\mathrm{AE}}(k/\alpha)$ where $R_0^{\mathrm{AE}}(x)=\frac{-1-\mathrm{e}^x}{h}$, while for $n>0$ $R_n^{\mathrm{AE}}(x)$ are polynomials determined using (39); the first four are: $R_1^{\mathrm{PE}}(x)=1$ $R_2^{\mathrm{PE}}(x)=1-\frac{1}{3}x$ $R_3^{\mathrm{PE}}(x)=1-\frac{9}{15}x+\frac{1}{15}x^2$ $R_4^{\mathrm{PE}}(x)=1-\frac{87}{105}x+\frac{18}{105}x^2-\frac{1}{105}x^3$ $\varphi_{\mathrm{C}}(k)=\frac{\varphi(k)\gamma(k)-\varphi(2k)\gamma(2k)}{\gamma(k)-\gamma(2k)}$	$\varphi(0)=1$ $\varphi_{\mathrm{C}}(0)=3/2$	$\varphi(\infty)=\varphi_{\mathrm{C}}(\infty)=1/2,\ n=0$ $\varphi(\infty)=\varphi_{\mathrm{C}}(\infty)=0,\ n>0$	(A100)
Autocovariance function	$c(h)=\lambda\mathrm{e}^{-h/\alpha}Q_n^{\mathrm{A}}(h/\alpha)$ where $Q_n^{\mathrm{A}}(x)$ is a polynomial determined by the power spectrum (see below) using (22); the first five are: $Q_0^{\mathrm{A}}(x)=1$ $Q_1^{\mathrm{A}}(x)=1-x$ $Q_2^{\mathrm{A}}(x)=1-\frac{5}{3}x+\frac{1}{3}x^2$ $Q_3^{\mathrm{A}}(x)=1-\frac{33}{15}x+\frac{12}{15}x^2-\frac{1}{15}x^3$ $Q_4^{\mathrm{A}}(x)=1-\frac{279}{105}x+\frac{141}{105}x^2-\frac{22}{105}x^3-\frac{1}{105}x^4$	$c^{\#}(0)=0$	$c^{\#}(\infty)=-\infty$	(A101)
Power spectrum	$s(w)=\frac{4^{n+1}(n!)^2}{(2n)!}\frac{\lambda\alpha(2\pi\alpha w)^{2n}}{\left(1+(2\pi\alpha w)^2\right)^{n+1}}$	$-s^{\#}(\infty)=2$	$-s^{\#}(0)=-2n$	(A102)

Table A7. Mathematical expressions of the second order characteristics of the Generalized Planck (GP) process and their asymptotic properties.

Characteristic	Mathematical Expression	Local Asymptotic Properties	Global Asymptotic Properties	Equation
Climacogram	$\gamma(k) = \frac{\gamma}{(\pi k/\alpha)} P_n^{\mathrm{GP}}(\pi k/\alpha)$ where $n = 0, 1, 2, \ldots$, $P_n^{\mathrm{GP}}(x)$ is a function determined using (23) the first three are: $P_0^{GP}(x) = 3\left(\ln \frac{e^{2x}-1}{2x} - x\right)$ $P_1^{\mathrm{GP}}(x) = \frac{5}{2}\left(1 - \frac{3}{x^2} + \frac{3}{\sinh^2 x}\right)$ $P_2^{\mathrm{GP}}(x) = \frac{21}{40}\left(1 - \frac{45}{x^4} - \frac{30-2\cosh 2x}{\sinh^4 x}\right)$	$\gamma^{\#}(0) = 0$	$\gamma^{\#}(\infty) = -1$ $n = 0$ $\gamma^{\#}(\infty) = -2$ $n > 0$	(A103)
Climacospectrum	$\zeta(k) := \frac{k(\gamma(k)-\gamma(2k))}{\ln 2}$	$\zeta^{\#}(0) = 3$	$\zeta^{\#}(\infty) = 0$ $n = 0$ $\zeta^{\#}(\infty) = -1$ $n > 0$	(A104)
Entropy production	$\phi(k)$ is determined from (39); the first three terms are: $\varphi_0(k) = \frac{1-(\pi k/\alpha)\coth(\pi k/\alpha)}{2\pi k/\alpha - 2\ln((e^{2\pi k/\alpha}-1)/(2\pi k/\alpha))}$ $\varphi_1(k) = \frac{3-3(\pi k/\alpha)^3\coth(\pi k/\alpha)}{3(\pi k/\alpha)^2+((\pi k/\alpha)^2-3)\sinh^2(\pi k/\alpha)}$ $\varphi_2(k) = \frac{180-15(\pi k/\alpha)^5(11\cosh(\pi k/\alpha)+\cosh(3\pi k/\alpha))}{30(\pi k/\alpha)^4(2+\cosh(2\pi k/\alpha))\sinh(\pi k/\alpha)-(2(\pi k/\alpha)^4+90)\sinh^5(\pi k/\alpha)}$ $\varphi_C(k) = \frac{\varphi(k)\gamma(k)-\varphi(2k)\gamma(2k)}{\gamma(k)-\gamma(2k)}$	$\varphi(0) = 1$ $\varphi_C(0) = 2$	$\varphi(\infty) = \varphi_C(\infty) = 1/2, n = 0$ $\varphi(\infty) = \varphi_C(\infty) = 0, n > 0$	(A105)
Autocovariance function	$c(h) = \frac{\lambda}{(\pi k/\alpha)^2} Q_n^{\mathrm{GP}}(\pi k/\alpha)$ where $Q_n^{\mathrm{A}}(x)$ is a function determined from (22); the first three are: $Q_0^{\mathrm{GP}}(x) = 3\left(1 - \frac{x^2}{\sinh^2 x}\right)$ $Q_1^{\mathrm{GP}}(x) = -\frac{45}{x^2} + \frac{15x^2(2+\cosh 2x)}{\sinh^4 x}$ $Q_2^{\mathrm{GP}}(x) = \frac{945}{2x^4} - \frac{x^2(2079+1648\cosh 2x++63\cosh 4x))}{8\sinh^6 x}$	$c^{\#}(0) = 0$	$c^{\#}(\infty) = -2(n+1)$	(A106)
Power spectrum	$s(w) = \frac{c_n \lambda \alpha (\alpha w)^{2n+1}}{e^{2\pi\alpha w}-1}$ where c_n is a normalizing constant making the area of the power spectrum $\gamma_0 = \lambda$; the first three are $c_0 = 24$, $c_1 = 240$, $c_2 = 504$	$-s^{\#}(\infty) = \infty$	$-s^{\#}(0) = -2n$	(A107)

Appendix A.2.8. Harmonic Oscillation Treated as a Stochastic Process

A harmonic oscillation, here expressed as

$$x(t) = \sqrt{2\lambda}\ \cos\left(2\pi(t+b)/T\right),\ 0 \le b \le T \tag{A108}$$

is very easily modelled as a deterministic process but, when it is superimposed to a stochastic process, the resulting process is a stochastic process, too. Obviously the second-order characteristics of this composite process are affected by periodic components and therefore we need to know which equations should be superimposed to those of the stochastic component (see also [40,41]). These equations are given in Table A8. Remarkably, the climacogram retains information about the phase ($2\pi b/T$) of the process (through the spikes appearing at $k = (m + 1/2)T$, $m \in \mathrm{N}_0$, which depend on the phase). Note that this information is completely lost in the autocovariance function and the power spectrum.

An application is depicted in Figure A10. The example process is the average of a FHK with $H = M = 0.8$ and a harmonic oscillation. It is difficult to identify the presence of the oscillation from visual inspection of the time series, but the detailed graph of any of the second-order characteristics, plotted in Figure A10, captures the periodic behaviour. Note that for the application, the generation was done by the SMA model for the entire (composite) process and not by separating the generation of the stochastic and the deterministic component. This is an indication of the flexibility and generality of the generation algorithm.

Table A8. Mathematical expressions of the second order characteristics of a harmonic oscillation treated as a stochastic process and their asymptotic properties.

Characteristic	Mathematical Expression	Local Asymptotic Properties	Global Asymptotic Properties	Equation
Climacogram	$\gamma(k) = \begin{cases} \frac{\lambda T^2}{\pi^2 k^2}\sin^2\left(\frac{\pi k}{T}\right), & k \neq \left(m+\frac{1}{2}\right)T \\ \frac{2\lambda T^2}{\pi^2 k^2}\sin^2\left(\frac{2\pi b}{T}\right), & k = \left(m+\frac{1}{2}\right)T \end{cases}$ where T is the period, $m \in \mathrm{N}_0$ and the harmonic oscillation is described by $x(t) = \sqrt{2\lambda}\ \cos(2\pi(t+b)/T), 0 \le b \le T$	$\gamma^{\#}(0) = 0$	$\gamma^{\#}(\infty) = -2$	(A109)
Climacospectrum	$\zeta(k) := \frac{k(\gamma(k)-\gamma(2k))}{\ln 2}$	$\zeta^{\#}(0) = 3$	$\zeta^{\#}(k)$ does not converge as $k \to \infty$	(A110)
Entropy production	$\varphi(k) = (\pi k/T)\cot(\pi k/T)$ $\varphi_C(k) = (2\pi k/T)\cot(\pi k/T)$ (for $k \neq \left(m+\frac{1}{2}\right)T$)	$\varphi(0) = 1$ $\varphi_C(0) = 2$	$\varphi(k)$ and $\varphi_C(k)$ do not converge as $k \to \infty$	(A111)
Autocovariance function	$c(h) = \lambda\cos(2\pi h/T)$	$c^{\#}(0) = 0$	$c^{\#}(k)$ does not converge as $k \to \infty$	(A112)
Power spectrum	$s(w) = \lambda T\delta(Tw - 1)$	(not applicable)		(A113)

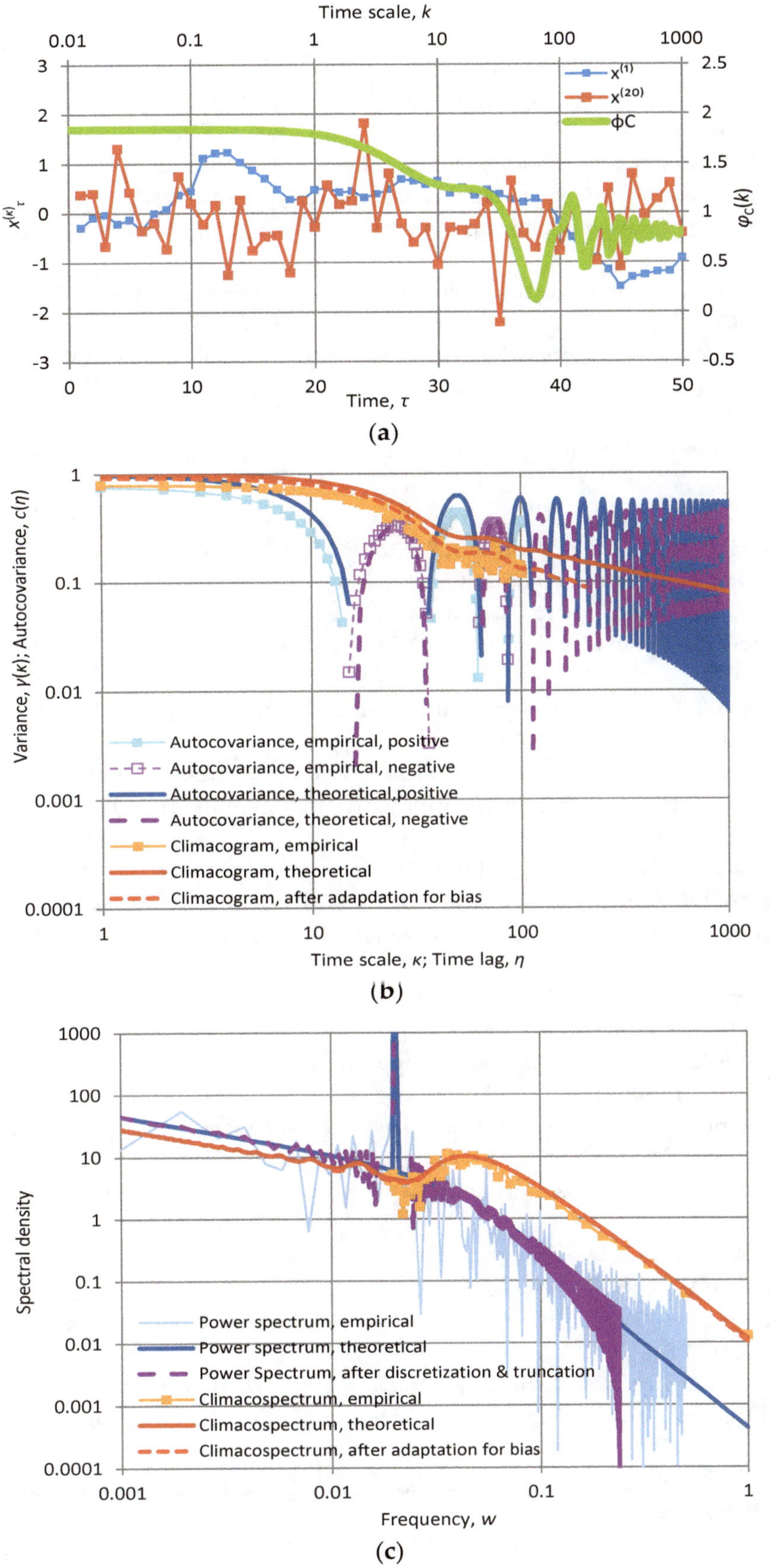

Figure A10. Characteristic graphs as in Figure A2 but for a time series generated by the FHK-C model with $\lambda = 1$, $\alpha = 10$, and $H = M = 0.8$ in which a harmonic oscillation with $T = 5\alpha = 50$ is superimposed.

References

1. Boltzmann, L. Über die Beziehung zwischen dem zweiten Hauptsatze der mechanischen Wärmetheorie und der Wahrscheinlichkeitsrechnung respektive den Sätzen über das Wärmegleichgewicht. *Wien. Ber.* **1877**, *76*, 373–435. (In German)
2. Shannon, C.E. The mathematical theory of communication. *Bell Syst. Tech. J.* **1948**, *27*, 379–423. [CrossRef]
3. Jaynes, E.T. Information theory and statistical mechanics. *Phys. Rev.* **1957**, *106*, 620–630. [CrossRef]
4. Sonuga, J.O. Principle of maximum entropy in hydrologic frequency analysis. *J. Hydrol.* **1972**, *17*, 177–191. [CrossRef]
5. Jackson, D.R.; Aron, G. Parameter estimation in hydrology: The state of the art. *J. Am. Water Resour. Assoc.* **1971**, *7*, 457–472. [CrossRef]
6. Singh, V.P.; Fiorentino, M. A Historical Perspective of Entropy Applications in Water Resources. In *Entropy and Energy Dissipation in Water Resources*; Singh, V.P., Fiorentino, M., Eds.; Springer: Dordrecht, The Netherlands, 1992.
7. Singh, V.P. The use of entropy in hydrology and water resources. *Hydrol. Process.* **1997**, *11*, 587–626. [CrossRef]
8. Singh, V.P. Hydrologic synthesis using entropy theory: review. *J. Hydrol. Eng.* **2011**, *16*, 421–433. [CrossRef]
9. Singh, V.P.; Cui, H. Entropy Theory for Groundwater Modeling. Available online: http://www.aggs.in/Issues/JGWR-2015101.pdf (accessed on 24 October 2017).
10. Singh, V.P.; Cui, H. Entropy theory for streamflow forecasting. *Environ. Process.* **2015**, *2*, 449–460. [CrossRef]
11. Singh, V.P. Entropy theory. In *Handbook of Applied Hydrology*, 2nd ed.; Singh, V.P., Ed.; McGraw Hill: New York, NY, USA, 2017.
12. Porporato, A.; Rigby, J.R.; Daly, E. Irreversibility and fluctuation theorem in stationary time series. *Phys. Rev. Lett.* **2007**, *98*, 94101. [CrossRef] [PubMed]
13. Kondepudi, D.; Prigogine, I. *Modern Thermodynamics*; Wiley: Chichester, UK, 1998.
14. Pujol, T.; Llebot, J.E. Extremal principle of entropy production in the climate system. *Q. J. R. Meteorol. Soc.* **1999**, *125*, 79–90. [CrossRef]
15. Ozawa, H.; Ohmura, A.; Lorenz, R.D.; Pujol, T. The second law of thermodynamics and the global climate system: A review of the maximum entropy production principle. *Rev. Geophys.* **2003**, *41*. [CrossRef]
16. Prigogine, I. Moderation et transformations irreversibles des systemes ouverts. *Bull. Classe Sci. Acad. R. Belg.* **1945**, *31*, 600–606.
17. Ziegler, H. *Progress in Solid Mechanics*; Sneddon, I.N., Hill, R., Eds.; North-Holland: Amsterdam, The Netherlands, 1963; Volume 4.
18. Martyushev, L.M.; Seleznev, V.D. Maximum entropy production principle in physics, chemistry and biology. *Phys. Rep.* **2006**, *426*. [CrossRef]
19. Niven, R. Entropy production extremum principles. In *Handbook of Applied Hydrology*, 2nd ed.; Singh, V.P., Ed.; McGraw Hill: New York, NY, USA, 2017.
20. Koutsoyiannis, D. HESS Opinions: A random walk on water. *Hydrol. Earth Syst. Sci.* **2010**, *14*, 585–601. [CrossRef]
21. Montanari, A.; Koutsoyiannis, D. A blueprint for process-based modeling of uncertain hydrological systems. *Water Resour. Res.* **2012**, *48*, W09555. [CrossRef]
22. Koutsoyiannis, D. Hurst-Kolmogorov dynamics as a result of extremal entropy production. *Phys. A Stat. Mech. Appl.* **2011**, *390*, 1424–1432. [CrossRef]
23. Hemelrijk, J. Underlining random variables. *Stat. Neerl.* **1966**, *20*. [CrossRef]
24. Koutsoyiannis, D. Generic and parsimonious stochastic modelling for hydrology and beyond. *Hydrol. Sci. J.* **2016**, *61*, 225–244. [CrossRef]
25. Koutsoyiannis, D.; Montanari, A. Negligent killing of scientific concepts: The stationarity case. *Hydrol. Sci. J.* **2015**, *60*, 1174–1183. [CrossRef]
26. Kolmogorov, A.N. A simplified proof of the Birkhoff-Khinchin ergodic theorem. *Uspekhi Mat. Nauk* **1938**, *5*, 52–56.
27. Papoulis, A. *Probability, Random Variables and Stochastic Processes*, 3rd ed.; McGraw-Hill: New York, NY, USA, 1991.

28. Dimitriadis, P.; Koutsoyiannis, D. Climacogram versus autocovariance and power spectrum in stochastic modelling for Markovian and Hurst–Kolmogorov processes. *Stoch. Environ. Res. Risk Assess.* **2015**, *29*, 1649–1669. [CrossRef]
29. Koutsoyiannis, D. Physics of uncertainty, the Gibbs paradox and indistinguishable particles. *Stud. Hist. Philos. Mod. Phys.* **2013**, *44*, 480–489. [CrossRef]
30. Koutsoyiannis, D. Entropy: from thermodynamics to hydrology. *Entropy* **2014**, *16*, 1287–1314. [CrossRef]
31. Koutsoyiannis, D.; Dimitriadis, P.; Lombardo, F.; Stevens, S. From fractals to stochastics: Seeking theoretical consistency in analysis of geophysical data. In *Advances in Nonlinear Geosciences*; Tsonis, A., Ed.; Springer: New York, NY, USA, 2018.
32. Dechant, A.; Lutz, E. Wiener-Khinchin theorem for nonstationary scale-invariant processes. *Phys. Rev. Lett.* **2015**, *115*, 80603. [CrossRef] [PubMed]
33. Mandelbrot, B. *Gaussian Self-Affinity and Fractals: Globality, the Earth, 1/f Noise, and R/S (Vol. 8)*; Springer: New York, NY, USA, 2002.
34. Kitanidis, P.K. Geostatistics. In *Handbook of Hydrology*; Maidment, D.R., Ed.; McGraw-Hill: New York, NY, USA, 1993.
35. Gaetan, C.; Guyon, X. *Spatial Statistics and Modeling*; Springer: New York, NY, USA, 2010.
36. Clark, I. Statistics or geostatistics? Sampling error or nugget effect? *J. South. Afr. Inst. Min. Metall.* **2010**, *110*, 307–312.
37. Gneiting, T.; Schlather, M. Stochastic models that separate fractal dimension and the Hurst effect. *SIAM Rev.* **2004**, *46*, 269–282. [CrossRef]
38. Keshner, M.S. 1/f noise. *Proc. IEEE* **1982**, *70*, 212–218. [CrossRef]
39. Wornell, G.W. Wavelet-based representations for the 1/f family of fractal processes. *Proc. IEEE* **1993**, *81*, 1428–1450. [CrossRef]
40. Markonis, Y.; Koutsoyiannis, D. Climatic variability over time scales spanning nine orders of magnitude: Connecting Milankovitch cycles with Hurst–Kolmogorov dynamics. *Surv. Geophys.* **2013**, *34*, 181–207. [CrossRef]
41. Pappas, C.; Mahecha, M.D.; Frank, D.C.; Babst, F.; Koutsoyiannis, D. Ecosystem functioning is enveloped by hydrometeorological variability. *Nat. Ecol. Evol.* **2017**, *1*, 1263–1270. [CrossRef] [PubMed]
42. Kang, H.S.; Chester, S.; Meneveau, C. Decaying turbulence in an active-grid-generated flow and comparisons with large-eddy simulation. *J. Fluid Mech.* **2003**, *480*, 129–160. [CrossRef]
43. Dimitriadis, P.; Koutsoyiannis, D. Stochastic synthesis approximating any process dependence and distribution. *Stoch. Environ. Res. Risk Assess.* **2017**. submitted.
44. Jaynes, E.T. *Probability Theory: The Logic of Science*; Cambridge University Press: Cambridge, UK, 2003; p. 728.
45. Swendsen, R.H. How physicists disagree on the meaning of entropy. *Am. J. Phys.* **2011**, *79*, 342–348. [CrossRef]
46. Stein, M.L. *Interpolation of Spatial Data*; Springer: New York, NY, USA, 1999.
47. Beran, J. *Statistics for Long-Memory Processes*; Chapman and Hall: New York, NY, USA, 1994.
48. Taqqu, M.S. Fractional Brownian motion and long-range dependence. In *Theory and Applications of Long-Range Dependence*; Doukhan, P., Oppenheim, G., Taqqu, M.S., Eds.; Birkhäuser: Boston, MA, USA, 2003.

12

An Entropy-Based Generalized Gamma Distribution for Flood Frequency Analysis

Lu Chen [1], Vijay P. Singh [2,3] and Feng Xiong [4,*]

1 College of Hydropower and Information Engineering, Huazhong University of Science and Technology, Wuhan 430074, China; chen_lu@hust.edu.cn

2 Department of Biological and Agricultural Engineering, Texas A&M University, College Station, TX 77843, USA; vsingh@tamu.edu

3 Zachry Department of Civil Engineering, Texas A&M University, College Station, TX 77843, USA

4 State Key Laboratory of Water Resources and Hydropower Engineering Science, Wuhan University, Wuhan 430072, China

* Correspondence: fxiong07@whu.edu.cn

Academic Editor: Kevin Knuth

Abstract: Flood frequency analysis (FFA) is needed for the design of water engineering and hydraulic structures. The choice of an appropriate frequency distribution is one of the most important issues in FFA. A key problem in FFA is that no single distribution has been accepted as a global standard. The common practice is to try some candidate distributions and select the one best fitting the data, based on a goodness of fit criterion. However, this practice entails much calculation. Sometimes generalized distributions, which can specialize into several simpler distributions, are fitted, for they may provide a better fit to data. Therefore, the generalized gamma (GG) distribution was employed for FFA in this study. The principle of maximum entropy (POME) was used to estimate GG parameters. Monte Carlo simulation was carried out to evaluate the performance of the GG distribution and to compare with widely used distributions. Finally, the T-year design flood was calculated using the GG and compared with that with other distributions. Results show that the GG distribution is either superior or comparable to other distributions.

Keywords: flood frequency analysis; generalized gamma (GG) distribution; principle of maximum entropy (POME)

---

1. Introduction

Flood frequency analysis (FFA) is needed for the design of water engineering and hydraulic structures. The sizing of bridges, culverts and other facilities; the design capacities of levees, spillways and other control structures; and reservoir operation or management all depend upon the estimated magnitude of various design flood values [1–3]. In FFA, flow data, such as the annual maximum data, are fitted using a theoretical frequency distribution, which is usually selected from a set of candidate distributions [4]. For example, the Pearson type three distribution (P3) has been recommended in China [5]. In the US, since 1967 the Log-Pearson type 3 distribution (LP3) has been the official distribution for all catchments which are fitted for planning and insurance purposes [6]. The UK has endorsed the GEV distribution [7,8] for FFA.

The choice of the appropriate model is one of the most important issues for FFA. The method commonly practiced is to try different distributions for the data at hand and choose the best fitted distribution using some particular goodness-of-fit measure [9]. One of the disadvantages of this method is that too many different distributions need to be tried and the selected distribution may be the best based on one goodness of fit criterion, but not based on another criterion. In order to overcome this disadvantage, some generalized frequency distributions have been recently used for FFA. The generalized gamma (GG) distribution is discussed in this study. It is a generalization of the two-parameter gamma distribution. The GG distribution includes as special cases the exponential distribution, the two-parameter gamma distribution, and the Weibull distribution, which provide sufficient flexibility to fit a large variety of data sets.

After deciding the distribution, the second issues is to estimate the parameters associated with the GG distribution. The popular techniques for parameter estimation include the methods of maximum likelihood (ML) [7], moments (MM) [10] and L-moments [11]. In addition, entropy theory can be used to derive more generalized distributions using different constraints [12]. The theory involves entropy maximizing in accord with the principle of maximum entropy (POME), in which the distribution parameter are determined, given the observed data and a set of constraints. Singh [12] indicated that the entropy method was reasonable and efficient for parameter estimation.

The objective of this study was therefore to propose an entropy based generalized gamma distribution for flood frequency analysis. The GG distribution parameters were estimated using POME. The GG distribution was tested using observed data sets. Also, Monte Carlo simulation was carried out to evaluate the predictive ability of the GG distribution and it was compared with some widely accepted distributions. Finally, the *T*-year design flood values were calculated and compared based on different FFA distributions.

2. Methodology

2.1. Generalized Gamma Distribution

Let X be a random variable and x be its specific value. The probability density function (PDF) of the generalized gamma (GG) distribution can be expressed as:

$$f(x) = \frac{r_2}{\beta\Gamma(\frac{r_1}{r_2})}(\frac{x}{\beta})^{(r_1-1)} \exp(-(\frac{x}{\beta})^{r_2}) \tag{1}$$

where $\Gamma(\bullet)$ is the gamma function, and r_1, r_2 are the shape parameters, and β is the scale parameter.

2.2. Estimation of Parameters of GB2 Distribution by POME

The GG distribution parameters were determined using the principle of maximum entropy (POME). The POME method involves the following steps: (1) specification of constraints; (2) maximization of entropy using the method of Lagrange multipliers; (3) derivation of the relation between Lagrange multipliers and constraints; (4) derivation of the relation between Lagrange multipliers and distribution parameters; and (5) derivation of the relation between distribution parameters and constraints. A flow chart showing the estimation procedure is shown in Figure 1.

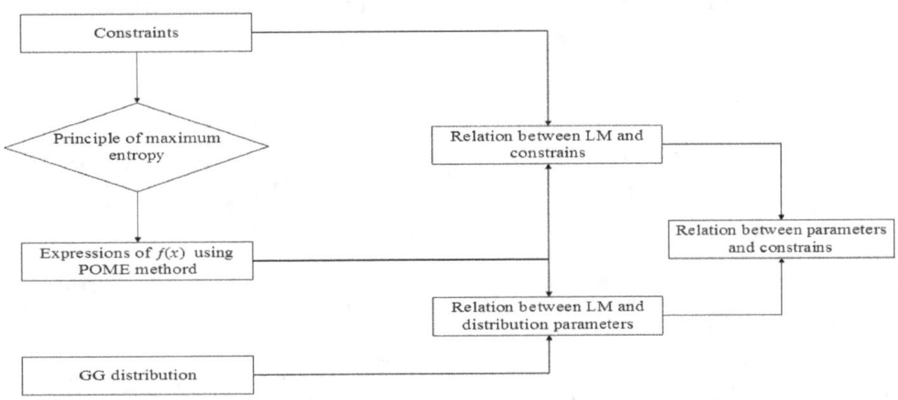

Figure 1. Flow chart of POME method for parameter estimation.

2.2.1. Specification of Constraints

Flood discharge is considered as a random variable X, which ranges from 0 to infinity. Its probability distribution function (PDF) and cumulative distribution function (CDF) are denoted as $f(x)$ and $F(x)$, respectively, where x is a specific value of X. Since constraints encode the information that can be given for the random variable, following Singh [12], the constraints for the GG distribution can be expressed as:

$$\int_0^\infty f(x)\mathrm{d}x = 1 \tag{2a}$$

$$\int_0^\infty f(x)\ln x\mathrm{d}x = E(\ln x) \tag{2b}$$

$$\int_0^\infty f(x)x^q\mathrm{d}x = E(x^q) \tag{2c}$$

The first constraint is the total probability law, the second constraint is the mean of log values or the geometric mean, and the third constraint is the mean of values raised to a power q or log of scaled values raised to a power and then shifted by unity.

2.2.2. Maximization of Entropy Using the Method of Lagrange Multipliers

The Shannon entropy of X, $H(X)$, can be expressed as [13]:

$$H(X) = -\int_0^\infty f(x)\log f(x)\mathrm{d}x \tag{3}$$

The $f(x)$ can be obtained by maximizing the Shannon entropy subject to given constraints in accord with the principle of maximum entropy (POME). Following Singh [14,15], maximization of Equation (3), subject to Equation (2a) to (2c), using the method of Lagrange multipliers leads to:

$$f(x) = \exp(-\lambda_0 - \lambda_1\ln(x) - \lambda_2 x^q) \tag{4}$$

where λ_0, λ_1, λ_2 are the Lagrange multipliers that are not known.

2.2.3. Relation between Lagrange multipliers and parameters

Substitution of Equation (4) in Equation (2a) yields:

$$\int_0^\infty f(x)\mathrm{d}x = \int_0^\infty \exp(-\lambda_0 - \lambda_1\ln(x) - \lambda_2 x^q)\mathrm{d}x = 1 \tag{5}$$

Equation (5) can be expressed as:

$$\exp(\lambda_0) = \int_0^\infty x^{(-\lambda_1)}\exp(-\lambda_2 x^q)\mathrm{d}x \tag{6}$$

Let $t = \lambda_2 x^q$. Then $x = (\frac{t}{\lambda_2})^{\frac{1}{q}}$, and $\mathrm{d}x = \frac{1}{q\lambda_2}(\frac{t}{\lambda_2})^{\frac{1}{q}-1}\mathrm{d}t$. Then Equation (6) can be expressed as:

$$\exp(\lambda_0) = \int_0^\infty \frac{1}{q}(\frac{1}{\lambda_2})^{\frac{-\lambda_1+1}{q}} t^{\frac{-\lambda_1+1}{q}-1}\exp(-t)\mathrm{d}t = \frac{1}{q}(\frac{1}{\lambda_2})^{\frac{-\lambda_1+1}{q}}\Gamma(\frac{-\lambda_1+1}{q}) \tag{7}$$

Substitution of Equation (7) in Equation (4) yields:

$$f(x) = \exp(-\lambda_0 - \lambda_1 \ln(x) - \lambda_2 x^q) = \frac{q\lambda_2^{\frac{-\lambda_1+1}{q}}}{\Gamma(\frac{-\lambda_1+1}{q})} x^{(-\lambda_1)} \exp(-\lambda_2 x^q) \tag{8}$$

Let $\frac{-\lambda_1+1}{q} = \frac{r_1}{r_2}$, and $\lambda_2 = (\frac{1}{\beta})^{r_2}$. Then, $r_2 = q$, and $r_1 = -\lambda_1 + 1$. Equation (8) can now be written as:

$$f(x) = \frac{r_2(\frac{1}{\beta})^{r_2 \frac{r_1}{r_2}}}{\Gamma(\frac{r_1}{r_2})} x^{(r_1-1)} \exp(-(\frac{1}{\beta})^{r_2} x^{r_2}) = \frac{r_2}{\beta\Gamma(\frac{r_1}{r_2})} (\frac{x}{\beta})^{(r_1-1)} \exp(-(\frac{x}{\beta})^{r_2}) \tag{9}$$

Equation (9) is the same as the generalized gamma distribution given by Equation (1). Hence, the relation between Lagrange multipliers and distribution parameters are given by:

$$\begin{cases} q = r_2 \\ \lambda_1 = 1 - r_1 \\ \lambda_2 = \beta^{-r_2} \end{cases} \tag{10}$$

2.2.4. Relation between Lagrange Multipliers and Constraints

Since the Lagrange multiplier λ_0 can be expressed by Equations (6) and (7), the set of equations can be used to obtain λ_0:

$$\lambda_0 = \ln(\int_0^\infty \exp(-\lambda_1 \ln(x) - \lambda_2 x^q) dx) \tag{11a}$$

$$\lambda_0 = -\ln q + \frac{\lambda_1 - 1}{q} \ln(\lambda_2) + \ln\Gamma(\frac{1-\lambda_1}{q}) \tag{11b}$$

Differentiation of Equation (11a) with respect to λ_1 and λ_2 yields:

$$\begin{cases} \frac{\partial\lambda_0}{\partial\lambda_1} = \frac{\int_0^\infty \ln x \exp(-\lambda_1 \ln(x) - \lambda_2 x^q) dx}{\int_0^\infty \exp(-\lambda_1 \ln(x) - \lambda_2 x^q) dx} = -E(\ln x) \\ \frac{\partial\lambda_0}{\partial\lambda_2} = \frac{\int_0^\infty x^q \exp(-\lambda_1 \ln(x) - \lambda_2 x^q) dx}{\int_0^\infty \exp(-\lambda_1 \ln(x) - \lambda_2 x^q) dx} = -E(x^q) \end{cases} \tag{12}$$

Defining $b = \frac{1-\lambda_1}{q}$, and differentiating Equation (11b) with respect to λ_1 and λ_2, we obtain:

$$\begin{cases} \frac{\partial\lambda_0}{\partial\lambda_1} = \frac{1}{q}\ln\lambda_2 - \frac{1}{q}\partial\frac{\ln\Gamma(b)}{\partial b} = \frac{1}{q}\ln\lambda_2 - \frac{1}{q}\varphi(b) \\ \frac{\partial\lambda_0}{\partial\lambda_2} = \frac{1}{\lambda_2}\frac{\lambda_1-1}{q} = \frac{-b}{\lambda_2} \end{cases} \tag{13}$$

where $\varphi(\bullet)$is a digamma function.

Based on Equations (12) and (13), the relation between Lagrange multipliers and constraints can be expressed as:

$$\begin{cases} \frac{\partial\lambda_0}{\partial\lambda_1} = \frac{1}{q}\ln\lambda_2 - \frac{1}{q}\varphi(b) = -E(\ln x) \\ \frac{\partial\lambda_0}{\partial\lambda_2} = \frac{-b}{\lambda_2} = -E(x^q) \end{cases} \tag{14}$$

Since there are three parameters, Equations (13) and (14) are not sufficient for calculating all the parameters, and one additional equation is therefore needed which is given as:

$$\frac{\partial^2 \lambda_0}{\partial^2 \lambda_1} = \frac{1}{q^2}\varphi'(b) = \mathrm{var}(\ln x) \tag{15}$$

2.2.5. Relation between Parameters and Constraints

Based on the relation between parameters and constraints and between parameters and Lagrange multipliers, the relation between parameters and constraints can be expressed as:

$$\begin{cases} \frac{1}{r_2}\ln(\beta^{-r_2}) - \frac{1}{r_2}\varphi(\frac{r_1}{r_2}) = -E(\ln x) \\ \beta^{r_2}\frac{r_1}{r_2} = E(x^{r_2}) \\ \frac{1}{r_2{}^2}\varphi'(\frac{r_1}{r_2}) = \mathrm{var}(\ln x) \end{cases} \tag{16}$$

where $\varphi(\bullet)$ is the digamma function; $\varphi'(\bullet)$ is the tri-gamma function. For a given data set X, the $E(\ln x)$ and $\mathrm{var}(\ln x)$ can be calculated directly. There are three parameters and three equations in Equation (16). Therefore, this set of nonlinear functions can be solved by the widely used Newton iteration method (Deuflhard, [16]) for parameter estimation. The initial value of the three parameters are set to (1, 1, 1). After multiple iterations, the optimal parameters can be obtained.

3. The Descriptive Ability of GG Distribution

Annual maximum (AM) flood peak data from 10 gauging stations, namely sites 1 to 10, were selected (Table 1). These ten stations are selected due to their diversity of statistical properties and climate types (arid, semi-arid and humid).

Table 1. Statistics of annual maximum flood data series for 10 sites.

Site No.	Gauging Station	Period	C_v	C_s	C_k
1	Rogue River at Raygold near Central Point, US	1906–2001	0.67	1.94	5.66
2	Quinault River at Quinault Lake, US	1912–2001	0.40	0.52	−0.6
3	Eel R A Scotla, US	1911–2001	0.51	0.61	−0.44
4	White River Near Meeker, US	1910–2001	0.34	0.65	0.77
5	Yellowstone River at Corwin Springs, US	1890–2001	0.30	0.66	0.59
6	Genesee River at Portageville, US	1909–2001	0.48	2.93	15.67
7	White River Near Meeker, US	1910–2001	0.34	0.65	0.77
8	Brokenstraw Creek at Younsville, US	1910–2001	0.33	0.75	0.65
9	Danjiangkou reservior at Danjiangkou, China	1929–2014	0.56	0.95	1.57
10	Geheyan reservior at Changyang, China	1951–2005	0.42	1.34	3.34

Besides AM series, partial-duration series can be also employed for the POME method. In this study, the AM series was considered since it is more widely used. The GG distribution was employed to fit the AM series of the 10 sites. The distribution parameters were estimated using Equations (16). The fitted GG distribution and the empirical frequency distribution of the AM series from sites 1, 5, 6 and 8 are shown in Figures 2–5. These four sites are selected because sites 5 and 8 have low skews, site 1 has moderate skew and site 6 has high skew, the cumulative distributions and histograms of AM series fitted by GG distribution for these sites can be representative. The line represents the fitted distribution and point represents the empirical frequencies of observations. Results show that the GG distribution fitted the empirical data well. Histograms of the AM flood peak series fitted by the GG distribution for the four sites are also shown in Figures 2–5 which also show that the GG distribution fitted the empirical histograms well. The skewness coefficient of AM series of sites 1, 5, 6 and 8 was 1.94, 0.66, 2.93 and 0.75, respectively, which showed that the GG distribution described both low and high skewed data well.

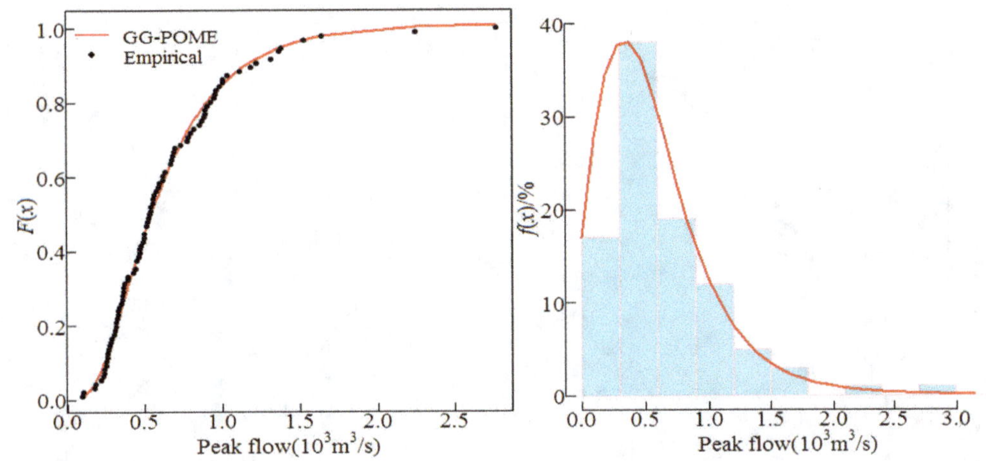

Figure 2. Cumulative distribution and histogram of AM flood peak series fitted by the GG distribution for site 1.

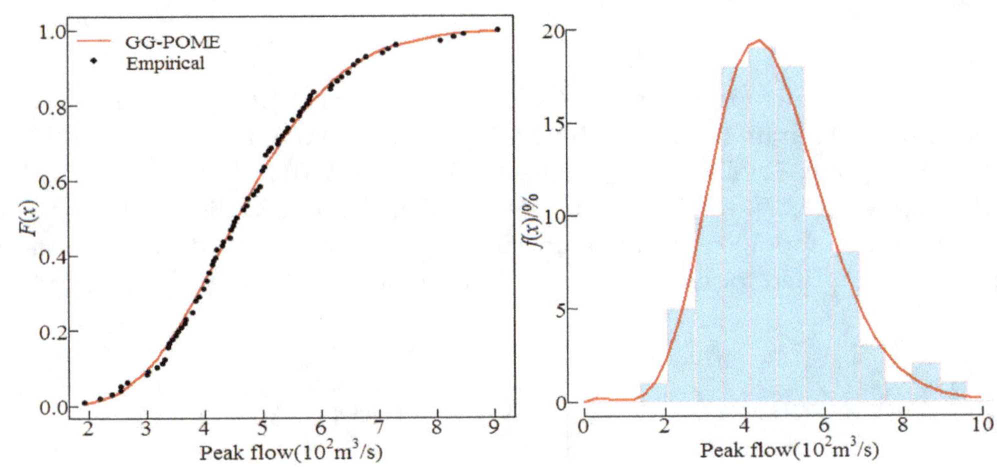

Figure 3. Cumulative distribution and histogram of AM flood peak series fitted by the GG distribution for site 5.

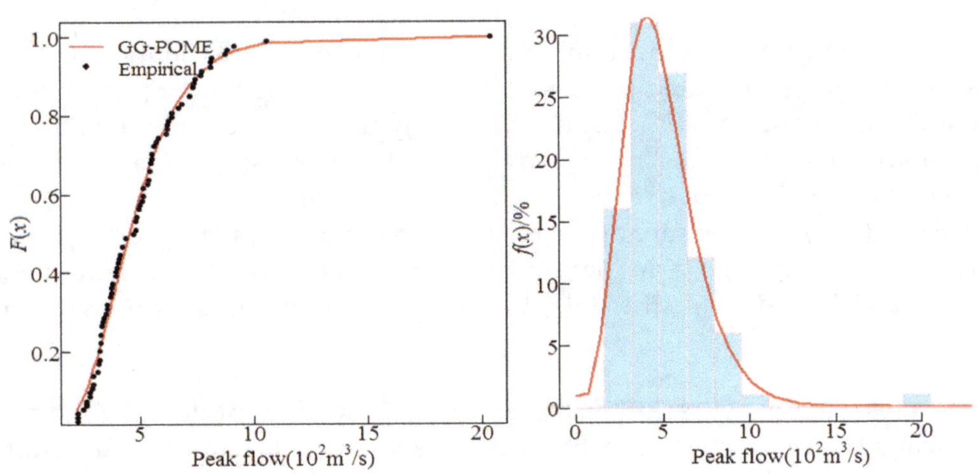

Figure 4. Cumulative distribution and histogram of AM flood peak series fitted by GG-POME model for site 6.

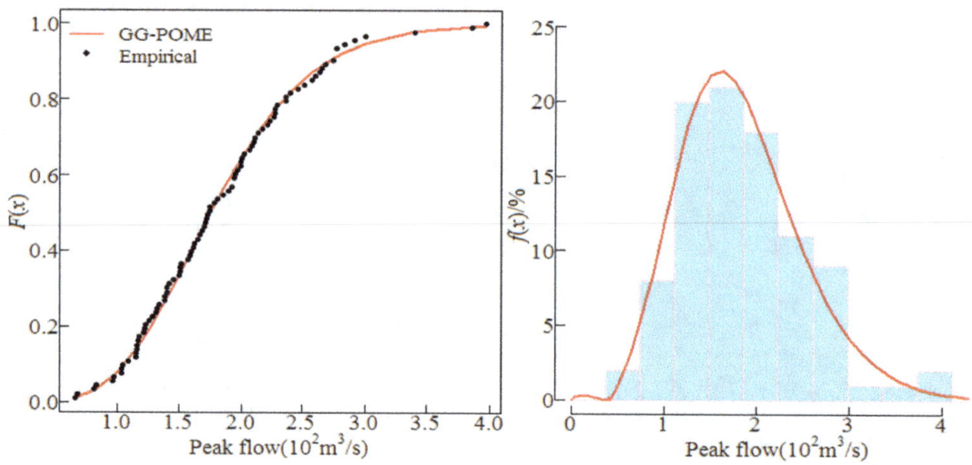

Figure 5. Cumulative distribution and histogram of AM flood peak series fitted by the GG distribution for site 8.

Several distributions, including normal (NM), exponential (EXP), generalized logistic (GLO), gamma (GM), generalized Pareto (GPA), Gumbel (GB), Weibull (WB), P3, GEV and LP3 distributions, in which the parameters of EXP, GLO, GM, GPA, GB, WB, P3, GEV distributions were estimated by the L-moment method (LM) [11,17], while the parameters of NM and LP3 distributions were estimated by MM [18,19]. These FFA models were also fitted to the AM series for the 10 sites and the values of *RMSE* and *AIC* were computed for each model using Equations (17) and (18) and listed in Table 2.

$$RMSE = \sqrt{\frac{\sum_{i=1}^{n}\left(\hat{P}(i) - P(i)\right)^2}{n}} \tag{17}$$

$$AIC = n\left(\ln\left(\frac{1}{n}\sum_{i=1}^{n}\left(\hat{P}(i) - P(i)\right)^2\right)\right) + 2K \tag{18}$$

where n denotes the sample size, K is the number of parameters of the distribution, $\hat{P}$ is the theoretical non-exceedance probability calculated by the distribution, and P is the empirical non-exceedance probability. Root mean square error (*RMSE*) is a frequently used measure of the differences between values (sample and population values) predicted by a model or an estimator and the values actually observed. The smaller *RMSE* values represent the better performance of the model. The Akaike information criterion (*AIC*) is a measure of the relative quality of statistical models for a given set of data. It also includes a penalty that is an increasing function of the number of estimated parameters. Given a set of candidate models for the data, the preferred model is the one with the minimum *AIC* value.

Table 2 illustrates that for sites 1, 2, 4, 5, 8, 9 and 10, the GG distribution had the smallest *RMSE* values, which means the GG distribution fitted the observed AM data best. In addition, the GG distribution had the smallest *AIC* values for sites 2, 5, 8, and 10. Table 2 also indicates that the average *RMSE* and *AIC* values of GG distribution are the smallest among all the compared distributions. Thus, the GG distribution performs better than other distributions.

Table 2. The *RMSE* and *AIC* values of different flood frequency analysis distributions.

Model	Site 1		Site 2		Site 3		Site 4		Site 5		Mean Value
	RMSE	*AIC*	*RMSE*	*AIC*	*RMSE*	*AIC*	*RMSE*	*AIC*	*RMSE*	*AIC*	*RMSE*
GG-POME	0.016	−508.12	0.028	−417.67	0.023	−430.63	0.021	−462.6	0.013	−578.63	0.020
NM-MM	0.078	−258.88	0.047	−306.32	0.048	−316.43	0.037	−394.91	0.038	−407.87	0.053
EXP-LM	0.021	−468.96	0.056	−273.3	0.054	−218.65	0.069	−244.39	0.058	−286.44	0.054
GLO-LM	0.025	−446.68	0.046	−303.81	0.035	−362.47	0.021	−474.16	0.015	−560.19	0.027
GM-LM	0.029	−447.1	0.032	−362.45	0.021	−447.48	0.021	−467.06	0.017	−550.61	0.027
GPA-LM	0.019	−504.1	0.029	−413.67	0.021	−444.15	0.042	−346.33	0.036	−389.21	0.030
GB-LM	0.039	−386.64	0.036	−343.98	0.025	−421.83	0.024	−430.95	0.015	−530.05	0.031
WB-LM	0.017	−522.29	0.029	−372.34	0.019	−459.48	0.025	−441.45	0.019	−515.48	0.024
P3-LM	0.016	−533.04	0.033	−351.64	0.023	−434.97	0.021	−468.95	0.015	−561.44	0.024
GEV-LM	0.019	−495.44	0.034	−347.79	0.025	−419.06	0.021	−472.43	0.014	−572.29	0.022
LP3-MM	0.016	−528.15	0.028	−376.18	0.021	−450.45	0.024	−451.6	0.016	−559.41	0.021
	Site 6		Site 7		Site 8		Site 9		Site 10		Mean Value
	RMSE	*AIC*	*RMSE*	*AIC*	*RMSE*	*AIC*	*RMSE*	*AIC*	*RMSE*	*AIC*	*AIC*
GG-POME	0.024	−427.53	0.021	−447.25	0.013	−562.19	0.021	−421.51	0.023	−269.84	−452.60
NM-MM	0.038	−407.87	0.113	−198.3	0.035	−400.26	0.046	−313.47	0.051	−210.33	−321.46
EXP-LM	0.058	−286.44	0.04	−373.78	0.059	−271.25	0.064	−233.3	0.061	−150.65	−280.72
GLO-LM	0.033	−377.21	0.014	−520.86	0.023	−449.16	0.031	−368.25	0.023	−267.35	−413.01
GM-LM	0.038	−393.42	0.046	−374.75	0.014	−541.17	0.023	−404.7	0.027	−264.86	−425.36
GPA-LM	0.024	−444.17	0.032	−359.81	0.025	−464.21	0.027	−384.59	0.044	−194.88	−394.51
GB-LM	0.033	−404.93	0.072	−282.68	0.019	−475.75	0.024	−406.35	0.024	−262.96	−394.61
WB-LM	0.022	−454.57	0.042	−312.87	0.013	−559.08	0.021	−429.65	0.030	−237.83	−430.50
P3-LM	0.022	−452.78	0.051	−275.86	0.014	−538.15	0.023	−419.04	0.027	−249.79	−428.57
GEV-LM	0.027	−411.21	0.021	−438.69	0.015	−525.05	0.024	−409.68	0.024	−260.26	−435.19
LP3-MM	0.026	−418.17	0.022	−450.08	0.014	−536.12	0.021	−426.08	0.026	−262.16	−445.84

Table 2 also shows that the GG, P3, GEV, LP3 distributions gave quite similar performances for most of the selected sites. However, it was observed that the GG distribution performed better at several sites. For site 2, the *RMSE* values for the GG, P3, and GEV distributions were 0.028, 0.033 and 0.034, respectively. The *AIC* values were −417.67, −351.64 and −347.79, respectively. Thus, the GG distribution performed much better than the P3 and GEV distributions for site 2. Compared with the LP3 distribution, the GG distribution was more appropriate for sites 5 and 7. For site 5, the *RMSE* and *AIC* values for the LP3 distribution (GG distribution) were 0.016(0.013) and −559.41 (−578.63), respectively. For site 7, the *RMSE* and *AIC* values for the LP3 distribution (GG distribution) were 0.019(0.016) and −545.85 (−571.71), respectively. Thus, the GG distribution outperformed the LP3 distribution for those two sites. The above discussions shows that the GG distribution is either superior or comparable to the commonly used distributions.

The maximum likelihood (ML) method was also employed for GG distribution and compared with the proposed GG-POME model for site 5 (low skew) and site 6 (high skew). Figure 6 gives comparisons of their probability density functions and indicates that the GG-POME model gives a better performance. The *RMSE* and *AIC* values of GG-ML model for sites 5 and 6 were also calculated. The *RMSE* and *AIC* values for the GG-ML (GG-POME) model are 0.023 (0.013) and −497.54 (−578.63), respectively for site 5. And the *RMSE* and *AIC* values for the GG-ML (GG-POME) model are 0.032 (0.024) and −379.24 (−427.53), respectively for site 6. Therefore it may imply that GG-POME model outperforms GG-ML model.

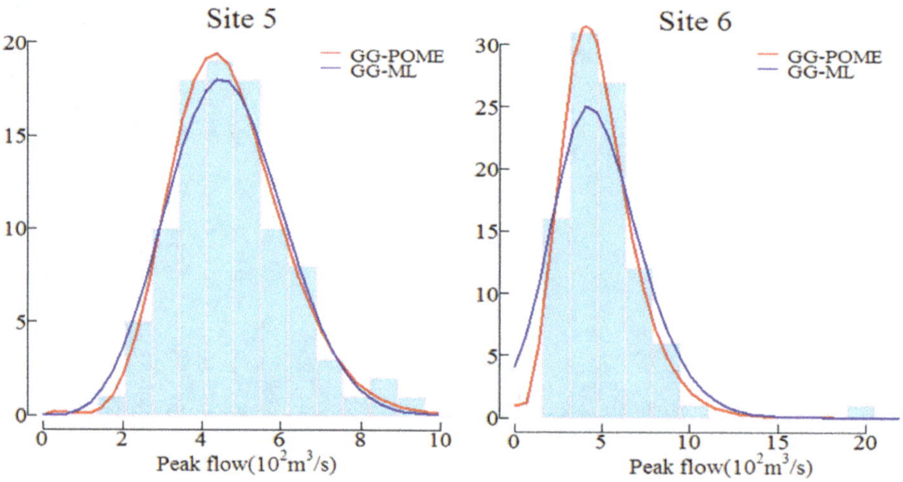

Figure 6. Comparisons of probability density functions of GG-POME and GG-ML models for sites 5 and 6.

4. Monte Carlo Simulation

The predictive ability of the GG distribution was evaluated using Monte Carlo simulation and compared with that of the P3, GEV, and LP3 distributions. To test how well a candidate distribution estimated the magnitude-return period relationship, a parent distribution which was not identical to any of the candidate distributions was chosen. Cunnane [20] recommended that such a parent distribution should be a Wakeby distribution with certain parameters. In this study, three kinds of data sets were generated from the Wakeby distribution with parameters as shown in Table 3. The Wakeby distribution has quantile function given as [21]:

$$x(F) = \xi + \frac{\alpha}{\beta}(1 - (1 - F)^{\beta}) - \frac{\gamma}{\delta}(1 - (1 - F)^{-\delta}) \tag{19}$$

where F is the uniform (0, 1) variate; and $\xi, \alpha, \beta, \gamma, \delta$ are the parameters.

Table 3. Monte Carlo simulation data sets generated from the Wakeby distribution.

	ξ	α	β	γ	δ	C_v	C_s
Case 1	30.4	114.2	11.3	19.2	−0.5	0.2	0.16
Case 2	15.4	308.8	10.25	38.5	−0.3	0.36	0.48
Case 3	23.5	198.6	3.7	109.2	−0.2	0.55	0.95

Then, the real quantile value Q_T was computed. $S = 1000$ samples with size n ($n = 20, 50, 100$) were generated from each Wakeby distribution and fitted by the four distributions to estimate the events of $T = 10$, 100 and 1000-year return periods. Table 4 lists the *RB* and *RRMSE* values computed by each distribution using Equations (20) and (21):

$$RB = \frac{1}{S}\sum_{i=1}^{S}\frac{(\hat{Q}_T)_i - Q_T}{Q_T} \tag{20}$$

$$RRMSE = \sqrt{\frac{1}{S-1}\sum_{i=1}^{S}\left(\frac{(\hat{Q}_T)_i - Q_T}{Q_T}\right)^2} \tag{21}$$

where Q_T is a given parent quantile, $(\hat{Q}_T)_1 \ldots (\hat{Q}_T)_S$ are the estimators for the samples generated from the Wakeby distribution, and S is the number of Monte Carlo trials. The relative bias (*RB*) and the relative root mean square errors (*RRMSE*) were used to evaluate the accuracy and efficiency of a candidate model, respectively.

Table 4. Calculated *RB* and *RRMSE* values for different FFA distributions.

			GG-POME		P3-LM		GEV-LM		LP3-MM	
			RB	*RRMSE*	*RB*	*RRMSE*	*RB*	*RRMSE*	*RB*	*RRMSE*
Case 1	$n = 20$	$T = 10$	−1.71	4.54	−1.21	3.64	−0.87	3.78	−0.89	4.07
		$T = 100$	6.11	7.95	4.6	9.26	4.89	9.31	2.56	8.84
		$T = 1000$	2.02	6.46	2.29	9.08	5.11	10.84	4.15	10.03
	$n = 50$	$T = 10$	−1.32	2.73	−0.87	2.45	−0.75	2.43	−0.81	2.48
		$T = 100$	5.53	6.5	4.38	6.28	6.12	8.02	4.25	6.77
		$T = 1000$	1.85	4.9	0.02	6.73	2.87	8.1	0.65	7.31
	$n = 100$	$T = 10$	−1.06	1.82	−1.05	1.86	−0.72	1.84	−0.71	1.82
		$T = 100$	6.02	6.55	4.97	5.72	5.47	6.34	4.74	5.82
		$T = 1000$	1.46	3.42	0.21	4.29	2.93	6.03	0.55	5.46
Case 2	$n = 20$	$T = 10$	−2.65	9.78	−0.82	9.78	−0.58	9.34	−0.35	14.78
		$T = 100$	1.93	10.23	2.25	16.85	3.64	19.23	5.97	39.86
		$T = 1000$	3.59	14.18	8.69	24.29	16.48	34.45	11.62	44.85
	$n = 50$	$T = 10$	−1.93	8.44	−1.57	6.76	−1.76	6.21	−1.26	6.63
		$T = 100$	1.66	7.55	1.76	10.27	1.37	10.74	−1.67	12.54
		$T = 1000$	3.27	10.33	8.64	16.46	6.67	18.96	4.32	21.43
	$n = 100$	$T = 10$	−1.21	5.86	−1.18	4.73	−0.58	4.34	−0.86	4.52
		$T = 100$	1.94	5.87	1.37	6.24	0.87	7.52	−1.24	7.88
		$T = 1000$	4.22	8.4	9.49	12.78	8.72	16.35	0.29	13.51
Case 3	$n = 20$	$T = 10$	−0.42	11.46	−0.58	13.43	−2.34	12.57	2.64	14.79
		$T = 100$	5.1	14	7.86	21.95	8.54	23.35	4.96	31.38
		$T = 1000$	17.67	28.16	13.57	29.87	21.82	42.72	26.58	57.28
	$n = 50$	$T = 10$	-0.41	8.78	0.86	7.96	−1.25	7.38	1.78	8.67
		$T = 100$	6.34	12.38	4.62	13.94	6.44	15.86	4.41	16.34
		$T = 1000$	16.86	22.85	14.53	24.63	21.02	33.68	8.29	27.32
	$n = 100$	$T = 10$	−0.42	5.72	−0.87	5.87	−1.15	5.46	0.26	5.74
		$T = 100$	6.37	8.54	4.75	8.98	6.65	11.87	3.54	12.64
		$T = 1000$	17.52	19.32	15.67	19.78	17.98	23.92	7.66	20.67

From Table 4, generally for all distributions and for all cases, it was observed that the *RB* and *RRMSE* values increased with the return period *T*. For a small return period ($T = 10$), the selected four distributions exhibited very similar behaviors regardless of the sample size. For moderate and large return periods ($T = 100$ and 1000), notable differences of *RB* and *RRMSE* values were observed. Thus, in the latter discussion, we would mainly focus on moderate and high return period quantile estimators.

For case 1 ($C_v = 0.2$, $C_s = 0.16$), it was observed that the GG and P3 distributions were superior to the GEV and LP3 distributions. When the sample size equaled 100 or 50, the P3 distribution quantile estimators had the smallest *RB* values for both moderate and large return periods ($T = 100$ and 1000). But the GG distribution quantile estimators had smaller *RRMSE* values for $T = 1000$ than other distributions. For a small sample size ($n = 20$), the GG distribution had the smallest *RB* and *RRMSE* values for both moderate and large return periods ($T = 100$ and 1000). For $T = 1000$, the *RRMSE* values of the GG, P3, GEV and LP3 distributions were 6.46, 9.08, 10.84 and 10.03, respectively. Apparently, the GG distribution performed much better when the sample size was small. This indicates that the GG distribution was more robust. Thus, for case 1, the P3 distribution was preferable when the sample size was large than 50, while the GG distribution was more appropriate when sample size did not exceed 50.

For case 2 ($C_v = 0.36$, $C_s = 0.48$), results indicated that for sample size $n = 50$ and $n = 100$, the GEV distribution quantile estimators had the smallest *RB* values for $T = 100$ and the LP3 distribution quantile estimators had the smallest *RB* values for $T = 1000$. However, their *RRMSE* values were quite large and increased significantly when the sample sizes decreased. For $T = 1000$, when the sample size decreased from 100 to 20, the *RRMSE* values of the GEV distribution rose from 16.35 to 34.45, and the *RRMSE* values of the LP3 distribution rose from 13.51 to 44.85. While the *RRMSE* values of the GG distribution rose slightly from 4.8 to 14.18. This was due to the poor accuracy of the GEV and LP3 distributions parameter estimators which had high variance for small sample sizes. In this case, the GG distribution performed significantly better than the other three distributions. Its *RB* values were quite small, and its *RRMSE* values were the smallest for all sample sizes and return periods. This was a good indication of the robustness of the GG distribution for this case.

For case 3 ($C_v = 0.55$, $C_s = 0.97$), all distribution quantile estimators had quite large *RB* and *RRMSE* values. For $n = 50$ and $n = 100$, *RB* and *RRMSE* of the GEV distribution were the highest, which amounted to 21.02 and 33.68, respectively, for $n = 50$, $T = 1000$, while the GG distribution yielded 16.86 and 22.85, respectively. Also for $n = 50$ and $n = 100$, the LP3 distribution quantile estimators had the smallest *RB* values for both $T = 100$ and $T = 1000$, and the other three distributions had similar *RB* values. But the LP3 distribution gave the worst performance for small sample sizes ($n = 20$). Its *RB* and *RRMSE* values were 26.58 and 57.28, respectively, for $T = 1000$, whereas the GG distribution yielded 17.67 and 28.16, respectively. In this case, the *RB* values of the GG distribution were comparable to the P3 and GEV distributions, and were a little larger than the LP3 distribution for $n = 50$ and $n = 100$, the *RRMSE* values of the GG distribution were the smallest for both moderate and large return periods ($T = 100$ and 1000) regardless of the sample size. Also, when the sample size decreased from 100 to 20, the *RB* and *RRMSE* values of the GG distribution rose from 17.52 and 19.32 to 17.67 and 28.16, respectively. This might imply that the distribution was less affected by sample size. Thus, the GG distribution was superior to other distributions for this case. Therefore, the predictive ability of the GG distribution was found to be comparable or superior to that of the other distributions, and it was more robust since it was less affected by sample size, and therefore, estimated the magnitude-return period relationships better.

5. *T*-Year Design Flood Calculation

The Danjiangkou reservoir lies in the upper Hanjiang basin and is the source of water for the Middle Route Project under the South-to-North Water Transfer Scheme in China [22]. The Geheyan reservoir, with a volume of 3.12 billion m^3, plays an important role in management of Qingjiang River [23]. Flood frequency analysis for these two sites was therefore considered in this study. The

T-year design flood calculated by different FFA distributions at Danjiangkou Reservoir and Geheyan Reservoir are listed in Table 5. Figures 7 and 8 compare frequency curves of different distributions at these two reservoir sites.

Table 5 indicates that design flood for small return periods was similar for these four distributions. However, significant differences were observed for large return periods. The 1000-year design flood calculated by the GG and LP3 distributions at Danjiangkou Reservoir were 55,234 m^3/s and 48,822 m^3/s, respectively. And the 1000-year design flood calculated by the GEV and LP3 distributions at Geheyan Reservoir were 15,746 m^3/s and 13,877 m^3/s, respectively.

Figure 7 indicates that the GG, P3, and GEV distributions had quite similar flood quantile estimators for large return periods at Danjiangkou Reservoir. However, the 1000-year design flood calculated by the LP3 distribution was smaller than by the other three distributions. Figure 8 indicates that the 1000-year design flood calculated by the GEV distribution at Geheyan Reservoir was the largest, and was the smallest for the LP3 distribution.

Table 5. Comparison of T-year design floods calculated by different FFA distributions at Danjiangkou and Geheyan sites.

Site	Model	Return Period (Year)				
		1000	500	100	50	10
Danjiangkou (m^3/s)	GG-POME	55,234	51,432	42,204	38,803	27,398
	P3-LM	53,838	50,202	41,407	37,411	27,311
	GEV-LM	55,369	51,490	42,054	37,785	27,217
	LP3-MM	48,822	46,561	40,261	36,999	27,692
Geheyan (m^3/s)	GG-POME	13,992	13,745	11,186	10,277	7957
	P3-LM	14,896	13,941	11,648	10,616	8039
	GEV-LM	15,746	14,594	11,910	10,746	7991
	LP3-MM	13,877	13,099	11,171	10,276	7963

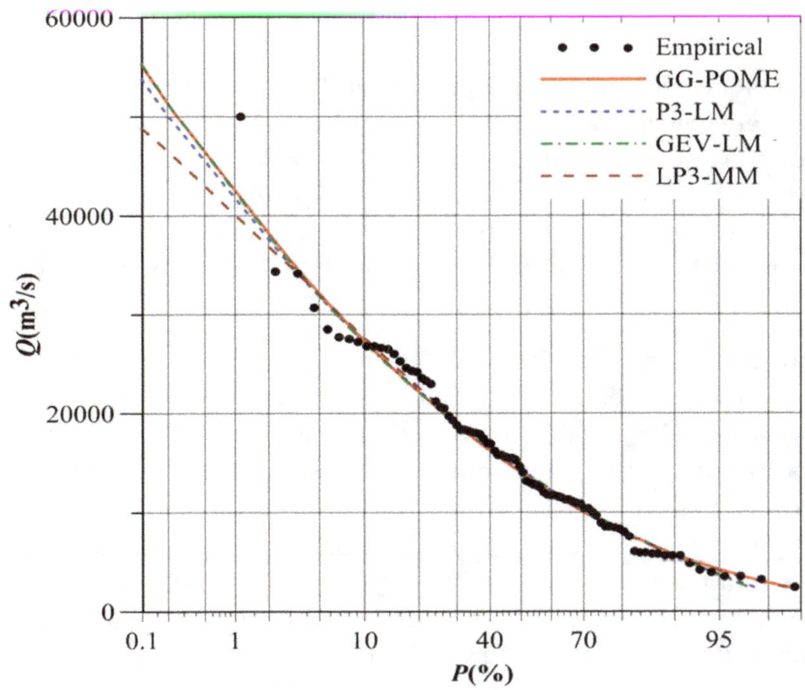

Figure 7. Frequency curves of different FFA distributions at Danjiangkou Reservoir.

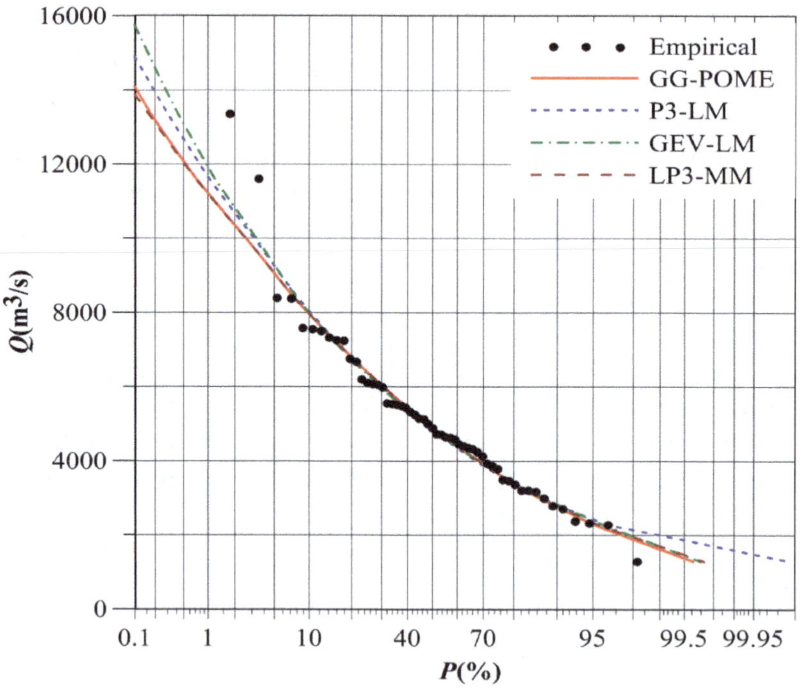

Figure 8. Frequency curves of different FFA distributions at Geheyan Reservoir.

The design flood calculated by the GG distribution was quite close to that by the LP3 distribution. Besides, the P3 distribution has been adopted in China as a uniform procedure for FFA [24,25]. Table 2 shows that *RMSE* and *AIC* values for the P3 distribution at Danjiangkou Reservoir were 0.023 and −419.04, respectively, and the GG distribution yielded 0.021 and −421.51, respectively. The *RMSE* and *AIC* values for the P3 distribution at Geheyan Reservoir were 0.027 and −249.79, respectively, and the GG distribution yielded 0.023 and −269.84, respectively. Thus, the performance of the GG distribution was better than that of the P3 distribution. Therefore, the design flood estimated by the GG distribution would be preferable in practice.

6. Conclusions

In this study, the GG distribution with parameters estimated by POME was applied for FFA. Ten gauging stations were selected as a case study to test the GG distribution. Frequency estimates from the GG distribution were also compared with those of commonly used distributions. A Monte Carlo simulation study was carried out to evaluate the predictive ability of the GG distribution and compare it with other distributions. In addition, some characteristics of frequency curves at Danjiangkou Reservoir and Geheyan Reservoir were evaluated. The following conclusions are drawn from this study:

(1) The GG distribution is appealing for FFA. The cumulative distributions and histograms show that the GG distribution can fit both low and high skewed data well.

(2) The parameters estimated by POME are found reasonable. Both the marginal distributions and histograms indicates that the GG distribution with so estimated parameters can successfully be fitted to empirical values.

(3) The performance of the GG distribution is comparable or superior to that of the other distributions. Results illustrate that for sites 1, 2, 4, 5, 8, 9 and 10, the GG distribution has the smallest *RMSE* values. In addition, the GG distribution has the smallest *AIC* values for sites 2, 5, 8, and 10. Thus, the GG distribution is preferred to other distributions for those sites. Furthermore, the GG, P3, GEV, and LP3 distributions give similar performance for most of the selected sites. However, the GG distribution fits better than them for a few sites.

(4) The predictive ability of the GG distribution is found to be comparable or superior to widely accepted distributions. The GG distribution performs significantly better than the other three distributions when sample sizes are small. Thus it is less effected by sample size and is more robust.

Acknowledgments: This study was supported by the National Natural Science Foundation of China (51679094; 51509273; 91547208) and Fundamental Research Funds for the Central Universities (2016YXZD048).

Author Contributions: All of the authors read and approved the final manuscript. Lu Chen and Feng Xiong conceived and designed the experiments; Vijay P. Singh analyzed the data and contributed materials; Lu Chen and Feng Xiong wrote the paper.

References

1. American Society of Civil Engineers. *Hydrology Handbook*, 2nd ed.; ACSE: New York, NY, USA, 1996.
2. Chen, L.; Guo, S.; Yan, B.; Liu, P.; Fang, B. A new seasonal design flood method based on bivariate joint distribution of flood magnitude and date of occurrence. *Hydrol. Sci. J.* **2010**, *55*, 1264–1280. [CrossRef]
3. Chen, L.; Singh, V.P.; Guo, S.; Hao, Z.; Li, T. Flood coincidence risk analysis using multivariate copula functions. *J. Hydrol. Eng.* **2012**, *17*, 742–755. [CrossRef]
4. Kasiviswanathan, K.S.; He, J.X.; Tay, J.H. Flood frequency analysis using multi-objective optimization based interval estimation approach. *J. Hydrol.* **2017**, *545*, 251–262. [CrossRef]
5. Chen, L.; Singh, V.P.; Guo, S.; Zhou, J.; Zhang, J. Copula-based method for multisite monthly and daily streamflow simulation. *J. Hydrol.* **2015**, *528*, 369–384. [CrossRef]
6. Benameur, S.; Benkhaled, A.; Meraghni, D.; Chebana, F.; Necir, A. Complete flood frequency analysis in Abiod watershed, Biskra (Algeria). *Nat. Hazards* **2017**, *86*, 519–534. [CrossRef]
7. Natural Environment Research Council. *Flood Studies Report*; Institute of Hydrology: London, UK, 1975.
8. Natural Environment Research Council. *Flood Studies Report*; Institute of Hydrology: Wallingford, UK, 1999.
9. Papalexiou, S.M.; Koutsoyiannis, D. Entropy based derivation of probability distributions: A case study to daily rainfall. *Adv. Water Resour.* **2012**, *45*, 51–57. [CrossRef]
10. Chebana, F.; Adlouni, S.; Bobée, B. Mixed estimation methods for Halphen distributions with applications in extreme hydrologic events. *Stoch. Environ. Res. Risk. Assess.* **2010**, *24*, 359–376. [CrossRef]
11. Hosking, J.R.M. L-moments: analysis and estimation of distributions using linear combinations of order statistics. *J. R. Stat. Soc.* **1990**, *52*, 105–124.
12. Singh, V.P. *Entropy-Based Parameter Estimation in Hydrology*; Springer: Dordrecht, The Netherlands, 1998.
13. Shannon, C.E. A mathematical theory of communication. *Bell Syst. Tech. J.* **1948**, *27*, 379–423. [CrossRef]
14. Singh, V.P. *Entropy Theory and Its Application in Environmental and Water Engineering*; John Wiley & Sons: Chichester, UK, 2013.
15. Singh, V.P. *Entropy Theory in Hydrologic Science and Engineering*; McGraw-Hill Education: New York, NY, USA, 2014.
16. Deuflhard, P. *Newton Methods for Nonlinear Problems: Affine Invariance and Adaptive Algorithms*; Springer: Berlin, Germany, 2011.
17. Liang, Z.; Hu, Y.; Li, B.; Yu, Z. A modified weighted function method for parameter estimation of Pearson type three distribution. *Water Resour. Res.* **2014**, *50*, 3216–3228. [CrossRef]
18. Rao, A.R.; Hamed, K.H. *Flood Frequency Analysis*; CRC Press: New York, NY, USA, 2000.
19. Griffis, V.W.; Stedinger, J.R. Log-Pearson type 3 distribution and its application in flood frequency analysis. II: Parameter estimation methods. *J. Hydrol. Eng.* **2007**, *12*, 482–491. [CrossRef]
20. Cunnane, C. Statistical distribution for flood frequency analysis. *Oper. Hydrol. Rep.* **1989**, *44*, 369.
21. Houghton, J.C. Birth of a parent: The Wakeby distribution for modeling flood flows. *Water Resour. Res.* **1978**, *14*, 1105–1109. [CrossRef]
22. Li, S.; Xu, Z.; Cheng, X.; Zhang, Q. Dissolved trace elements and heavy metals in the Danjiangkou Reservoir, China. *Environ. Geol.* **2008**, *55*, 977–983. [CrossRef]
23. Qi, S.; Yan, F.; Wang, S.; Xu, R. Characteristics, mechanism and development tendency of deformation of Maoping landslide after commission of Geheyan reservoir on the Qingjiang River, Hubei Province, China. *Eng. Geol.* **2006**, *86*, 37–51. [CrossRef]

24. Ji, X.; Ding, J.; Shen, H.; Salas, J.D. Plotting positions for Pearson type-III distribution. *J. Hydrol.* **1984**, *74*, 1–29.
25. Shao, Q.; Wong, H.; Xia, J.; Ip, W. Models for extremes using the extended three-parameter Burr XII system with application to flood frequency analysis. *Hydrol. Sci. J.* **2004**, *49*, 685–702. [CrossRef]

PERMISSIONS

All chapters in this book were first published in MDPI; hereby published with permission under the Creative Commons Attribution License or equivalent. Every chapter published in this book has been scrutinized by our experts. Their significance has been extensively debated. The topics covered herein carry significant findings which will fuel the growth of the discipline. They may even be implemented as practical applications or may be referred to as a beginning point for another development.

The contributors of this book come from diverse backgrounds, making this book a truly international effort. This book will bring forth new frontiers with its revolutionizing research information and detailed analysis of the nascent developments around the world.

We would like to thank all the contributing authors for lending their expertise to make the book truly unique. They have played a crucial role in the development of this book. Without their invaluable contributions this book wouldn't have been possible. They have made vital efforts to compile up to date information on the varied aspects of this subject to make this book a valuable addition to the collection of many professionals and students.

This book was conceptualized with the vision of imparting up-to-date information and advanced data in this field. To ensure the same, a matchless editorial board was set up. Every individual on the board went through rigorous rounds of assessment to prove their worth. After which they invested a large part of their time researching and compiling the most relevant data for our readers.

The editorial board has been involved in producing this book since its inception. They have spent rigorous hours researching and exploring the diverse topics which have resulted in the successful publishing of this book. They have passed on their knowledge of decades through this book. To expedite this challenging task, the publisher supported the team at every step. A small team of assistant editors was also appointed to further simplify the editing procedure and attain best results for the readers.

Apart from the editorial board, the designing team has also invested a significant amount of their time in understanding the subject and creating the most relevant covers. They scrutinized every image to scout for the most suitable representation of the subject and create an appropriate cover for the book.

The publishing team has been an ardent support to the editorial, designing and production team. Their endless efforts to recruit the best for this project, has resulted in the accomplishment of this book. They are a veteran in the field of academics and their pool of knowledge is as vast as their experience in printing. Their expertise and guidance has proved useful at every step. Their uncompromising quality standards have made this book an exceptional effort. Their encouragement from time to time has been an inspiration for everyone.

The publisher and the editorial board hope that this book will prove to be a valuable piece of knowledge for researchers, students, practitioners and scholars across the globe.

LIST OF CONTRIBUTORS

Santiago Zazo and José-Luis Molina
IGA Research Group, Higher Polytechnic School of Ávila, University of Salamanca, Av. de los Hornos Caleros, 50, 05003 Ávila, Spain

Verónica Ruiz-Ortiz
Department of Industrial Engineering and Civil Engineering, University of Cadiz, Campus Bay of Algeciras, Avda. Ramon Puyol s/n, 11202 Algeciras, Spain

Mercedes Vélez-Nicolás and Santiago García-López
Department of Earth Sciences, University of Cadiz, Campus Rio San Pedro, s/n, 11510 Puerto Real, Spain

Vlasta Ondrejka Harbulakova, Martina Zelenakova and Adrian Olejnik
Department of Environmental Engineering, Faculty of Civil Engineering, Technical University of Kosice, Vysokoskolska 4, 042 00 Kosice, Slovakia

Pavol Purcz
Department of Applied Mathematics, Faculty of Civil Engineering, Technical University of Kosice, Vysokoskolska 4, 042 00 Kosice, Slovakia

Baohui Men, Yangsong Li, Huanlong Liu and Zhijian Wu
Beijing Key Laboratory of Energy Safety and Clean Utilization, North China Electric Power University, Renewable Energy Institute, Beijing 102206, China

Rishang Long
State Key Laboratory of New Energy Power System, North China Electric Power University, Beijing 102206, China

Wei Tian
Key Laboratory of Water Cycle and Related Land Surface Process, Institute of Geographic Sciences and Natural Resources Research, Chinese Academy of Sciences, Beijing 100101, China
University of Chinese Academy of Sciences, Beijing 100049, China

Gengxi Zhang and Xiaoling Su
College of Water Resources and Architectural Engineering, Northwest A & F University, Yangling 712100, China

Olusola O. Ayantobo
College of Water Resources and Architectural Engineering, Northwest A & F University, Yangling 712100, China
Department of Water Resources Management and Agricultural-Meteorology, Federal University of Agriculture, PMB 2240, Abeokuta 110282, Nigeria

Lu Chen and Kangdi Huang
College of Hydropower & Information Engineering, Huazhong University of Science & Technology, Wuhan 430074, China

Liang Cheng, Jun Niu and Dehai Liao
Center for Agricultural Water Research in China, China Agricultural University, Beijing 100083, China

Lina Hao
College of Water Resources and Architectural Engineering, Northwest A & F University, Yangling 712100, China

Vijay P. Singh
Department of Biological and Agricultural Engineering, Texas A&M University, College Station, TX 77843-2117, USA
Zachry Department of Civil Engineering, Texas A&M University, College Station, TX 77843-2117, USA

Bellie Sivakumar
School of Civil and Environmental Engineering, The University of New South Wales, Sydney, NSW 2052, Australia
Department of Land, Air and Water Resources, University of California, Davis, CA 95616, USA

Huijuan Cui
Key Laboratory of Land Surface Pattern and Simulation, Institute of Geographic Sciences and Natural Resources Research, Chinese Academy of Sciences, Beijing 100101, China

Sufen Wang
Center for Agricultural Water Research in China, China Agricultural University, Beijing 100083, China

Rongxing Zhou
School of Civil Engineering and Environmental Engineering, Anhui Xinhua University, Hefei 230088, China

Zhengwei Pan
School of Civil Engineering and Environmental Engineering, Anhui Xinhua University, Hefei 230088, China
Institute of Safety and Environmental Assessment, Anhui Xinhua University, Hefei 230088, China

Juliang Jin and Shaowei Ning
School of Civil Engineering, Hefei University of Technology, Hefei 230009, China

Chunhui Li
Ministry of Education Key Lab of Water and Sand Science, School of Environment, Beijing Normal University, Beijing 100875, China

Demetris Koutsoyiannis
Department of Water Resources, School of Civil Engineering, National Technical University of Athens, 15780 Athina, Greece

Lu Chen
College of Hydropower and Information Engineering, Huazhong University of Science and Technology, Wuhan 430074, China

Feng Xiong
State Key Laboratory of Water Resources and Hydropower Engineering Science, Wuhan University, Wuhan 430072, China

Index

www.ingramcontent.com/pod-product-compliance
Lightning Source LLC
LaVergne TN
LVHW080949210926
848485LV00004B/808